1st Siliwangi International Conference on Innovation in Research (SICIR 2018)

IOP Conference Series: Materials Science and Engineering
Volume 550

Bandung, Indonesia
14 August 2018

Editors:

Aripin
I. Made Joni
Camellia Panatarani
Supratman
Arief Cahyanto
Tantry Maulina
Reginawanti Hindersah

ISBN: 978-1-5108-9384-9
ISSN: 1757-8981

Printed from e-media with permission by:

Curran Associates, Inc.
57 Morehouse Lane
Red Hook, NY 12571

CURRAN ASSOCIATES INC. proceedings.com

Some format issues inherent in the e-media version may also appear in this print version.

Printed by Curran Associates, Inc. (2019)

Institute of Physics
Dirac House, Temple Back
Bristol BS1 6BE UK

Phone: 44 1 17 929 7481
Fax: 44 1 17 920 0979

techtracking@iop.org

Additional copies of this publication are available from:

Curran Associates, Inc.
57 Morehouse Lane
Red Hook, NY 12571 USA
Phone: 845-758-0400
Fax: 845-758-2633
Email: curran@proceedings.com
Web: www.proceedings.com

Table of contents

Volume 550

The 1st Siliwangi International Conference on Innovation in Research 2018 (SICIR)

14 August 2018, Bandung, Indonesia

Accepted papers received: 13 May 2019
Published online: 23 August 2019

Preface

Papers

Energy

Material

IOP Conf. Series: Materials Science and Engineering **550** (2019) 011001 doi:10.1088/1757-899X/550/1/011001

LIST OF COMMITTEES

Editorials Team

7. Dr. A. Marthin Kalay
Universitas Pattimura
Jln. Ir. M. Putuhena, Kampus Poka, Ambon, 97233, Maluku, Indonesia
marthinkalay@gmail.com

8. Assoc. Prof. Aranya Chaowalit, RN, PhD
Prince of Songkla University
15 Karnjanavanich Rd., Hat Yai, Songkhla 90110
mailto:mAranya.c@psu.ac.th

9. Assc. Prof. Dr. Khatijah Binti Abdullah Lim Geok Lim,
University of Malaya
Jalan Universiti, 50603 Kuala Lumpur, Wilayah Persekutuan Kuala Lumpur, Malaysia
mailto:katlim@um.edu.my

10. Assc. Prof. Kusman Ibrahim, MNS., Ph.D (INA)
Universitas Padjadjaran
Jl. Raya Bandung Sumedang KM 21, Jatinangor, 45363, Jawa Barat, Indonesia
k.ibrahim@unpad.ac.id

11. Dr. Kathy Nelson, PhD, MA, RN,
Victoria University of Wellington,
Kelburn, Wellington 6012, New Zealand
mailto:Kathy.nelson@vuw.ac.nz

12. Prof. Alison Hutton,
New Castle University
University Dr, Callaghan NSW 2308, Australia
Alison.hutton@newcastle.edu.au

13. Sandra K. Cesario, PhD, RNC, FAAN,
Texas Woman's University
304 Administration Dr, Denton, TX 76204, USA
scesario@twu.edu.

14. Ass. Prof. Esther Ching-Lan Lin
National Cheng Kung University of Taiwan
No. 1號, Dasyue Rd, East District, Tainan City, Taiwan 701
mailto:chinglan@mail.ncku.edu.tw

IOP Conf. Series: Materials Science and Engineering **550** (2019) 011001 doi:10.1088/1757-899X/550/1/011001

Advisory Boards

IOP Conf. Series: Materials Science and Engineering **550** (2019) 011001 doi:10.1088/1757-899X/550/1/011001

Organizing Committee

1. **Assoc. Prof. Dr. Supratman – (Chairperson)**
   (Universitas Siliwangi, Indonesia)
2. **Dr. Nani Ratnaningsih**
   (Universitas Siliwangi, Indonesia)
3. **Yuyus Saputra**
   (Universitas Siliwangi, Indonesia)
4. **Husni Mubarok, S.TP., MT.**
   (Universitas Siliwangi, Indonesia)
5. **Rohmat Gunawan, ST., MT.**
   (Universitas Siliwangi, Indonesia)
6. **Rianto, ST.,MT**
   (Universitas Siliwangi, Indonesia)
7. **Nurul Hiron, ST., MT.**
   (Universitas Siliwangi, Indonesia)
8. **Asri Siti Fatimah**
   (Universitas Siliwangi, Indonesia)
9. **Yuni Ertinawati**
   (Universitas Siliwangi, Indonesia)
10. **Nita Sari Narulita Dewi**
    (Universitas Siliwangi, Indonesia)
11. **Ratu Sarah P**
    (Universitas Siliwangi, Indonesia)
12. **Rima Hermawati**
    (Universitas Siliwangi, Indonesia)
13. **Fera Sulastri**
    (Universitas Siliwangi, Indonesia)
14. **Ai Siti Nurjamilah**
    (Universitas Siliwangi, Indonesia)
15. **Vita Meilani**
    (Universitas Siliwangi, Indonesia)
16. **Zulfi Miftahudin**
    (Universitas Siliwangi, Indonesia)
17. **Agi Ahmad Ginanjar**
    (Universitas Siliwangi, Indonesia)
18. **Fuad Abdullah**
    (Universitas Siliwangi, Indonesia)
19. **M. Adi Khairul**
    (Universitas Siliwangi, Indonesia

SICIR 2018 Partners

Partner University

http://www.unpad.ac.id/

Conference Management

http://jpfoundation.or.id/

Ikhsan Saputro, M.T.

M. Abdul Al Mujahid, S.Si.

Angga Nugraha, S.Pt

IOP Conf. Series: Materials Science and Engineering **550** (2019) 011001 doi:10.1088/1757-899X/550/1/011001

SICIR 2018

Organized by:

Universitas Siliwangi

https://unsil.ac.id/

Managed by:

JP Global Transtech

http://jpfoundation.or.id/

Partner University:

Universitas Padjadjaran

http://www.unpad.ac.id/

IOP Conf. Series: Materials Science and Engineering **550** (2019) 011002 doi:10.1088/1757-899X/550/1/011002

Preface: The 1st Siliwangi International Conference on Innovation in Research 2018 (SICIR 2018)

We are proud to present to you The 1st Siliwangi International Conference on Innovation in Research 2018, located in HARRIS Hotel & Conventions, Bandung on 14 August 2018. This conference covers a wide range of innovations in the field of education and renewable engineering. Innovations in the field of education include the problems of teaching and learning as well as the solution. Engineering field includes the discovery and development of renewable technologies.

SICIR 2018, organized by Directorate of Research and Community Service (LP2M) of Universitas Siliwangi (UNSIL) and managed by JP Global Transtech. The objectives of this conference are to disseminate the research results of researchers in University and R & D Institutions and to improve collaboration between researchers with community education and engineering.

There were 4 plenary speakers, 121 scientific participants from countries like Indonesia, Malaysia, and Japan, contributing to the success of the conference. The articles were considered under 8 categories, Agrotechnology, Education, Energy, Health Care, Health Care Engineering, Instrumentation & control and Information Technology, Materials, and Regional Development.

On behalf of the organizing committee, we would like to thanks Rector of Universitas Siliwangi, advisory committee, participants for supporting of this scientific meeting. We would also like to thank all the members of the organizing committee, it is a pride to be able to work with all of you.

Corresponding Editors

Prof. Dr. Eng. Aripin
Prof. Dr. Eng. I Made Joni
Prof. Dr. Eng. Camellia Panatarani
Assoc. Prof. Dr. Supratman

Peer review statement

All papers published in this volume of *IOP Conference Series: Materials Science and Engineering* have been peer reviewed through processes administered by the proceedings Editors. Reviews were conducted by expert referees to the professional and scientific standards expected of a proceedings journal published by IOP Publishing.

Fabrication and characterization of *Intermediate-Temperature Solid Oxide Fuel Cell* (IT-SOFC) single cell using Indonesia's resources

P Widiatmoko, H Devianto, I Nurdin, F Yusupandi, Kevino and E N Ovani

Institut Teknologi Bandung, Ganesha 10 Indonesia

E-mail: hardev@che.itb.ac.id

Abstract. Solid Oxide Fuel Cell (SOFC) is one of the promising developments in the fuel cell system. SOFC can be applied to micro-scale combined heat and power generation (micro CHP) as a power plant in housing. Recently, intermediate-temperature SOFC (500-700°C) is being widely developed since cheaper than the conventional SOFC (~1000°C). Intermediate-temperature and conventional SOFC are less developed in Indonesia due to expensive electrode and electrolyte materials, such as Yttria and Lanthanum, which are not available in Indonesia. This study aims to develop a novel disk-shaped single cell SOFC based on materials from Indonesia's resources. Calcia Cobalt Zinc Oxide (CCZO) is an alternative cathode to replace conventional cathode such as Lanthanum Strontium Manganite (LSM). CCZO cathode was made using spray coating. Calcia can also substitute Ytrria as a stabilizer of zirconia in electrolyte fabrication. Calcia-stabilized Zirconia (CSZ) electrolyte was produced using spray coating. NiO anode was fabricated through hydraulic pressing. The single cell was operated using hydrogen (125 sccm) as a fuel and ambient air (500 sccm) as an oxidant. The results showed an open-circuit voltage (OCV) of 0.577, 0.558, and 0.513 V and produced the maximum power densities of 1.4, 2.8, and 10 mW m^{-2} at 500, 600, and 700°C, respectively. Measurement of the cell impedance indicated that the overall resistance decreased with increasing temperature.

1. Introduction

Electricity generations in Indonesia are still dominated by fossil fuels such as coal, oil and gas which produce greenhouse gases. Meanwhile, Indonesia's electricity needs in 2050 will rise around 193 million tons of equivalent (TOE) according to business as usual (BAU) scenario [1]. To reduce Indonesia's dependency of fossil fuels, utilization of new and renewable energy (NRE) is very important. Hydrogen, produced through biomass gasification, is a new energy source which can be converted into electricity. One of the methods to convert hydrogen into electricity is by using Fuel cell. Fuel cell is an electrochemical device which converts chemical energy into electricity directly without involving the process of combustion. The advantages of fuel cell are higher efficiency (40 – 60%) than heat engine (20 – 30%), quite operation, and low or zero emission [2].

Solid Oxide Fuel Cell (SOFC), a fuel cell used solid oxide electrolyte, is one of the promising developments in the fuel cell system. The advantages of SOFC are higher electrical efficiency (50 – 60%) than other types of fuel cell, higher resilience to corrosion of electrode, wider fuel flexibility and higher reduction of CO_2 emission from 90% to 35% when using hydrocarbon fuels [3]. SOFC is usually operated at 850 – 1000°C, which is called as high temperature SOFC (HT-SOFC). The requirement of high heat and complex material such as ceramic and high temperature metal alloys make HT-SOFC expensive. The purpose of decreasing temperature is to cut material cost and decrease heat requirement. The intermediate-temperature SOFC (IT-SOFC) which operates at 500 – 800°C has been developed to be used in small scale applications such as micro combined heat and power (micro-CHP) and auxiliary power unit (APUs). IT-SOFC gives a wide range of material, rapid start-up and shut down, low corrosion rate of metallic component, improved durability and more robust construction [4].

Electrolyte material of IT-SOFC commonly used is yttria-stabilized zirconia (YSZ). Anode and cathode material of IT-SOFC commonly used is NiO cermet such as NiO-YSZ and lanthanum strontium manganite (LSM), respectively. However, these materials such as ytrria, strontium and lanthanum are rare in Indonesia thus IT-SOFC is less developed in Indonesia. To replace those rare materials, cheaper and common material such as calcia (CaO) can substitute ytrria as a stabilizer of zirconia in electrolyte fabrication. Indonesia has an abundant zirconia and calcia resources. Zirconia is a by-product from tin production in Bangka Belitung Island [5]. Calcia can be produced from limestone which has huge ore reserves in Indonesia, around 28.7 billion tons. [6]. Calcia-stabilized zirconia (CSZ) ionic conductivity has been investigated and it is suitable to be used for IT-SOFC electrolyte [7]. An alternative material for cathode is calcium cobalt zinc oxide (CCZO). Indonesia's cobalt reserves in 2017 are around 180,344 tons [8]. Cobalt is a by-product from nickel production. An alternative material, NiO-CSZ, can be used as Anode material because of high electrocatalytic activity and electrically conductive [9]. Indonesia's nickel reserves in 2017 are around 4 million tons and Indonesia is the seventh largest country of nickel production [8].

In this study, an IT-SOFC single cell was fabricated using Indonesia's resources then characterized and evaluated its performance. NiO-CSZ anode was produced through hydraulic pressing method. In other hand, CSZ electrolyte and CCZO cathode were fabricated using spray coating method. The design of single cell was anode-supported. The single cell was evaluated on power density and impedance with different operating temperature, respectively.

2. Experimental

2.1. Material preparation

Material compositions of anode are nickel oxide (NiO), zirconia (ZrO_2), maizena flour, CaO and polyvinyl alcohol (PVA). The electrolyte was formed by ZrO_2, CaO, PVA, isopropyl alcohol (IPA) and polyethylene glycol (PEG). Meanwhile, cathode's materials are cobalt oxide (Co_3O_4), zinc oxide (ZnO), CaO, PVA, ZrO_2, IPA and PEG. Maizena flour, PVA, IPA and PEG act as pore former, binder, dispersion agent and plasticizer, respectively. All materials except PEG and IPA was milled using ball mill for 12 hours. After ball milling, the material powders was sieved for 325 mesh.

2.2. Cell fabrication

First of all, NiO-CSZ anode as support was fabricated using hydraulic pressing method. The mass ratio between NiO and CSZ was 65:35 wt%. The function of CSZ addition is to expand an active surface area. The concentration of PVA and maizena was 3 and 15 wt%. All powders were mixed and pressed at 60 kg/cm^2 for 1 hour to achieve 1.5 mm in thickness. The sintering temperature of anode production was 1000°C for 1 hour. For electrolyte fabrication, ZrO2, CaO 3 wt%, PVA 1 wt% and PEG 1 wt% were mixed and dispersed using IPA. The function of PEG addition was to increase flexibility and durability of electrolyte. A spray coating method in electrolyte production was intended to reach 20 μm in thickness. The sintering temperature of electrolyte production was 1000°C for 3 hours. In cathode fabrication, Co3O4, ZnO, CaO (CCZO) were mixed with CSZ. The mass ratio between CCZO and CSZ was 65:35 wt%. The concentration of PVA and PEG was 1 wt%, respectively. The method of cathode production was spray coating to achieve 50 μm in thickness. The sintering temperature of cathode production was 800°C for 5 hours. The single cell was fabricated with a disk shape of 4.3 cm in anode's and electrolyte's diameter, 4.2 cm in cathode's diameter and 1.8 mm in thickness. The process flow diagram of single cell production as shown in figure 1.

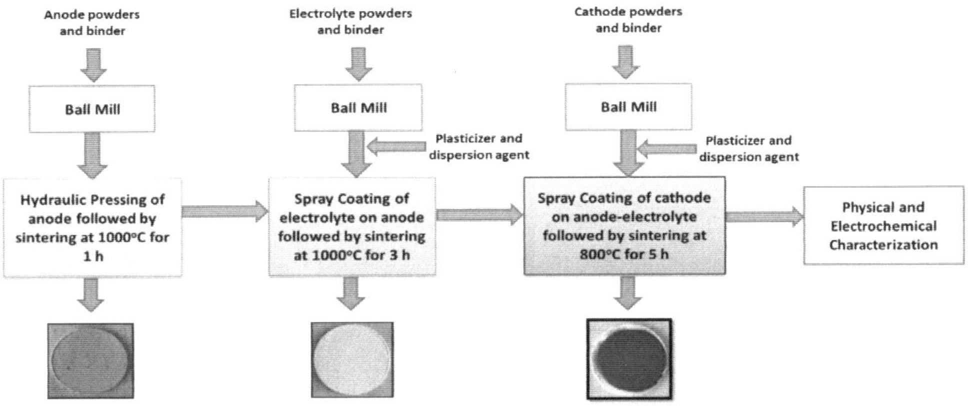

Figure 1. Schematic flow diagram for fabrication of IT-SOFC single cell.

2.3. Microstructure and electrochemical characterization

The single cell microstructure was examined with scanning electron microscope (SEM). This characterization aim to determine thickness and structure of each component. The analysis was done before electrochemical characterization. To observe electrochemical measurement, SS mesh was used as current collector on the anodic and cathodic side. The electrochemical performance of single cell was tested by I-V-P curve and electrochemical impedance spectroscopy (EIS). The electrochemical measurement was carried out by Gamry V3000. The performance of single cell was observed at 500, 600 and 700°C using hydrogen as fuel and ambient air as oxidant.

3. Results and discussion

3.1. Effect of maizena on anode's porosity and PVA concentration on anode morphology

The anode required a porous structure due to fuel diffusion to triple phase boundary (TPB). Maizena was commonly used as pore former. Testing of anode's porosity was done by ASTM C373-88. The result showed that the porosity of NiO-CSZ anode without pore former was 6%. When a 15 wt% of maizena was added, the porosity of NiO-CSZ anode increased to 31%. The requirement of anode's porosity was 20 – 40% because very large porosity weakened a bending strength of anode [10, 11]. A 15%wt of maizena as pore former is sufficient to reach anode's porosity requirement while maintaining a bending strength of anode as support [12]. PVA as a binder is used to increase compact strength of NiO-CSZ anode. PVA concentrations used in this experiment were 3 and 10 wt%. The 10 %wt of PVA made a fragile structure because there are too much mass loss of PVA during sintering. Reduction of PVA concentration is used to avoid high mass loss. The 3 wt% of PVA made a strong and uniform structure of NiO-CSZ anode with the diameter of the anode was 4.4 cm after sintering, as shown in figure 2.

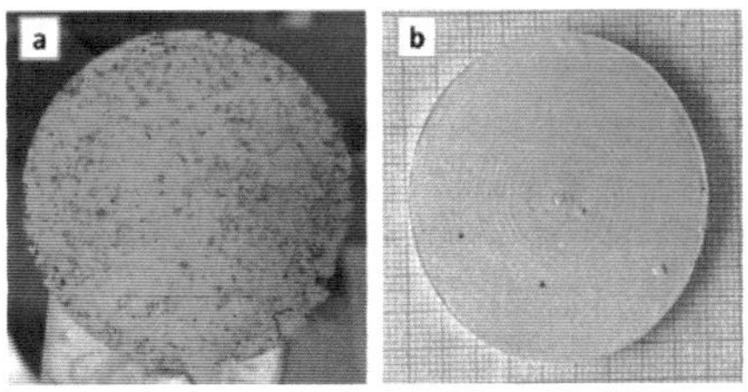

Figure 2. NiO-CSZ anode morphology with (a) 10 wt% (b) 3 wt% of PVA after sintering.

3.2. Effect of PVA phase to electrolyte and cathode morphology

The requirement of electrolyte is dense structure. Production of CSZ electrolyte used two phase of PVA; liquid and solid. The 1 wt% of PEG was added as plasticizer in each PVA. The 1 wt% of liquid PVA made microcrack on surface electrolyte after sintering. The microcrack was formed by a bumpy surface which caused unequal thermal expansion in electrolyte's surface. The 1 wt% of powder PVA and 1 wt% of PEG made the surface electrolyte uniform with the diameter of anode-electrolyte was 4.3 cm after sintering as shown in figure 3.

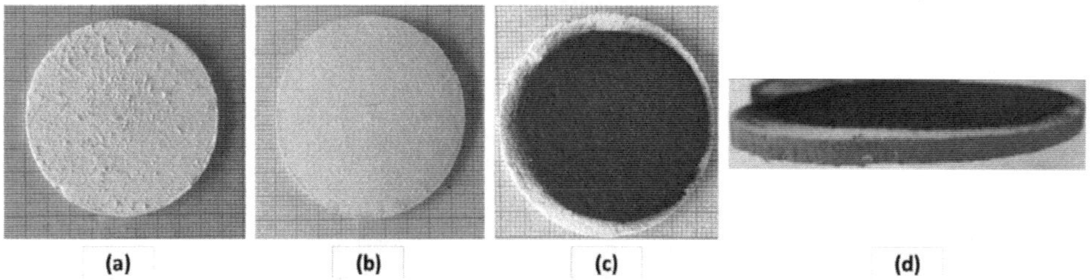

Figure 3. (a) liquid (b) powder PVA effect in CSZ electrolyte fabrication (c) CCZO cathode production on surface anode-electrolyte (d) the single cell.

The composition of PEG and PVA in electrolyte fabrication was used to fabricate CCZO cathode through spray coating. The diameter of CCZO cathode was 4.2 cm which is smaller than anode-electrolyte diameter as shown in figure 3. The purpose was to avoid short circuit between anode and cathode. The single cell was successfully fabricated with 13.2 cm^2 in active area and 1.8 mm in thickness.

3.3. Microstructural characterization

Microstructure of the single cell was shown in figure 4. In figure 4a, NiO-CSZ anode as a support was thicker than the electrolyte and cathode. The thickness of anode, electrolyte and cathode was 1.65 mm, 100 μm and 50 μm, respectively. The electrolyte's thickness was too large because the requirement of electrolyte thickness was 20 – 50 μm to reduce impedance in electrolyte [13]. Figure 4b showed the microstructure of CSZ electrolyte is less dense. It has potential to hydrogen crossover. The microstructure of CCZO cathode and NiO-CSZ anode, as shown in figure 4c and 4d, was porous structure, thus fuel and oxidant can diffuse to TPB in order to conduct oxidation and reduction reaction.

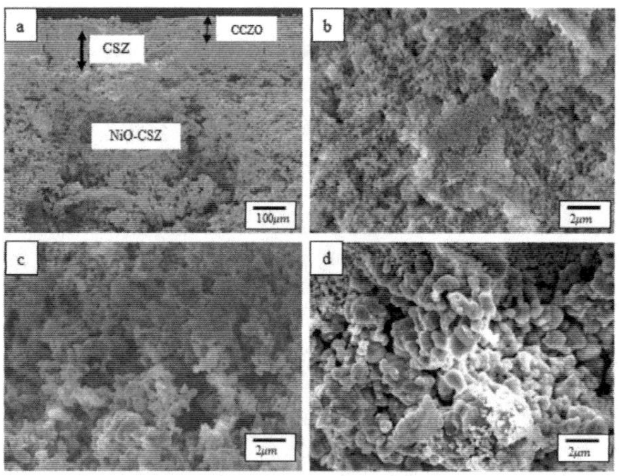

Figure 4. A cross-sectional SEM images of (a) single cell (b) electrolyte (c) cathode (d) anode.

3.4. Electrochemical characteristic of single cell

Electrochemical tests to determine I-V-P curve and the impedance of single cell were performed at 500, 600 and 700°C, respectively. The single cell was tested using hydrogen as fuel with flow rate of 125 sccm and ambient air as oxidant with flow rate of 500 sccm. Figure 5 showed I-V-P curve of IT-SOFC single cell. The single cell had open circuit voltage (OCV) of 0.577, 0.558 and 0.513 V at 500, 600 and 700°C, respectively. The low OCV indicated that structure of electrolyte was less dense as shown in figure 4b [14]. The maximum power density of single cell at 500, 600 and 700°C was 1.4, 2.8 and 10 mW m^{-2}, respectively. The low power density is caused by a thick electrolyte.

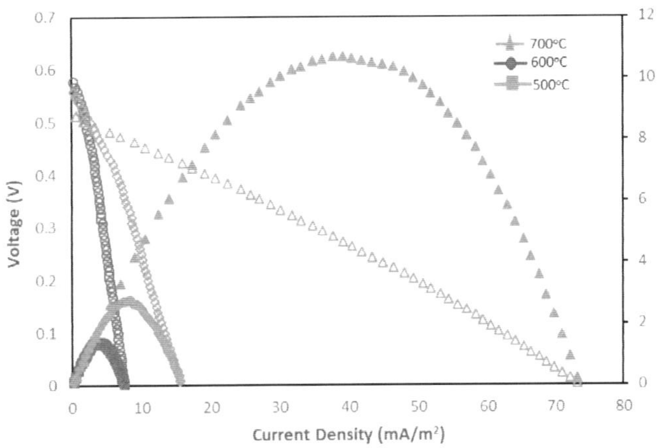

Figure 5. I-V-P curve of IT-SOFC single cell.

Process of charge and mass transfer in SOFC cell was showed by Nyquist plots from EIS measurement. Nyquist plots ilustrate the relationship between imaginary and real impedance. Figure 6 showed Nyquist plost of single cell was measured at OCV condition of 500, 600 and 700oC, respectively. An impedance cell was measured at frequency of 100 kHz – 0.1 Hz.

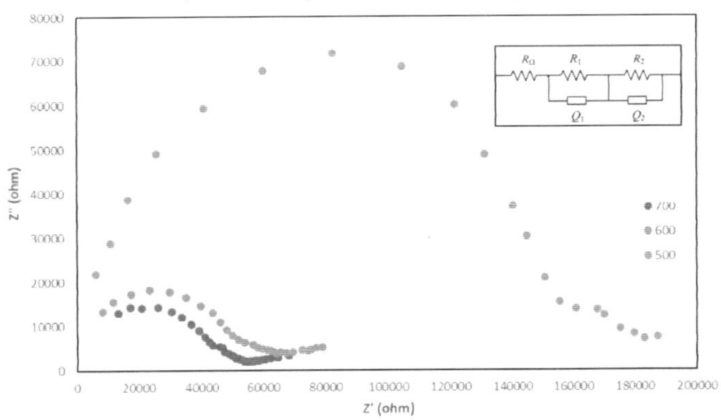

Figure 6. Nyquist plots and equivalent circuit of IT-SOFC single cell.

The first intercept on the real axis at high frequency represents the ohmic resistance (R_Ω) and the second intercept on the real axis at low frequency corresponds to the total resistance of the cell. Polarization resistance (R_p) was determined by two different arcs where one was at high frequency range (HF, ~10^5 Hz) and the other at medium frequency range (MF, ~10^2 – 10^3Hz). The Nyquist plots were fitted using equivalent circuit model of two parallel R and Q or CPE (constant phase element) and one resistor connected in series as shown in figure 6 [15]. The values from fitting was summarize in Table 1.

IOP Conf. Series: Materials Science and Engineering **550** (2019) 012001 doi:10.1088/1757-899X/550/1/012001

Table 1. Summary of values from equivalent circuit fitting of Nyquist plots.

T(°C)	R_Ω (ohm)	R_1 (ohm)	R_2 (ohm)	$R_p = R_1 + R_2$ (ohm)
500	3303	152300	104700	257000
600	3452	44170	38230	82400
700	3624	22750	31920	54670

R_1 correlated with charge transfer process and R_2 related with mass tranfer process [16]. Based on fitting result, the low performance of IT-SOFC single cell was at lower temperature (500°C). The high resistance of charge transfer happened because of a slow electrochemical kintetic of CCZO cathode when low operating temperature. The ohmic resistance was still high due to a thick electrolyte.

4. Conclusion

In this study, the IT-SOFC single cell using Indonesia's resources was successfully fabricated using hydraulic pressing and spray coating method. Maizena as pore former can increase a porosity of anode and 3 wt% of PVA powder made anode's structure strong and uniform surface. 1 wt% of PVA powder and 1 wt% of PEG made uniform surface of electrolyte and cathode even though the thickness of electrolyte was not fulfill a requirement of thin electrolyte fabrication. However, the process is cheap and easy to scale up. The IT-SOFC single cell exhibited the maximum power density of 10 mW m^{-2} at 700°C. The total resistance was still too high because the structure of electrolyte was less dense and too thick. Moreover, the slow electrochemical kinetic of cathode at low temperature caused a high resistance. Advancement of electrolyte thickness and structure will be researched to enhance the performance of IT-SOFC single cell fabricated using Indonesia's resources.

References

[1] Dewan Energi Nasional 2016 Outlook Energi Indonesia 2016 (Jakarta : DEN)
[2] Kirubakaran A, Jain S and Nema R K 2009 Renewable and Sustainable Energy Reviews 13 2430-2440
[3] Abdalla A M, Hossain S, Azad A T, Petra P M I, Begum F, Eriksson S G and Azad A K 2018 Renewable and Sustainable Energy Reviews 82 353-368
[4] Brett D J L, Atkinson A, Brandon N P and Skinner S J 2008 Chem. Soc. Rev 37 1568-1578
[5] Poernomo H, Biyantoro D and Purwani M V 2016 Eksplorium 37 73-88
[6] Habibie S, Wargadipura A H S, Gustiono D, Herdianto N, Riswoko A, Nikmatin S and Clarke S 2017 International Journal of Biomedical Engineering and Science 4 11-23
[7] Muccillo R, Netto R C B and Mucillo E N S 2001 Material Letters 49 197-201
[8] Direktorat Jenderal Mineral dan Batubara 2018 Laporan Kinerja 2017 (Jakarta : Kementerian ESDM)
[9] Goodenough J B and Huang Y-H 2007 J. Power Sources 173 1-10
[10] Wincewiz K C and Cooper J S 2005 J. Power Sources 140 280-296
[11] Yu H J, Park G W, Lee S and Woo S K 2007 J. Power Sources 163 926-932
[12] Yoshito W K, Matos J R, Ussui V, Lazar D R R and Paschoal J O A 2009 J Therm Anal Calorim 97 303-308
[13] Gao Z, Mogni L V, Miller E C, Railsback J G and Barnett S A 2016 Energy & Enviromental Science 9 1602-1644
[14] Kim K-J, Choi S-W, Kim M-Y, Lee M-S, Kim Y-S and Kim H-S 2016 Journal of Industrial and Engineering Chemistry 42 69-74
[15] Li Y, Wang S and Su P-C 2016 Scientific Reports 6 22369
[16] Singhal S C and Kendall K 2003 High Temperature Solid Oxide Fuel Cells : Fundamentals, Design and Applications (Oxford : Elsevier Ltd)

Community behavior analysis of electricity consumption in rural areas (a case study of Batang Regency)

A D Rarasati[1], M D Ardian[1], E A Setiawan[2], I H Mohammad[3]

[1]Civil Engineering Department, Faculty of Engineering, Universitas Indonesia, Indonesia
[2]Electrical Engineering Department, Faculty of Engineering, Universitas Indonesia, Indonesia
[3]Department of Geophysics, Faculty of Mathematics and Natural Sciences, Universitas Padjadjaran, Indonesia

Email: imran@geophys.unpad.ac.id

Abstract. The need for electricity linearly increases despite the lack of accessibility to electric power, especially in rural or downtown area. Since electricity has several positive impacts such as supporting human activities, increasing competitiveness and improving the economy, lack of access to it could lead to an area being underdeveloped. The government expected that procuring a 35,000 Mega Watt generator would overcome this problem. However, from the consumer's perspective, it is important to avoid over or under design during the infrastructure development lead. This study aims to analyze the energy reserve needs from the villagers' own perspectives by determining peak loads, electricity consumption, and the reliability of the power distribution systems. The research took place in Batang Regency, Central Java, Indonesia. It used interviews to deeply explore consumer behavior in energy consumption presented in commonly demanded factors, diversity factors, SAIDI and SAIFI values, and the respondent's understanding of energy saving. The results of this study revealed some unique characteristic behaviors of Batang electricity consumers based on their locations.

1. Introduction

The demand for electricity continuously rises every year as a consequence of economic growth [1] with the largest consumer being the household sector, followed by the industrial, business and public sectors [2] A very close relationship between economic growth and expanding electricity demand shows the energy needs as one of the main components for the development of a nation [3]. The very crucial role of electricity, as shown by Susila and Pribadi (2014), directly influences the constraints in the pace of development and the changes in social patterns such as the emergence of urbanization flows and the lack of opportunities to improve livelihoods as indicated by the low human development index [4].

The electric power indicators of availability are generally expressed in the form of electrification ratio, which is the ratio between the number of electrified consumers compared to the total number of energy consumers [4]. In Indonesia, statistics from the PT PLN (the National Electric Company shows that the outside electrification ratio of Java has reached about 90% of the total demand [5]. Since the provision of equal and sustainable electricity is the main foundation in advancing a nation's economy, the development of reliable electric infrastructure is therefore fundamentally necessary. On the other hand, a comprehensive analysis of electric needs is crucial to prevent under design or over design. Over design condition occurs when the developed infrastructure yields products exceeding its initial planning, while under design is the opposite [6]. Both these conditions should be avoided as they affect the effectiveness of project resource utilization [7].

Two alternative analysis can be conducted to estimate the electricity needs in order to avoid the aforementioned conditions. First, electricity requirement can be calculated based on centralized statistical data, generally provided by the Indonesian Central Statistic Agency and commonly used by policy makers. Another alternative is based on consumers' perception of their electricity needs and how effective such needs have been fulfilled. Since the consumers are directly involved in how electricity is efficiently consumed, their perception will accurately describe the community participation in actively utilizing electricity infrastructure. Both alternatives are ideally carried out simultaneously to produce an optimum model as a framework for the development of electricity infrastructure.

This paper describes the analysis of electricity needs around Batang Regency in Central Java. The main concern of the paper is to explore the behavior of the consumers in a rural area and to directly correlate the behavior to its regional development planning.

2. Research methodology

The survey area was divided into three main regions characterized by their altitude as follow: high altitude (350–700m above sea level/ASL), low altitude (100–350m ASL) and coastal area (below 100m ASL). The electricity infrastructure efficiency level was assessed from public perceptions through the following steps:

1. Initial data collection. Interviews were conducted to the corresponding samples with various socioeconomic backgrounds using a set of questions with statistically satisfying validity and reliability. The questions were designed to profile the respondents' electrical consumption, which is their use of electronic devices daily. The data collected were subsequently processed to determine the estimated electric load profile of the groups of interest (the highland area, the low altitude region and the coastal region). According to Skrotzki and Vopat (1960) [6], there are several factors as follow:
   a. Maximum demand: Maximum load of one group (Watt/Volt Ampere).
   b. Connected load: Load connected in one group (Watt/Volt Ampere).
   c. Demand factor: The ratio of maximum demand with the connected load.
   d. Group diversity factor: The ratio between the sums of individual maximum demands to the group's actual demand at a specific time interval.
2. Secondary data collection. This was obtained from public information provided by the Indonesian Central Statistic Agency to identify the socio historical and economic condition of the areas of interest. The socio historic backgrounds of three interest groups could be described and qualitatively correlated with the electrical consumption profiles.
3. Two other indicators widely accepted to describe the reliability of electricity supply are System Average Interruption Index (SAIDI), and System Average Interruption Frequency (SAIFI) [7]. SAIDI provides an average amount of time per year where energy supply to customers is disrupted, while SAIFI is the number of disrupted experiences per customer per year. SAIDI and SAIFI are calculated as follow:

$$SAIDI = \frac{\sum r_i N_i}{N_T}$$

$$SAIFI = \frac{\sum N_i}{N_T}$$

where

r_i : duration of electricity power off in hours
N_i : the number of consumers power interruption
N_T: total number of consumers

4. The village community's knowledge of electricity was investigated using descriptive statistics and chi square test for homogeneity. All the indicators are subsequently compared to identify the overall use of electricity through a community perspective, yielding a comprehensive and integral picture of the electrical condition of the area of interest and how much electricity needs must be fulfilled in this perspective.

3. Results and discussion

Table 1 shows the average of electricity indicators in all geographic altitude zones in a 24-hour period. The maximum demand is the highest demand of electrical needs in one specific interval time while the connected load is all the electric devices connected to the distribution system. In Table 1, the maximum demand and the total connected load refer to the average of highest demands and whole devices connected to an electrical installation in one house in the area of interest. The highest maximum demand was found in the high altitude area, followed by the low altitude area in second place and the coastal area having the lowest electricity demand on average. On the contrary, the highest connected load was found in the low altitude area and followed by the high altitude area, while the coastal area showed the lowest connected load with a value of almost half of the highest value in all areas of interest. The ratio of maximum demand and the connected load is defined as the factor that describes how much system capacity is required to supply the connected load. The highest demand factor was found in the coastal area with 0.62, followed by the high altitude area with 0.49 and the low altitude area with 0.43 (lowest value). The all connected loads did not operate simultaneously, only partial equipment connected were under operation in a specific time window during the 24-hour period. The average sum of individual time windows of one house is shown in Table 1. The high altitude area had the highest value with the total average sum of 1,567.69 Watt, followed by the low altitude area with 1,395.46 Watt and 1,312.14 Watt in the coastal area. Since the residences all use 900 Watt of power, and assuming only 95 percent of the power could be used while the other 5 percent is dissipated, the actual demand of the system is about 850 Watt. The ratio of sum of individual demand with the actual demand of the system is called Diversity Factor. Table 1 shows that the group diversity factor in the high altitude area is 1.84, followed by the low altitude area with 1.64 and the coastal area with 1.54.

Table 1. The electricity indicators of the research area.

Geographical Altitude	Maximum Demand (Watt)	Connected Load (Watt)	Sum of Individual Demand (Watt)	Actual Demand of the System (Watt)	Demand Factor	Diversity Factor
High	495.44	1,019.69	1,567.69	850	0.49	1.84
Low	460.67	1,071.69	1,395.46	850	0.43	1.64
Coastal	389.47	623.38	1,312.14	850	0.62	1.54

The indicators in table 1 can be further assessed with the parameters indicated in table 2, which shows that the income per capita per area of interest in the highland is the highest, almost one and a half times that of the low altitude area. The lowest income per capita was found in the coastal area with around IDR 13,500,000,- per year, about five sixth of that in the lowland area. Interestingly, the household number in the high area is lower than the number in the low altitude zone, while the income per capita is about one and a half times higher. The data clearly indicated that with a small number of population, the income per capita in the highland area is significantly higher than the two other areas. The coastal area has the lowest population with only about 20,000 households as well as the lowest income per capita.

A direct comparison of the income per capita with maximum demand and connected load from Table 1 reveals some facts around the areas under consideration. With the total load connected about 1,071 Watts, the residents in the low altitude zone have more electric equipment in general than the other two areas. On the other hand, the maximum demand in the lowland area is relatively lower than that in the high altitude area, whereas the demand factor in the high altitude area is relatively higher than that in the lowland area, indicating a more effective use of electric power. Similarly, the same conclusion is seen from the higher diversity factor which indicates the less cost of energy generated for the system. Table 1 electric indicator data revealed the unique facts in the comparison between people in the highland and low altitude areas: despite the relatively higher total connected load in the low land area, it is clear that the high altitude residents could use the equipment more efficiently. Based on per capita income as shown by table 2, the high income highland residents can afford more electric equipment but the data reveals the opposite: with an income of only two-thirds of the highland residents, the lowland residents own more equipment. The interview also explored the residents' knowledge of electricity efficiency and its results in Table 2 shows the same conclusion as that given by the electric parameters and income per capita: the highland residents have good knowledge of the importance of efficiently maintaining electricity, while the lowland residents showed lack of awareness.

The aforementioned approaches in describing people's characteristics in consuming electricity could not be implemented well in the coastal area. Instead of having the highest demand factor (about 0.62), the area's total connected load is almost about 40 percent less than the other two areas in general. This low connected loads signifies the area's economical condition: they only own very few electric equipment compared to other communities. This is in line with the finding that the income per capita of this area is also the lowest. The highest demand factor means that the lowland residents use their equipment effectively for fundamental needs only. Similarly, the knowledge of electricity efficiency is also poor in the coastal area as a direct result of the low income per capita and the low number of connected load. The number of households in the coastal region is about 20,000, and this is the lowest population in the area of interest. The region is less attractive since it has the lowest income per capita as well as the lowest connected load and the lowest diversity factor.

Table 2. The socio economic profiles of the research area.

	High Altitude Area (350-700m ASL)	Low Altitude Area (100-350m ASL)	Coastal Area (0-100m ASL)
Income per Capita per Year	IDR 23,647,059	IDR 16,200,000	IDR 13,571,429
Household number	87,347	130,705	20,140
Knowledge of Electricity Efficiency	Good	Poor	Poor

Table 3 shows the reliability of the electricity distribution in the area of interest. Table 3 clearly shows the coastal area as the area with the most frequent blackouts. Its SAIDI rate is estimated at 114-1,152 hours per customer per year and SAIFI > 48 disruptions per customer per year. A reduced scale time period of power off duration is about 3 hours a day, and based on Poland's category, this region is reaching the "disaster" level. The low altitude land occupies the second place in terms of SAIDI with about 12-36 hours per customer per year and SAIFI 12 disturbances per customer per year while the highland area has the lowest disturbances with SAIDI less than 12 hours per customer per year and SAIFI 12 disruptions per customer per year. The reliability clearly shows that the highland area has a robust electric supply, while the area of low altitude has relatively low stabilization and the coastal area has a very unpredictable power supply. A very interesting fact is clearly displayed here: the low population density of high altitude region shows a good level of electricity consumption, higher income per capita, high distribution reliability and high level of energy saving awareness. It seems like this awareness drives them to maintain more effective behavior in consuming electricity, which consequently leads to high power supply reliability. The residents in the low altitude zone have the highest population and high connected load but with lower demand factor, lower income per capita, and relatively less robust power supply stabilization. The coastal area displays less efficient profile with a small population, less income and high disturbances in power supply.

Table 3. The SAIDI and SAIFI of interest area.

Geographical Altitude	SAIDI (hours/consumers/years)	SAIFI (disruptions/consumers/years)	Duration of disruption (hours a day)	Category according to Poland's Standard
High	114-1,152	>48	>3	Very long-disaster
Low	12-36	12	1-3	Long
Coastal	<12	12	<1	Short-long

The low altitude community living near Batang's activity centers such as the Pantura Highway, which is the most important highway for national logistics, ideally has convenient access to information for productive activities including how to effectively maintain electric power. On the other hand, people in the high altitude area has more effective electrical consumption behavior which leads less power supply fewer disturbances. Lastly, the lack of electricity infrastructure in the coastal area (marked by the very high value of SAIFI and SAIDI) causes this region to be less attractive to live in (reflected by the low population) and results in an urgent need to build said electricity infrastructure.

4. Conclusions

This research has been successfully carried out up until the public perceptions data collection stage about electricity efficiency. The electricity usage pattern in Batang Regency shows unique characteristics where the highland residents showed a high level of electrical energy consumption accompanied by the highest level of distribution network reliability. These two factors affect one another: the high network distribution reliability (thus a relatively very few blackouts in a year) is directly related to electricity consumption in the area. The highland population is relatively lower in number than that in low altitude area but those living in low altitude zone frequently access information since the main highway of national transportation and logistics (the Pantura Highway) is located in this zone. This fact reinforces the assumption that existing power production has not been able to cover the electricity needs of the entire population of Batang. This assumption is also strengthened by the high value of SAIFI and SAIDI in the coastal and low altitude regions. As the low number of the highland population has a relatively higher income than those in other areas, the power supplied here can fulfill the electricity needs. On the contrary, the high number of lowland population and its high electricity demand cannot be adequately met by the current supply as shown by the relatively high SAIDI and SAIFI values. Lastly, the coastal area, with its less efficient profile compared to the others, should receive first priority with regards to building electricity infrastructures.

The high productivity of the highland population is accompanied by a relatively high level of awareness with regards to the importance of saving electricity. This fact is also supported by the high public access to information about energy saving efforts. At the same time, highland communities show good knowledge in utilizing electricity for agricultural productivity as their main livelihood. There is an imbalance in meeting the extraordinary electricity needs between coastal areas and areas with higher altitude. In the lowland area, the presence of Pantura Highway as the center of attraction causes many residents to choose to live there, leading to an increase in population and subsequently an increase in electricity consumption. Understandably, the existing power installation is unable to meet the demands of the area's entire population.

Meanwhile the coastal area is still lacking in electricity infrastructure, causing its lack of attractiveness as reflected by its very small population. This study managed to explore and show the current urgency of developing new electricity infrastructure in Batang district, especially to meet the needs of its densely populated areas at low altitudes and coastal areas. However, such infrastructure development must be accompanied by an increase of awareness regarding the importance of energy saving. The existence of Pantura Highway located lowland has indeed triggered a high level of public interest to living in the area, but this interest is not accompanied by a high level of awareness with regards to energy saving and the use of electricity towards productive activities.

Acknowledgment

The authors would like to express their gratitude to the Directorate General of Higher Education the Republic of Indonesia through PDUPT 2018 Grant for supporting this research and its dissemination.

References

[1] Estiri H *2014 The impacts of household behaviors and housing choice on residential energy consumption* (Seattle: University of Washington)
[2] Ministry of Energy and Mineral Resources 2017 *Statistik ketenagalistrikan 2016* Edition No. 3 Year 2017 (Jakarta: Indonesia Government Printing Office)
[3] Karanfil F and Li Y 2015 *Energy Policy* **82** 264–77
[4] Susila I M A D and Pribadi D R 2014 *Ketenagalistrikan dan Energi Terbarukan* **13** 61–68
[5] PT PLN (Persero) 2017 *Rencana Usaha Penyediaan Tenaga Listrik (RUPTL) PLN 2017-2026 (Ministry of Energy and Mineral Resources Decree No. 1415 K/20/MEM/2017)* (Jakarta: Indonesia Government Printing Office)

[6] Duan Y and Narendra N C 2015 *3rd International Conference on Applied Computing and Information Technology/2nd International Conference on Computational Science and Intelligence* 34–9
[7] Halpin D W and Senior B A 2011 *Construction Management* 4th Ed. (New York: John Wiley & Sons, Inc.)
[8] Skrotzki B GA and Vopat W A 1960 *Power station engineering and economy* (New York: McGraw-Hill)
[9] Kornatka M 2017 18th International Scientific Conference on Electric Power Engineering (EPE) (Ostrava: VŠB-Technical University of Ostrava and Brno University of Technology) p 1–4

IOP Conf. Series: Materials Science and Engineering **550** (2019) 012003 doi:10.1088/1757-899X/550/1/012003

Metal oxides, metal sulphides and hybrid cathode materials for aluminium ion batteries – a mini review

M Vanitha[1], I Made Joni[1,2], B M Wibawa[3] and C Panatarani[1,2]

[1] Nanotechnology and Graphene Research Center, Universitas Padjadjaran, Jl. Raya Bandung-Sumedang KM 21, Jatinangor, 45363, Jawa Barat, Indonesia.
[2] Department of Physics, Universitas Padjadjaran, Jl. Raya Bandung-Sumedang KM 21, Jatinangor, 45363, Jawa Barat, Indonesia.
[3]Departement of Electrical Engineering, Universitas Padjadjaran, Jl. Raya Bandung-Sumedang KM 21, Jatinangor, 45363, Jawa Barat, Indonesia.

E-mail: imadejoni@phys.unpad.ac.id

Abstract. Immense progress is required in the field of energy conversion and storage to achieve cost-effective high-performance batteries which could meet future energy demands. It also becomes mandatory to utilize the abundant materials from the earth's crust, along with safe conversion and storage of electrochemical reactions. In this perspective, aluminium ion batteries (AIBs) could be a viable alternative for the conventional lithium-ion batteries (LIBs). In this mini-review, we focus on the challenges and the latest growth related to the cathode active materials (metal oxides, metal sulfides and other hybrid systems) in AIBs. The development of highly efficient, low cost and safe energy storage systems based on aluminium ion batteries is much explored. Still, commercialization remains a theoretical one, hence this review highlights the significance of the developed materials and their shortcomings that need to be addressed for the commercialization of aluminium ion batteries in the future.

1. Introduction

The lithium-ion technology, with its high specific energy and power density, is the most extensively used energy storage systems in portable electronic devices over the past 20 years [1- 3]. Although LIBs have dominated the consumer electronics market, the unavailability of lithium, flammability and high cost lead to the transformation of the prevailing research focus towards an alternative technology [3, 4]. To accomplish long-term sustainability, lightweight, toxicity and high cost the exploration of an alternative energy storage system can be drifted to materials like aluminium, magnesium, sodium and potassium. Special attention is focussed on AIBs owing to the abundance of Al, low reactivity, easy handling with a high volumetric capacity of 8046 mAh cm^{-3} [4] and gravimetric capacity of 2980 mAh g^{-1}. Figure 1 compares the volumetric and gravimetric capacity of various (Al, Li, Mg, Ca, Zn, Na, K) battery systems. From the comparison, it can be ascertained that AIBs have superior capacity compared to the commercial LIBs and other battery systems.

In fact, rechargeable aluminium ion batteries are considered as a notable candidate for the future energy demands. Since the three electron transfer during the electrochemical charge-discharge process in AIBs provides an efficient storage capacity than LIBs [5]. Added to this the fabrication of electrochemical cell using aluminium can be done in an open environment which uplifts the safety of aluminium ion battery systems [6]. All of the mentioned batteries operates with the basic principle of intercalation/de-intercalation mechanism, where the cathode delivers the positive potential and the anode generates the cell voltage [7]. The first voltaic cell proposed by Sargent, employs Al anode, carbon as a cathode and aqueous caustic alkaline solution with dissolved zinc oxide [8]. The first secondary aluminium battery was reported by Holleck and Giner with the use of chlorine as a cathode, Al metal anode in an electrolyte composed of $AlCl_3$, KCl, NaCl [9].

Many improvements in the anode and cathode material, electrolytes and additives were investigated and numerous novel results have also emerged [7]. Progress in the cathode materials was stimulated in the last decade and several materials have been assessed to obtain AIBs with good efficiency, safety and low cost. Based on the results obtained with the development of unique materials and enhanced volumetric and gravimetric capacity relative to LIB systems, AIBs could be the future generation energy storage systems. However, the commercialization and development of AIBs are hindered by lack of efficient cathode materials to produce batteries with high specific capacity and stable cycle life [1, 7]. Several cathode materials have been reported so far, despite that still the commercialization of AIBs remains underdeveloped. The foremost reason being the complex issues associated with the cathode materials such as low discharge voltage [11, 12], poor cycle life [5, 13], low coulombic efficiency, unstable discharge capacity, volume expansion [14, 15] and unstable cycling ability [15].

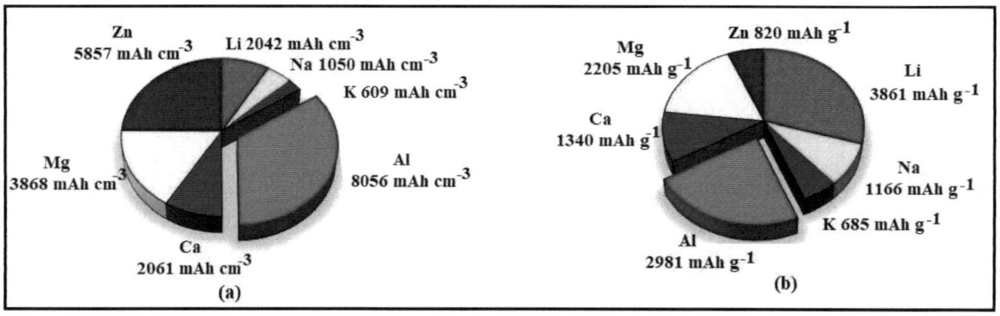

Figure 1. Comparison of the volumetric and gravimetric capacity of Al, Li, Mg, Ca, Zn, Na, K battery systems.

This paper is focused on the recent development of the cathode materials based on metal oxides, metal sulphides and hybrid materials with carbon, graphene, (carbon nanotubes) CNTs for AIBs. The cathode materials based on graphite is not discussed in this review since it is a vast topic and entails an extravagant study. In order to have a profound understanding of the disputes associated with the current cathode materials and suggested requirements of the future cathode materials for successful commercialization will also be addressed.

2. Cathode materials

The cathode materials for AIBs were developed by taking into consideration the materials employed in LIBs since the mechanism involved in both the cases are similar. The cathode active material in the battery system is one of the key parameters in building highly efficient battery systems with high power and energy density for commercial applications [7, 17]. The development of AIBs is hampered by the non-availability of cathode materials for practical applications and still, the outline and the principles of cathode materials remains un-determined [1]. Over the past decade, various cathode materials were proposed for AIBs which is discussed in detail.

2.1. Metal oxide-based cathode materials

Wang et al in 2013, for the first time, attempted the use of vanadium oxides as electrode materials for aluminium ion batteries. The tunnel structure of VO_2 and the suitable electrochemical potential window lead to its use in batteries as a cathode material. The VO_2 nanorods prepared by hydrothermal treatment has a specific area of about 37.3 $m^2 g^{-1}$, responsible for shorter diffusion length. The cyclic voltammetry studies were performed in the potential ranging from 0.0 to 0.9 V and at a current density of 50 mA g^{-1}, the discharge capacity of 165 mAh g^{-1} is achieved for an initial cycle which is declined to 116 mAh g^{-1} after 100 cycles. This is the initial work based on aluminium ion batteries sustaining the capacity for 100 cycles [17]. Hence, Wang et al. initiated the use of metal oxides as an electrode material in aluminium ion batteries which led to several other developments on metal oxides.

Other oxides of vanadium i.e., V_2O_5 is also considered to be a flexible host material for batteries which works on the principle of intercalation/de-intercalation mechanism such as lithium, sodium, aluminium and magnesium battery systems [18]. The reason for its use in batteries is the layered structure of V_2O_5, water absorption ability and high redox potential of V^{5+}. The electrochemical behavior of V_2O_5 is influenced by the intercalated water which has a profound repercussion in AIBs relative to the other battery systems [5, 20]. Gonzalez et al., in 2016, reported the electrochemical nature of V_2O_5 xerogel in aqueous $AlCl_3$ electrolyte using Swagelok type cells. The smaller water content present in V_2O_5 camouflaged the repulsion occurring between oxygen within the $Al_xV_2O_5$ during the intercalation process. The chemically intercalated aluminium ions with positive charge balance the negative charge on vanadium oxide, the results also suggest that V_2O_5 may suffer from limited intercalation with subsequent release of oxygen. It was also reiterated that there is no possibility for electrolyte decomposition and corrosion in the prescribed voltage range using V_2O_5 as an electrode material [19].

With all these different oxides of vanadium being reported as one of the efficient cathode material for aluminium ion batteries still, the controversy exists regarding the contribution of V_2O_5 to the capacity of those batteries [20]. This leads to a comprehensive investigation of the storage mechanism of Al^{3+} ions using metal oxide based electrode materials. This was addressed by Gu et al., in 2017, by preparing V_2O_5 nanowire with a diameter of about 80-120 nm by hydrothermal treatment [21]. To validate that the capacity obtained using V_2O_5 electrode doesn't arise from the current collector, Galvanostatic cycling was performed on bare nickel foam as a current collector without the addition of V_2O_5, which unveils zero capacity [10]. The specific capacity achieved using V_2O_5 nanowire is 107 mAh g^{-1}, quite low compared to the other reported results in V_2O_5 [21]. A two phase-transition reaction and intercalation of Al^{3+} ions into the metal oxide with the formation of amorphous layers was observed. This study provides a new pathway and detailed manifestation, substantiating the participation of V_2O_5 in the electrochemical reaction which gives rise to the obtained capacity.

The anatase form of TiO_2 being nontoxic and chemically stable is widely used in several energy storage devices [23, 24] and self-cleaning applications [25, 26]. Liu et al in 2012 synthesized TiO_2 nanotube array film with a thickness of about 14 μm by anodizing the metallic titanium foil. The nanotube array promotes an exceptional contact between the electrode and the electrolyte along with diminished ion diffusion with high electronic conduction. The intercalation process is highly dependent on the anion from the electrolyte accompanying Al^{3+} ions. The discharge capacity of 75 mAh g^{-1} was acquired at a current density of 4 mA cm^{-2} and TiO_2 nanotube offers a boosted discharge capacity compared to the anatase TiO_2. Another reason for the improved performance is the nanotube structure, facilitating fast transport of electrons and a short ion diffusion path. The aluminium storage is calculated from the highest discharge capacity and it was observed to be 0.074 mol per anatase. The reason beyond the reported discharge capacity is owing to the reversible electrochemical redox reaction of Ti^{4+}/Ti^{3+} during the intercalation process [12]. But TiO_2 nanotube suffers from the formation of oxide film on the Al anode surface and corrosion, hampering the redox reactions which lowers the cell potential [6].

Based on the previous reports on TiO_2 as an electrode material for LIBs, He et al, fabricated black anatase phase TiO_2 nano leaves for rechargeable aluminium ion battery and compared the results with the commercial white anatase phase TiO_2. The discharge capacity of commercial TiO_2 was 77.2 mAh g^{-1} and for the TiO_2 nano leaves initial discharge capacity of 278.1 mAh g^{-1} was reported, which is approximately 4 folds higher than the former. The irreversible capacity loss in TiO_2 nano leaves is only 2.84 % with excellent reproducibility and smaller capacity fading even after 300 cycles. The reason behind this superior performance of nano leaves is its shorter diffusion length, larger interfacial area and lower polarization. It is noteworthy to mention that the discharge capacity and the cycling stability obtained in this work is far superior when compared to other TiO_2 reports available for AIBs [5]. The summary of metal oxide based cathode materials and its initial discharge capacity is tabulated in table 1. Only a few reports are available on the development of metal oxide based electrode materials for AIBs unlike LIBs, since in the reported work its cell voltage, efficiency and cycle life were not much pronounced even though they show a higher capacity. Hence much focus is directed towards metal sulfides and other hybrid systems.

Table 1. Summary of metal oxide based cathode materials and its initial discharge capacity.

S. No.	Cathode Material	Electrolyte	Discharge capacity (mAh g-1)	Ref.
1.	V2O5 nanowire	AlCl3+ [EMIM]Cl	305	[21]
2.	VO2	AlCl3+ [BMIM]Cl	116	[17]
3.	V2O5 xerogel	AlCl3+ deionized water	130-250	[19]
4.	TiO2	Aqueous AlCl3	75	[12]
5.	TiO2 nanoleaves	Al(NO3)3	278	[5]
6.	Commercial TiO2	Al(NO3)3	77	[5]

[EMIM]Cl = 1-ethyl-3-methylimidazoliumchloride, [BMIM]Cl = 1-butyl-3-methylimidazoliumchloride

2.2. Metal sulfides based cathode materials

Among transitional metal sulphides, NiS is considered as an efficient electrode material for lithium ion batteries because of its good conductivity, theoretically high capacity and excellent catalytic activities [26]. Related to its use in lithium-ion batteries, hexagonal NiS nanobelts with length around 10-25 μm and size of 50–100 nm were synthesized by hydrothermal method and used as an electrode material for aluminium ion batteries. Pouch cells were assembled using NiS nanobelts as a cathode and high purity Al as an anode, the discharge capacity of 104.7 mAh g^{-1} was attained for the first cycle, NiS with nanobelt structure eases Al^{3+} ions diffusion. With further increase in the cycles, discharge capacity increases and at 100th cycle the discharge capacity of 104.4 mAh g^{-1} was accomplished at a current density of 200 mAh g^{-1}. Although, the synthesized NiS nanobelts was able to retain 90% capacity over 100 cycles it endeavors from low cell voltage hindering its use for the practical system [27]. Advancement is made by means of integrating graphene with Ni_3S_2 in 2016 by Wang et al. The discharge capacity of 295 mAh g^{-1} which slowly decreased for consecutive cycles with a coulombic efficiency of 84.3%. Although the coulombic efficiency is low, there exists an enhanced reversible charging and discharging process. The integration of Ni_3S_2 sheets with graphene matrix can be reaffirmed by the low internal resistance of about 10Ω [28].

TiS_2 and cubic $Cu_{0.31}Ti_2S_4$ are also encountered as cathode materials for AIBs. The electrochemical intercalation of Al in titanium sulfides as the electrode was experimented at both room temperature and at 50°C since in the latter temperature an advanced capacity and distinctive charge-discharge characteristics were observed. The total capacity of TiS_2 with a layered structure is 70 mAh g^{-1}, which was reported to be gradually uplift upon the consecutive cycles. Although the behavior of TiS_2 in room temperature and 50°C are similar, the capacity is more at higher temperature due to the better reaction kinetics. The author also hypothesized that with both the sulfides based cathode high capacity was not accomplished because of the strong coulombic attraction between aluminium and sulphide ions [29]. Hence, the future studies should focus on improving Al^{3+} ions transport by doping of ions in the crystal structures.

A new class of electrode material which is worth mentioning is based on chevrel phase molybdenum sulphide (Mo_6S_8), used in rechargeable magnesium-ion batteries as a cathode material for the first time by Aurbach and co-workers [30]. The unique structure of Mo_6S_8 is utilized initially by Linxiao et al. in

2015 as an electrode material for aluminium ion battery using $AlCl_3$-[BMIM]Cl electrolyte. The discharge capacity of 148 mAh g^{-1}, declined to 70 mAh g^{-1} for the 2nd cycle and stabilized after 50 cycles exhibiting promising cycle stability [31]. The low specific capacity may be due to the high molecular weight of Mo_6S_8. Recently, MoS_2 a chalcogenide from the transition metal family with microspherical morphology was synthesized by hydrothermal route and employed as a cathode material in AIBs. The discharge specific capacity of 253.6 mAh g^{-1} was reported for a current density of 20 mAg^{-1} and the charge is stored between the electrode and electrolyte interface [32]. Further, the detailed mechanism for the intercalation and de-intercalation of Al^{3+} ions to the electrode material during the charging/discharging process was elaborated in detail with the crystal structure. The coulombic efficiency of MoS_2 as cathode material is low, the reason being the phase transition of MoS_2 and formation of solid electrolyte interface (SEI) on the surface of the electrode [33]. Very recently $Al_{4/3}Mo_6S_8$ was developed by Agiorgousis et al., which could function as a promising material for energy generation and storage, i.e., act as a Janus material (dual function) [34].

Iron disulphide has gained consideration in LIBs as a cathode active material because of its high theoretical capacity and non-toxic nature [35] and hence it has been used extensively in lithium and sodium batteries [36]. In terms of sulfides, very recently the first attempt of employing FeS_2 as cathode active material was exemplified by Koura et al., operating aluminium rechargeable battery systems at a low temperature of about 25°C and an elevated temperature of about 240°C respectively [37]. Based on these studies FeS_2 was used as electrode material for rechargeable aluminium batteries operating at 55°C and the possible charge/discharge reaction mechanism was illustrated by Mori et al in 2016. The theoretical capacity of FeS_2 as an electrode material is 447 mAh g^{-1} and in this reported study about 85% was achieved, the smaller loss in the capacity may originate from the unreacted FeS_2 in the electrochemical cell. It is also suggested that a phase transition from FeS to FeS_2 phase may occur during the charging process [11]. The summary of metal sulfphide based cathode material and its initial discharge capacity is given in table 2.

Table 2. Summary of metal sulfide-based cathode materials and its initial discharge capacity.

S. No.	Cathode Material	Electrolyte	Discharge capacity (mAh g^{-1})	Ref.
1.	NiS nanobelts	AlCl3+ [EMIM]Cl	105	[27]
2.	TiS2	AlCl3+ [BMIM]Cl	70	[29]
3.	Cu 0.31Ti2S4	AlCl3+ [BMIM]Cl	25	[29]
4.	Mo6S8	AlCl3+ [BMIM]Cl	148	[31]
5.	MoS2	AlCl3+ deionized water	254	[33]
6.	FeS2	Aqueous AlCl3	390	[11]

2.3. Other hybrid systems

When the electrode material development for AIBs is focussed on metal oxides and metal sulfides, the first attempt in developing copper hexacyanoferrate (CuHCF) as a cathode material for aqueous Al-ion battery was attempted by Liu et al. in 2014 [38]. Prussian blue analogs (PBAs) have been much explored as battery electrodes and as an electrochromic material [39] and CuHCF and nickel hexacyanoferrate (NiHCF) with a structure similar to PBAs also establish electrochemical properties [40]. In 2013, Reed et al., developed CuHCF cathode and tested its electrochemical performance in an organic aluminium trifluoromethanesulfonate (aluminium triflate) dissolved in diethylene glycol dimethyl ether (diglyme) as the electrolyte. An initial discharge capacity of 60 mAh g^{-1} was attained and gradual fading of the discharge capacity was noticed after 10 cycles [41]. PBAs is expected to have the high-rate capability, long cycle life and good electrochemical behavior [42], the synthesized CuHCF also has the face-centered cubic Prussian blue structure and expected to display a similar electrochemical nature. The efficiency of this newly developed system is claimed to carry more charge in AIBs because of the trivalent Al ions compared to the other monovalent and divalent systems reported. The developed CuHCF electrode in aqueous aluminium sulfate reveals a discharge capacity of 63 mAh g^{-1} at a current density of 50 mA g^{-1}. The specific capacity depends mainly on the Fe^{2+}/Fe^{3+} redox couple in CuHCF electrode material [38]. The low discharge capacity using CuHCF as a cathode in the both aqueous and organic electrolyte may be owing to the cathodic and anodic passivation.

$Cu_{0.31}Ti_2S_4$ as cathode material with spinel structure delivered reversible capacity around 25 mAh g^{-1} substantiating that TiS_2 can function as a better cathode material than $Cu_{0.31}Ti_2S_4$. Upon subsequent cycles the capacity is found to diminish from the first cycle, the reason being the rigid and less open crystal structure of cubic $Cu_{0.31}Ti_2S_4$ thwarting the Al intercalation [29]. A novel electrode material $Cu_{2-x}Se$ nanorods were developed in an acidic ionic liquid electrolyte by Jiang et al. in 2018. The discharge capacity of 260 mAh g^{-1} was obtained at a current density of 50 mA g^{-1}, upon further cycles, the capacity sustained approximately to 245 mAh g^{-1} with 98% coulombic efficiency. It is speculated that the ultrahigh electronic conductivity of $Cu_{2-x}Se$ nanorods (340 S cm^{-1}) assist in stimulating the capacity with high retention in coulombic efficiency [17, 44]. Added to this, the mechanism pertaining to the intercalation/de-intercalation of Al^{3+} ions during charging and discharging process was also validated. The redox reaction depends on the interconversion between $Al_2Cl_7^-$ and $AlCl_4^-$ ions, where $Al_2Cl_7^-$ ions decompose to $AlCl_4^-$ and Al^{3+} during the charging process and the reverse occurs during discharge. The structural change of $Cu_{2-x}Se$ nanorods in each cycle with the intercalation and de-intercalation of $AlCl_4^-$ ions is a quasi-reversible process [16].

A new kind of hybrid system by combining the chemistry of lithium ($LiFePO_4$) and aluminium in a single device was developed by and could be used for grid and stationary applications [44]. $LiFePO_4$ is one of the most widely used cathode materials till date in LIBs, owing to its stability, low cost, safety and durability [45]. Based on the works reported in LIBs with $LiFePO_4$ as cathode active material, hybrid battery with an aluminium anode, $LiFePO_4$ as a cathode and acidic ionic liquid (EMIMCl) was attempted which gives an initial capacity of 160 mA h g^{-1}. A special note to be mentioned in this hybrid system is there is no requirement of lithium salts since $LiFePO_4$ will preserve the required electrochemical reactions during the charging process for the successive cycles. The reversible capacity gradually decreases to 144 mA h g^{-1} after 10 cycles which is further declined to 99 mA h g^{-1} after 20 cycles. Lithium intercalation was confirmed as the foremost reaction occurring at the cathode in the course of the discharge process, except the observed voltage which is below the theoretical value challenging further insight into this [44]. Certain discrepancies were left for the present-day researchers, to have a better understanding of the reasons behind these and to develop electrode materials which could overcome these disputes.

Micro spherical CuS dispersed with acetylene black nanoparticles was synthesized by Wang et al, acetylene black will facilitate the charge transfer process. The integrated CuS@C based electrode was used as a cathode for aluminium ion batteries in $AlCl_3$/[EMIM]Cl ionic liquid electrolyte. The theoretical capacity of CuS@C as calculated from Faraday's law is 280.3 mAh $g^{-1}$and practically, the capacity of about 240 mAh g^{-1} was achieved [46]. On subsequent cycles, the discharge capacity declined which is due to the formation of solid electrolyte interface and irreversible capacity loss [47]. When the charge/discharge study was conducted for 20 consecutive cycles, the discharge capacity of 100 mAh g^{-1} was acquired with similar discharge curves for all the cycles. Nearly 100% coulombic efficiency was achieved using integrated CuS@C as cathode material in AIBs which could deliver a specific capacity of 90 mAh g^{-1} [46]. Another type of carbon coated cathode material for AIBs is reported by Jiang et al., Li_3VO_4 coated with carbon by the spray drying process. The initial discharge capacity of 137 mAh g^{-1} attained at a current density of 20 mA g^{-1}, added to the development of a new material as cathode the authors have also targeted on figuring out the charge-discharge mechanism. Briefly, in the anode, the battery undergoes deposition and dissolution of metallic aluminium and insertion/de-insertion of Al^{3+} ions in the cathode. The amorphous V_2O_5/C composite was prepared by a zero-emission technique which reveal an initial discharge capacity of 150 mAh g^{-1} when tested as a cathode material for AIBs. It is noteworthy to mention that the valence state of vanadium changes during the charging/discharging process [4]. Although the incorporation of carbon is expected to enhance the conductivity and in turn the capacity of the electrode material, the results observed in this case is less compared to the other oxides of vanadium as discussed in table 1.

Graphene and graphene-based materials have attracted attention in the field of energy storage materials owing to their unique properties, making it a promising candidate for electrochemical energy devices [47, 48]. There are many reports available with the use of graphite as an electrode material for energy storage in ALBs and even offers high charge/discharge voltage [50], the role of graphite as an electrode material is not dealt in the present review as mentioned already. Based on the earlier studies on graphene/graphene-based materials, Ni_3S_2/graphene microflakes was developed and fabricated to study its electrochemical properties. The discharge capacity of 350 mAh g^{-1} was achieved which declined down to 60 mAh g^{-1} after 100 cycles sustaining a coulombic efficiency of 99% [51]. An alternative material for the current collector (tantalum) was also employed to overcome the poor electrochemical stability of metallic current collectors used so far [52]. Ni_3S_2/graphene electrode exhibits a stable capacity from 10th to 30th cycle which may be due to the stabilization of the other side reactions [53] and even at higher current density, the capacity lies between 235 to 300 mAh g^{-1} [51]. Cobalt sulfide encapsulated with carbon nanotube (CNT) and carbon nanofiber film (CNF) (Co_9S_8@CNT-CNF) as a new electrode material was developed recently which reveal a high discharge capacity of 315 mAh g^{-1}. The composite with its porous structure enables more active sites leading to an enriched kinetic process in AIBs [54].

Hu et al., attempted a novel electrode material for AIBs by preparing freeze-dried reduced graphene oxide (rGO) supported SnS_2 as the cathode active material. Among the other hybrid materials developed so far, SnS_2/rGO offers the highest capacity, because of the larger surface area of SnS_2 and highly conductive rGO matrix which brings about improved ion-electron diffusion pathway. The initial discharge capacity of SnS_2/rGO was 392 mAh g^{-1} which is high compared to bare SnS_2 with a capacity of 246 mAh g^{-1}. This clearly indicates that graphene in the cathode material plays an imperative role in facilitating the transportation of ions [55]. Figure 2 shows the comparison of initial discharge capacity with discharge capacity after cycles for the various hybrid systems. It is clear from the figure that hybrid systems developed using graphene, CNT depicts initial discharge capacity over 300 mAh g^{-1} which has declined to 87 mAh g^{-1} and 60 mAh g^{-1} for CoS_2/CNT and Ni_3S_2/graphene respectively. In the case of Co_9S_8@CNT-CNF hybrid system even after 200 cycles the discharge capacity of 297 mAh g^{-1} maintaining efficiency of 94% from the initial discharge capacity. It was reported that even after 6000 cycles the discharge capacity of 87 mAh g^{-1} [54]was achieved, which is high compared to the discharge capacity of graphene/CNT based hybrid systems for 100 cycles.

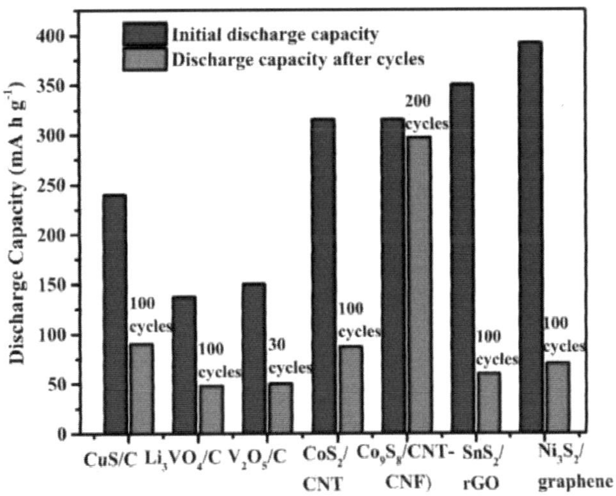

Figure 2. Comparison of discharge capacity initial and after cycles for various hybrid systems.

3. Future perspectives

In this mini-review, the cathode materials based on metal oxides, metal sulfides and hybrid systems based on the integration of metal oxides/metal sulfides with carbon-based materials (carbon, graphene, CNTs) and CuHCF used in AIBs have been summarized. It can be witnessed that the discharge capacity value reported in this review were diverse in all the cases, which establish that capacity depends on the cathode material engaged in each case. Based on all these studies it is suggested that researchers need to develop a cathode material with long cycle life, enhanced capacity, no side reactions, long stability, good coulombic efficiency and absence of structural disintegration during the intercalation and de-intercalation process. Novel graphene-based composite materials can overcome the mentioned aspects together with progress in the electrolyte and aid in the commercialization of AIBs which is still in the inception. Detailed analysis of the materials developed based on graphene its intercalation/de-intercalation mechanism, reaction kinetics, diffusion process should be addressed which should make up with the limitations of the already existing cathode materials.

References

[1] Zafar Z A, Imtiaz S, Razaq R, Ji S, Huang T, Zhang Z, Huang Y, Anderson J A 2017 *J. Mater. Chem. A.* **5** 5646–60
[2] Bruce P G, Scrosati B, Tarascon J-M 2008 *Angew Chemie Int Ed* **47** 2930–46
[3] Li Y, Dai H 2014 *Chem. Soc. Rev.* **43** 5257–75
[4] Chiku M, Takeda, Matsumura S, Higuchi E, Inoue H 2015 *ACS Appl Mater Interfaces* **7** 24385–9
[5] He YJ, Peng J F, Chu W, Li Y Z and Tong D G 2014 *J. Mater. Chem. A.* **2** 1721–31
[6] Elia G A, Marquardt K, Hoeppner K, Fantini S, Lin R, Knipping E, Peters W, Drillet J F, Passerini S and Hahn R 2016 Adv. Mater. **28** 7564–79
[7] Ambroz F, Macdonald T J, Nann T 2017 *Adv. Energy Mater.* **7**(15) 1602093
[8] Voltaic cell, 1949. https://patents.google.com/patent/US2554447A/en (accessed August 27, 2018)
[9] Holleck GL and Giner J 1972 *J. Electrochem. Soc.* **119** 1161
[10] Wang H, Bai Y, Chen S, Luo X, Wu C, Wu F, Lu J, Amine K 2015 *ACS Appl. Mater. Interfaces* **7** 80–84
[11] Mori T, Orikasa Y, Nakanishi K, Kezheng C, Hattori M, Ohta T and Uchimoto Y 2016*J. Power Sources* **313** 9–14
[12] Liu S, Hu J J, Yan N F, Pan G L, Li G R and Gao X P 2012 *Energy Environ. Sci.* **5** 9743–6
[13] Sun H, Wang W, Yu Z, Yuan Y, Wang S and Jiao S, 2015 *Chem. Commun.* **51** pp 11892–5
[14] Lin M C, Gong M, Lu B, Wu Y, Wang D Y, Guan M, Angell M, Chen C, Yang J, Hwang B J, Dai H 2015 *Nature* **520** 325–8
[15] Stadie N P, Wang S, Kravchyk K V and Kovalenko M V 2017 *ACS Nano.* **11** 1911–9
[16] Jiang J, Li H, Fu T, Hwang BJ, Li X and Zhao J 2018 *ACS Appl. Mater. Interfaces.* **10** 17942–9
[17] Wang W, Jiang B, Xiong W, Sun H, Lin Z, Hu L, Tu J, Hou J, Zhu H and Jiao 2013 S *Sci. Rep.* **3** 3383
[18] Kristoffersen H H and Metiu H 2016 *J. Phys. Chem. C.* **120** 3986–92
[19] González J R, Nacimiento F, Cabello M, Alcántara R, Lavela P and Tirado J L 2016 *RSC Adv.* **6** 62157–64
[20] Reed L D and Menke E 2013 *J. Electrochem. Soc.* **160** A915–7
[21] Gu S, Wang H, Wu C, Bai Y, Li H and Wu F 2017 *Energy Storage Mater.* **6** 9–17
[22] Giordano F, Abate A, Correa Baena J P, Saliba M, Matsui T, Im S H, Zakeeruddin S M, Nazeeruddin M K, Hagfeldt A and Graetzel M 2016 *Nat. Commun.* **7** 10379
[23] Batmunkh M, Macdonald T J, Shearer C J, Bat-Erdene M, Wang Y, Biggs M J, Parkin I P, Nann T and Shapter J G 2017 *Adv. Sci.* **4** 1600504
[24] Lu Y, Sathasivam S, Song J, Crick C R, Carmalt C J, Parkin I P 2015 *Science* **347** 1132–5
[25] Sotelo-Vazquez C, Noor N, Kafizas A, Quesada-Cabrera R, Scanlon D O, Taylor A, Durrant J R

and Parkin IP 2015 *Chem. Mater.* **27** 3234–42
[26] Kostov S, denBoer M, Strauss E, Golodnitsky D, Greenbaum S and E. Peled 1999 *J. Power Sources* **81** 709–14
[27] Yu Z, Kang Z, Hu Z, Lu J, Zhou Z and Jiao S 2016 *Chem. Commun.* **52** 10427–30
[28] Wang S, Yu Z, Tu J, Wang J, Tian D, Liu Y and S Jiao 2016 *Adv. Energy Mater.* **6** 2–11
[29] Geng L, Scheifers J P, Fu C, Zhang J, Fokwa B P T and J Guo 2017 *ACS Appl. Mater. Interfaces.* **9** 21251–7
[30] Aurbach D, Lu Z, Schechter A, Gofer Y, Gizbar H, Turgeman R, Cohen Y, Moshkovich M and Levi E 2000 *Nature* **407** 724–7
[31] Geng L, Lv G, Xing X and Guo J 2015 *Chem. Mater.* **27** 4926–9
[32] Chao D, Liang P, Chen Z, Bai L, Shen H, Liu X, Xia X, Zhao Y, Savilov S V, Lin J and Shen ZX 2016 *ACS Nano* **10** 10211–19
[33] Li Z, Niu B, Liu J, Li J and Kang F 2018 *ACS Appl. Mater. Interfaces.* **10** 9451–9
[34] Agiorgousis M L, Sun Y Y, West D J and Zhang S 2018 *ACS Appl. Energy Mater.* **1**(2) 440-6
[35] Kitajou A, Yamaguchi J, Hara S, Okada S 2014 *J. Power Sources.* **247** 391–5
[36] Kim T B, Choi J W, Ryu H S, Cho G B, Kim K W, Ahn J H, Cho K K and Ahn H J 2007 *J. Power Sources.* **174** 1275–8
[37] Koura N, Yui T, Takahashi S 1985 *J. Japan Inst. Light Met.* **35** 203–8
[38] Liu S, Pan G L, Li G R and Gao XP 2015 *J. Mater. Chem. A.* **3** 959–62
[39] Rosi N L, Eckert J, Eddaoudi M, Vodak D T, Kim J, O'Keeffe M and Yaghi OM 2003 *Science* **300** 1127–9
[40] Chang Z, Wang X, Yang Y, Gao J, Li M, Liu L and Wu Y 2014 *J. Mater. Chem. A.* **2** 19444–50
[41] Reed L D, Ortiz SN, Xiong M and Menke EJ 2015 *Chem. Commun.* **51** 14397–400
[42] Wessells C D, Huggins R A and Cui Y 2011 *Nat. Commun.* **2** 550
[43] Yue J L, Sun Q and Fu Z W 2013 *Chem. Commun.* **49** 5868
[44] Sun X -G, Bi Z, Liu H, Fang Y, Bridges C A, Paranthaman M P, Dai S and Brown G M 2016 *Chem. Commun.* **52** 1713–6
[45] Wang Y, He P, Zhou H and Olivine 2011 *Energy Environ. Sci.* **4** 805–17
[46] Wang S, Jiao S, Wang J, Chen H S, Tian D, Lei H and Fang D N 2017 *ACS Nano.* **11** 469–77
[47] Zou R, Zhang Z, Yuen M F, Sun M, Hu J, Lee C S and Zhang W 2015 *NPG Asia Mater.* **7** e195
[48] Zhu J, Yang D, Yin Z, Yan Q, Zhang H 2014, *Small* **10** 3480–98
[49] Hou J, Shao Y, Ellis M W, Moore R B, Yi B 2011 *Phys. Chem. Chem. Phys.* **13** 15384 –402
[50] Li Z, Niu B, Liu Y, Li J and Kang F 2018 *Electrochim. Acta.* **263** 68–75
[51] Wang S, Yu Z, Tu J, Wang J, Tian D, Liu Y and Jiao S 2016 *Adv. Energy Mater.* **6** 1600137
[52] Ni S, Lv X, Li T, Yang X, Zhang L 2013 *J. Mater. Chem. A.* **1** 1544–7
[53] Feng N, Hu D, Wang P, Sun X, Li X and He D 2013*Phys. Chem. Chem. Phys.* **15** 9924
[54] Hu Y, Ye D, Luo B, Hu H, Zhu X, Wang S, Li L, Peng S and Wang L 2018 *Adv. Mater.* **30** 1–6
[55] Hu Y, Luo B, Ye D, Zhu X, Lyu M and Wang L 2017 *Adv. Mater.* **29** 1606132

Acknowledgement
The first author acknowledges the financial support from Riset Hilirisasi Produk Unggulan, Universitas Padjadjaran, Bandung, Indonesia.

Effective thermal conductivity of U-shaped heat pipe

M H Kusuma[1], U Setiorini[2], T A R Putri[2], G, A R Antariksawan[3] and M Juarsa[1]

[1]Centre for Nuclear Reactor Technology and Safety, National Nuclear Energy Agency of Indonesia (BATAN), Kawasan Puspiptek Serpong, Tangerang Selatan 15314, Indonesia
[2]Department of Physics, Padjadjaran University, Jatinangor, Bandung, Indonesia
[3]Center for Science and Accelerator Technology, National Nuclear Energy Agency of Indonesia (BATAN), Jl. Babarsari Kotak Pos 6101 ykbb Yogyakarta 55281, Indonesia

E-mail: luluikal@batan.go.id

Abstract. The nuclear power station severe accident in Fukushima Daiichi became an important lesson learn to involve the passive cooling system in order to enhance the safety in a nuclear power station when prolonged station blackout occur. A U-shaped heat pipe is proposed as one of alternative passive cooling technology to absorb the decay heat generation in the cooling tank that connected to the reactor cavity cooling system. This research objective is to know the effective thermal conductivity of the U-shaped heat pipe. The effective thermal conductivity is an important factor to know the effect of the U-shaped heat pipe to remove the decay heat generation to the heat sink. An experiment was conducted to know the effective thermal conductivity of U-shaped heat pipe. The air cooling velocities on a bonded-fin heat sink in the condenser section were used as a parameter to obtain the effective thermal conductivity of U-shaped heat pipe. The air cooling velocities were varied at 0.65, 0.74, 1.02, and 1.21 m/s. The experiment results showed that the effective thermal conductivity of 35.37 was obtained at the highest variation of air cooling velocity on a bonded-fin heat sink in the condenser section. The result obtained shows that U-shaped heat pipe as a heat transfer device has thermal conductivity 35.37 times greater than the solid U-shaped Copper with the same geometry and dimension. This investigation results can be used to justify the use of heat pipes as a good passive cooling system in cooling tanks connected to Reaktor Daya Eksperimen reactor cavity cooling system.

Nomenclatures:

A	:	Surface area of U-shaped heat pipe (m^2)
C_p	:	Specific heat of coolant (J/kg. K)
k_{eff}	:	Effective thermal conductivity of U-shaped heat pipe (W/m.°C)
La	:	Length of adiabatic section (m)
Lc	:	Length of condenser section (m)
Le	:	Length of evaporator section (m)
L_{eff}	:	Effective length of U-shaped heat pipe (m)
$\dot{m}$	:	Mass flow rate of coolant (kg/s)

Q	:	Heat transfer rate (W)
Q_{out}	:	Amount of heat transport at condenser (W)
ρ	:	Density of water (kg/m^3)
R_T	:	Thermal resistance (°C/W)
T_i	:	Inlet temperature of coolant (°C)
T_o	:	Outlet temperature of coolant (°C)
ΔT	:	Temperature different between evaporator and condenser section (°C)
$\bar{v}$	:	Volumetric flow rate of air coolant (L/min)

1. Introduction

Since the occurrence of severe accident at Japan's Fukuhisma Dai-ichi nuclear power plant, passive cooling systems (PCS) has begun to be highly considered as a heat dissipation systems of decay heat generation in a nuclear reactor. Although it has less effective cooling capability than an active cooling system, PCS can be relied on to remove the residual heat generation when an accident occurs in a nuclear reactor. Nuclear reactor accidents can be caused by loss of coolant accident that caused by station blackout (SBO), or caused by the failure of the active cooling system.

In the design of nuclear reactor, PCS will be implemented into the existing nuclear safety system. The PCS must be simple, increase the heat dissipation capability, and easy to operate in normal nuclear reactor conditions or when an accident occurs.

One of the passive safety system technologies commonly used in the electronics field is heat pipe. The heat absorbed by the heat pipe is discharged into the heat sink, thus there is no heat accumulation in the heat source chamber. Heat pipe is working in two phases, no electricity needed to operate and it has a simple geometry [1–3].

The investigation of heat pipe concept as a passive residual heat removal system in a nuclear reactor has also been carried out by the other researchers, for example in spent fuel storage pool [4–7], nuclear for space [8], nuclear core [9], control rod [10], etc. In addition, heat pipes can also be used for nuclear seawater desalination [11]. The investigation result stated that the heat pipe had good thermal performance, and it significantly dissipated the heat produced in normal nuclear reactor operating conditions or in accident conditions. The importance of decay heat generation dissipation in nuclear reactor result in many nuclear reactor designs that involve passive cooling systems in their safety systems. For generation 3 and 4 of NPP design, the concept of passive cooling system has to be implemented [12–14].

In the design concept of the Reaktor Daya Eksperimen (RDE) which is being developed by Indonesia, a U-shaped heat pipe is proposed as one of alternative passive cooling technology in order to absorb the decay heat generation in the cooling tank that connected to the reactor cavity cooling system (RCCS). The novelty of this research is the U-shaped heat pipe design that never used as passive cooling system in the cooling tank that connected to RCCS.

This research objective is to know the effective thermal conductivity of the U-shaped heat pipe. The effective thermal conductivity is an important factor to know the effect of the U-shaped heat pipe to remove the decay heat generation to the heat sink. An experiment was conducted to know the effective thermal conductivity of U-shaped heat pipe. The air cooling velocities on a bonded-fin heat sink in the condenser section were used as a parameter to obtain the effective thermal conductivity of U-shaped heat pipe. The air cooling velocities were varied at 0.65, 0.74, 1.02, and 1.21 m/s.

2. Methodology

2.1. Experimental setup

The experimental setup of U-shaped heat pipe as a passive cooling technology to absorb the decay heat generation can be seen in figure 1.

IOP Conf. Series: Materials Science and Engineering **550** (2019) 012004 doi:10.1088/1757-899X/550/1/012004

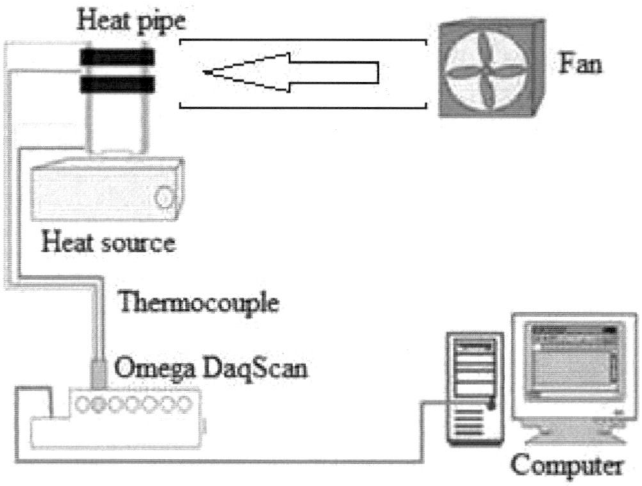

Figure 1. The experimental setup of U-shaped heat pipe.

U-shaped heat pipe consists of 4 copper tube with bonded-fin heat sink in the condenser section. Each heat pipe has outer diameter of 6 mm and inner diameter of 5.8 mm. Four of evaporator sections are placed on the bottom of heat pipe, and four of condenser sections are placed on the top of heat pipe. Evaporator and condenser section have length of 132.7 mm and 131 mm, respectively. On the outer side of the condenser section is placed 57 pieces of horizontal fins. Fins are made of Aluminum and have a rectangular shape. Each fin has a length, width, and thickness of 116.3 mm, 45.1 mm, and 0.4 mm, respectively. Fins are placed in a stacked horizontal position with a distance of 0.6 mm between the fins. The heat pipe is vacuumed for eliminate non-condensable gas and to decrease the boiling point of working fluid inside the heat pipe. De-mineral water is charged into the heat pipe with filling ratio of 100%. Filling ratio is the ratio between fluid volumes inside the evaporator compared with evaporator volume.

The function of evaporator is to absorb heat from the heat source, and the function of condenser is to release heat to heat sink. As a heat source, a heater with maximal power of 3000 W is placed on the evaporator section. Heat transfer by radiation from the heater is absorbed by the outer wall of the evaporator. The heat causes the working fluid in the heat pipe boils and change the phase of the working fluid from liquid to vapour. The heat provided by the heater to the evaporator is given at constant heat load of 900 W.

The steam then transfers the heat to the condenser. Latent heat is transferred to the outer walls of the condenser and fins. To accelerate the release of heat to the environment, air with ambient temperature of 29°C is exhaled to the fin with variations in speed at 0.65, 0.74, 1.02, and 1.21 m/s. Air velocity measurement uses a digital anemometer which has an uncertainty of $\pm$ 0.005 m/s. A chimney is used to direct the air to the fins.

To measure the temperatures data, the ethernet-based Omega DaqScan/2000 series is used as data acquisition system. Sixteen K type thermocouples with uncertainty of $\pm$ 0.1°C are placed in the test section. In the heat pipe, thermocouples are placed on the outside wall of Copper tube. Three thermocouples on evaporator section, 8 on condenser section, 1 thermocouple on heater room, 1 thermocouple on adiabatic section, 1 thermocouple for air inlet, 1 thermocouple for air outlet, and 1 thermocouple for measuring ambient temperature.

2.2. Data calculation

2.2.1. Thermal resistance calculation. The thermal resistance of U-shaped heat pipe is calculated using equation (1).

$$R_T = \frac{\overline{T_e} - \overline{T_c}}{Q_{out}} \quad (1)$$

The amount of heat removed across the U-shaped heat pipe to the environment is defined as:

$$Q_{out} = \dot{m}.c_p.(T_o - T_i) \quad (2)$$

Additionally, the mass flow rate of air coolant that absorbed the latent heat on condenser section can be obtained using:

$$\dot{m} = \rho.\tilde{v} \quad (3)$$

2.2.2. Effective thermal conductivity calculation. The effective thermal conductivity was used to compare the effectivity between heat pipe and a cylinderical copper with the same geometry. It assumed that the heat pipe as a cylindrical copper which absorb and transport the heat with conduction way.
The effective thermal conductivity calculation is used the Fourier heat conduction formulation as follow:

$$k_{eff} = Q\frac{L_{eff}}{A\,\Delta T} \quad (4)$$

Additionally, the effective length of U-shaped heat pipe can be obtained using:

$$L_{eff} = \left(\frac{L_e + L_c}{2}\right) + L_a \quad (5)$$

3. Results and discussion

3.1. Temperatures distribution

The experiment transient temperatures distribution obtained with variations on the air cooling flow rate and heat load of 900 W can be seen in figure 2.

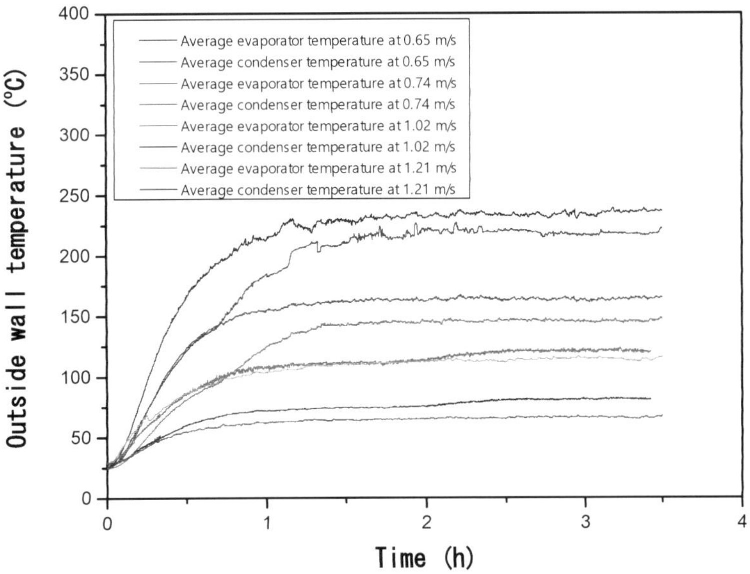

Figure 2. Transient temperature distributions at variation of air flow rate.

Figure 2 showed that the beginning of heating process on heat pipe, the evaporator temperatures were increase significantly because of heat absorbed continuously from heat source and it is effect on fluid to boil rapidly. The steam that resulted from the boiling process rises to the adiabatic and condenser section

because it has lighter density than the fluid, thus affecting to increase the adiabatic and condenser temperature significantly. In this state, there has not been a natural circulation process in the heat pipe because the condensate has not formed enough in the condenser due to incomplete absorption of latent heat by the exhaled air into the condenser section. This heat pipe phenomenon is called overshoot.
The overshoot phenomenon is finished after the latent heat in the condenser section has been discharged into the environment and the condensate is formed. The condensate which has a heavier density than vapour then drops gravitationally through the adiabatic section and returns to the evaporator section. In this state, the vapour that rises to the condenser and condensate which drops to the evaporator does not occur smoothly because of the unstable vapour and condensate that produced in the evaporator and condenser. The amount of vapour rises to condenser is sometimes more than the condensate that drops to evaporator, and vice versa. The collisions between vapour and condensate cause the fluctuations in temperature distribution in overall of heat pipe. In this situation the natural circulation process has occurred, but it has not reached the steady state. This heat pipe phenomenon is called zigzag.
After the overshoot and zigzag phenomena occurred, the last phenomenon that occurs is stable. In this state, the overall temperature distributions in the heat pipe has a steady pattern continuously. This indicates that the natural circulation process has occurred perfectly in the heat pipe. The latent heat which generated in the condenser can be discharged properly into the environment through the fins.
Figure 2 also showed that the temperature at the evaporator and condenser can reach overshoot, zigzag, and stable conditions in a faster time due to the higher air velocity that blown to absorb heat in the condenser. Zigzag conditions are more dominant when the condenser is blown with a low air velocity, while the zigzag is not dominant if the air is blown at a high air velocity.
The distribution of average heat pipe surface temperature after reaching the stable condition is shown in figure 3.

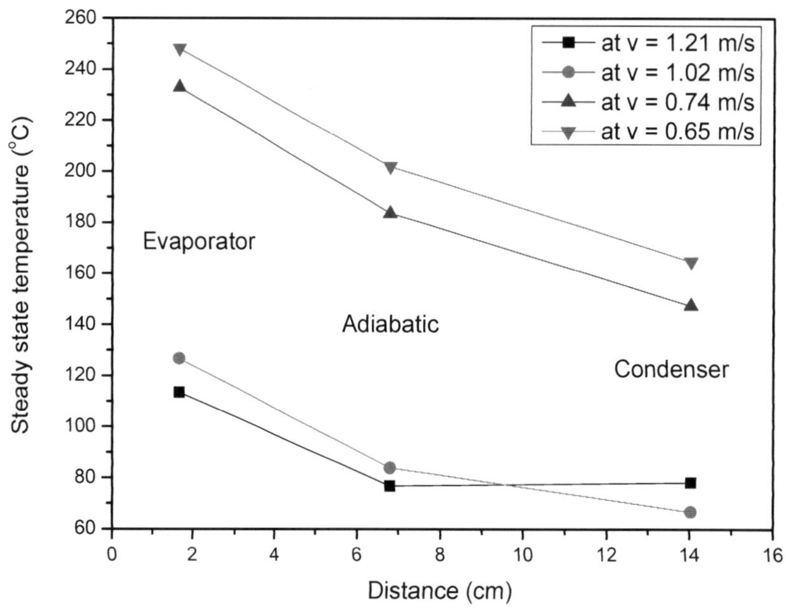

Figure 3. Steady state temperature distribution on U-shaped heat pipe.

Figure 3 showed that at a constant heat load, the lowest average temperatures of the evaporator and condenser were obtained when the condenser is cooled with air velocity of 1.21 m/s, and vice versa. The higher air velocity that used to absorb heat in the condenser causes the lower temperature of the evaporator and condenser, and the temperature difference between evaporator and condenser. The experiment showed that the highest air velocity of 1.21 m/s resulted in the smallest temperature difference between the evaporator and condenser. It can be seen that the air velocity has a significant effect on the temperature distribution in the U-shaped heat pipe. Increasing the air velocity is accelerate the steam condensation process in the condenser section, and it is resulted in the smallest temperature difference between the evaporator and condenser. Heat removal on the U-shaped heat pipe is strongly influenced by the air velocity that is used to take the heat in the condenser section.

3.2. Thermal resistance and effective thermal conductivity

3.2.1. Thermal resistance. The thermal resistance is an important parameter to know the heat pipe thermal performance. The heat pipe with the lowest thermal performance is indicated the best thermal performance of heat pipe to transport the heat to environment, and vice versa.
The thermal resistance of U-shaped heat pipe that obtained from experiment is shown in figure 4.

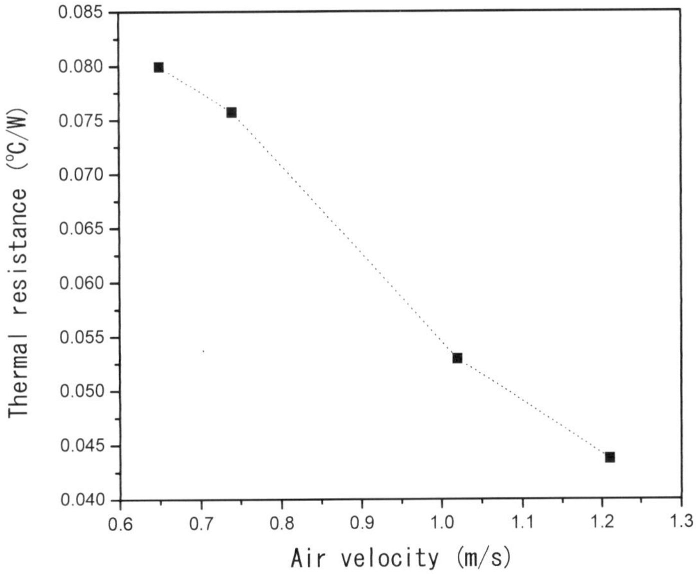

Figure 4. Thermal resistance of U-shaped heat pipe.

Figure 4 showed that the lowest thermal resistance of 0.044 °C/W is obtained when air was blown the condenser section at the velocity of 1.21 m/s. It is shown that the trend of thermal resistance decrease with the increase of air velocity because higher air velocity is absorb more latent heat in the condenser section and affected to decrease the condensate temperature produced in the condenser section. The low-temperature condensate that drop into the evaporator result in lower evaporator temperature than the condensate which drop at high temperature. Thus at the highest air velocity, the U-shaped heat pipe is resulting lower temperatures difference between evaporator and condenser section.
The experimental results show that the U-shaped heat pipe has the best thermal performance when it is operated at higher air velocity. The results obtained indicate the compatibility with t research that has been done before [4, 15].

3.2.2. Effective thermal conductivity. The highest effective thermal conductivity of the U-shaped heat pipe is 35.37, obtained when U-shaped heat pipe is operated with the largest air velocity of 1.21 m/s. The experimental results show that the effective thermal conductivity is increase with the increase of air velocity which used to take latent heat in the condenser section. The effective thermal conductivity obtained shows that the U-shaped heat pipe made from copper used in the experiment has the heat transport ability of 35.37 times greater than the copper solid rod with the same geometry.

4. Conclusion

The effective thermal conductivity of U-shaped heat pipe, which will be proposed as a passive cooling system technology to absorb the decay heat generation in the cooling tank that connected to the reactor cavity cooling system, was investigated experimentally. It was concluded that U-shaped heat pipe have an overshoot, zigzag, and stable temperature distribution. The lower thermal resistance of 0.044 °C/W and effective thermal conductivity of 35.37 was obtained when the condenser section of the U-shaped heat pipe was blown with the highest variation of air velocity of 1.21 m/s. The U-shaped heat pipe has the heat transport ability of 35.37 times greater than the copper solid rod with the same geometry. Based on the thermal resistance and effective thermal conductivy, the U-shaped heat pipe can be proposed as an alternative for passive cooling system technology to absorb the decay heat generation in the cooling tank that connected to the reactor cavity cooling system.

References

[1] Wang J-C 2014 Microelectron Reliab **54** 1344–54
[2] Kumar K R S, R Dinesh, Roseline A A, Kalaiselvam S 2017 *Microelectron Reliab* **73** 1–13
[3] Wang Y, Wang B, Zhu K, Li H, He W and Liu S 2018 *Appl Therm Eng* **143** 630–38
[4] Kusuma M H, Putra N, Antariksawan A R, Koestoer R A, Widodo S, Ismarwanti S and Verlambang B T 2018 *Int. J. Therm. Sci.* **126** 162–71
[5] Kim I G and Bang I C 2018 *Energy Convers Manag* **173** 233–43
[6] Lai K, Wang W, Yi C, Kuang Y and Ye C 2018 *Ann. Nucl. Energy* **111** 523–35
[7] Ye C, Zheng M G, Wang M L, Zhang R H and Xiong Z Q 2013 *Ann. Nucl. Energy* **58** 124–31
[8] Wang C, Chen J, Qiu S, Tian W, Zhang D and Su G H 2017 *Ann. Nucl. Energy* **103** 74–84
[9] Panda K K, Dulera I V and Basak A 2017 *Nucl. Eng. Des.* **323** 376–85
[10] Kim K M, Bang I C 2018 *Int. J. Heat Mass Transf.* **119** 472–83
[11] Jouhara H, Anastasov V and Khamis I 2009 *Desalination* **249** 1055–61
[12] Wang Y 2013 *Energy Procedia* **39** 240–47
[13] IAEA 2009 *IAEA-TECDOC*-1624
[14] Yu X-G, Park H-S, Kim Y-S, Kang K-H, Cho S and Choi K-Y 2015 *Nucl. Eng. Des.* **259** 205–20
[15] Kusuma M H, Putra N, Antariksawan A R, Imawan F A 2017 *Nucl. Eng. Technol.* **49** 476–83

Acknowledgment

The research funded by DIPA PTKRN BATAN 2018 and research grant from Programme INSINAS Ministry of Research, Technology and Higher Education.

The determination of geothermal potential area based on remote sensing, Micromine software, and land surface temperature calculation

N Sulaksana, D Gentana, P P Raditya and R A Sentosa

Faculty of Geological Engineering, Padjadjaran University

Email: n.sulaksana@unpad.ac.id

Abstract. Indonesia has a significant geothermal energy potential because it is in the Ring of Fire. However, only 7% of identified potential geothermal has been utilized until now. This research aims to determine and predict geothermal potential area by using Landsat 8 imagery, Micromine software, and Land Surface Temperature Calculation. Land Surface Temperature calculation can be carried out by thermal infrared bands which can be obtained from Landsat 8 imagery. Land Surface Temperature calculation is based on a Split-Window algorithm which providing a surface temperature value and its distribution. Furthermore, the method can support to generate the distribution of Land Surface Temperature in research area using Micromine software. The remote sensing analysis also being done to determine geological structures development in the research area which assumed to control manifestation occurrence. It also shows the appearances of geothermal surface manifestation. The result of Land Surface Temperature analyses could predict surface temperature zone in the research area. The area that has high surface temperature value tends to be an area that has a potential geothermal area. The application of this method is being done on a potential geothermal field around Mt. Patuha, Ciwidey, West Java to verify these approaches. This research shows that Land Surface Temperature method to determine geothermal potential has a positive result.

1. Introduction

Landsat satellite continuously takes imageries of the earth's surface from space and provide valuable data since 1972. One analysis that can be done with Landsat imagery is Land Surface Temperature (LST) analysis. LST is temperature felt when the land surface is touched with the hands or can be defined as the skin of the ground [1].

LST is one key energetic equilibrium parameters on the surface and is also a major climatological variable [2]. The utilization of this LST is used further in this research, that is to estimate geothermal potential areas based on the distribution of the LST obtained using remote sensing. The utilization of LST for geothermal potential area estimation was carried out around Mt. Patuha, Ciwidey, West Java as the research area to prove the validity of this method.

2. Methodology

Two methods are used in this research, i.e. lineament delineation and calculation of LST in the 5x5 km of the research area. Lineaments delineation is performed on a digital elevation model (DEM) represented from four directions of illumination which support each other to optimize the showing of lineaments in the delineation step [3]. Illumination angles of 0°, 45°, 90°, and 135° are used in this research. This is done with the aim of estimating fractures that can be assumed to be the pathway of thermal water to the surface.

The 8th version of Landsat imagery has a total of 11 bands with their respective uses and also two types of sensor, i.e., Operational Land Imager (OLI) and Thermal Infrared Sensor (TIRS) [4]. Land Surface Temperature (LST) utilizes some bands from those bands contained in a Landsat 8 imagery. However, the main parts of LST calculations are the 10th and 11th band.

LST calculations are broken down into several stages using the Split-Window algorithm [5,1].

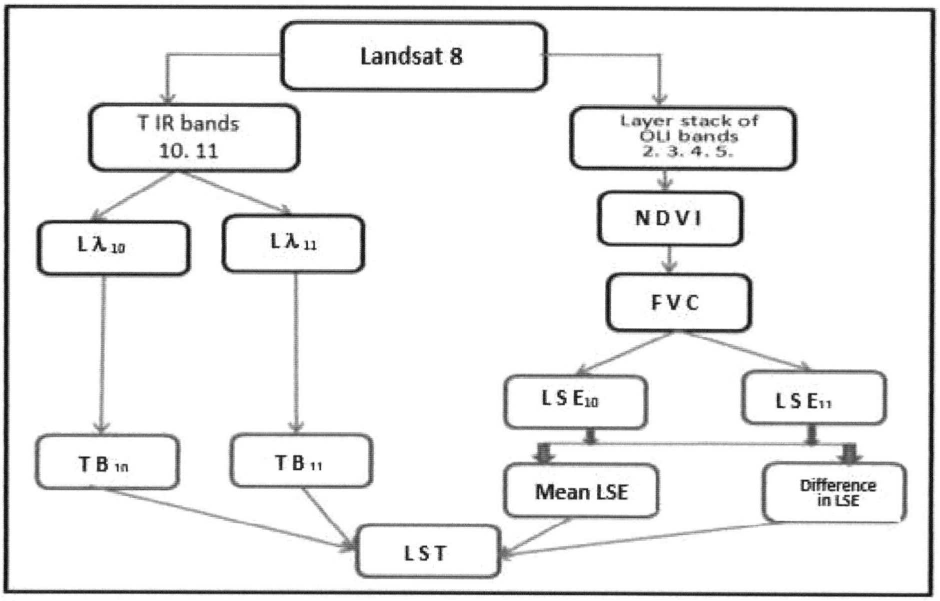

Figure 1. Flowchart of split-window algorithm [5,1].

Split-Window algorithm is an algorithm that is prepared to simplify the calculation of surface temperature using the following formula:

$$LST = TB_{10} + C_1 (TB_{10}-TB_{11}) + C_2 (TB_{10}-TB_{11})^2 + C_0 + (C_3+C_4W) (1- \varepsilon) + (C_5+C_6W) \Delta \varepsilon$$

where :

- LST: *Land Surface Temperature* (K)
- C_0-C_6: Split-window coefficient (table 1) (Rajeshwari & Mani, 2014; Skokovic et al., 2014).
- TB_{10} and TB_{11}: Brightness temperature of 10 and 11 bands (K). TB_{10} and TB_{11} can be determined by formula $TB = K2/Ln((K1/L\lambda) + 1)$
  where:
  - K1 and K2: thermal conversion constant of 10 and 11 bands.
  - $L\lambda$: Top of atmospheric spectral radiance (m^2.srad.μm). $L\lambda$ can be determined by $L\lambda = ML * Qcal + AL$
    where:
    - M_L: multiplicative rescaling factor of 10 and 11 bands. M_L can be known directly from Landsat attached note = 0.000342
    - Qcal: the value of 10 and 11 bands themselves.
    - A_L: additive rescaling factor of 10 and 11 bands. A_L can be known directly from Landsat attached note = 0.1

- ε: mean of Land Surface Emissivity of 10 and11 bands. ε can be determined from $\varepsilon : (\varepsilon_{10} - \varepsilon_{11}) / 2$
- $\Delta \varepsilon$: difference of Land Surface Emissivity of 10 and 11 bands. $\Delta \varepsilon$ can be determined from $\Delta \varepsilon : \varepsilon_{10} - \varepsilon_{11}$ where ε_{10} and ε_{11} are the value of Land Surface Emissivity. Each emissivity value of 10 and 11 bands can be determined from $\varepsilon = \varepsilon s (1 - FVC) + \varepsilon v * FVC$
where:
    - FVC: Fractional Vegetation Cover. FVC of 10 and 11 bands with formula $FVC = (NDVI - NDVIs) / (NDVIv - NDVIs)$
    where:
        - NDVI: Normalized Difference Vegetation Index. NDVI can be determined from $NDVI = ([B5] - [B2]) / ([B5] + [B2])$ where Bn is the value of the bands contained in Landsat.
        - NDVIs: soil NDVI. NDVIs can be determined from $NDVIs = ([B5] - [B4]) / ([B5] + [B4])$ where Bn is the value of the bands contained in Landsat.
        - NDVIv: vegetation NDVI. NDVIs can be determined from $NDVIv = ([B5] - [B3]) / ([B5] + [B3])$ where Bn is the value of the bands contained in Landsat.
    - εs : Soil Emmisivity [1,6].
    - εv : Vegetation emmisivity[1,6].

Table 1. Constant and coefficient value in LST calculation [1,6].

Parameter	Desc.	Value
C0	Split-Window Coefficient	-0.268
C1		1.378
C2		0.183
C3		54.300
C4		-2.238
C5		-129.200
C6		16.400
ML Band 10	Multiplicative Rescaling Factor of Band 10	0.000342
ML Band 11	Multiplicative Rescaling Factor of Band 11	0.000342
AL Band 10	Additive Rescaling Factor of Band 10	0.1
AL Band 11	Additive Rescaling Factor of Band 11	0.1
εs Band 10	Soil Emmisivity of Band 10	0.971
εs Band 11	Soil Emmisivity of Band 11	0.977
εv Band 10	Vegetation Emmisivity of Band 10	0.987
εv Band 11	Vegetation Emmisivity of Band 11	0.989
K1 Band 10	Thermal Constant K1 of Band 10	774.8853
K1 Band 11	Thermal Constant K1 of Band 11	1321.0789
K2 Band 10	Thermal Constant K2 of Band 10	480.8883
K2 Band 11	Thermal Constant K2 of Band 11	1201.1442

Landsat imagery used in this research is Landsat version 8 with 15 meter resolution captured by satellite on 24th September 2018 which is the latest research area image data at the time this research was conducted.

LST calculation results using the 11 bands from Landsat imagery will produce maps of LST distribution that are mapped using Micromine software. The LST distribution map will be compared with the results lineament delineation and manifestations location to find out the similarity of the analysis results of those elements.

3. Results and discussion

3.1. Lineament delineation

Delineation is carried out in the research area and produces a distribution of lineament with varying orientations. Lineaments are delineated from the direction of light 0, 45, 90, and 135 degrees are respectively distinguished by red, orange, yellow, and blue in the map below.

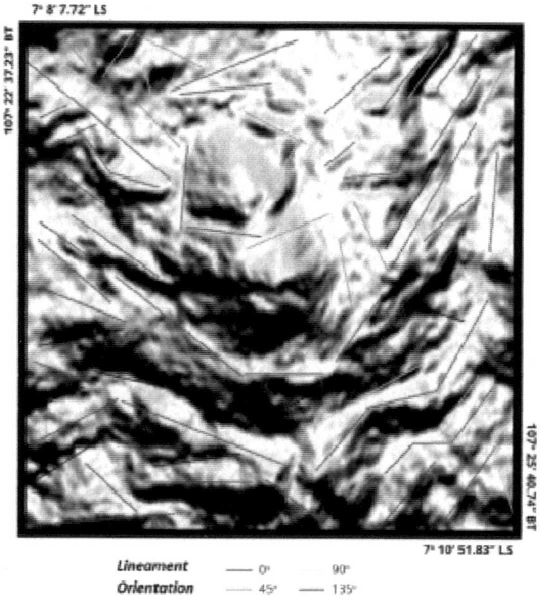

Figure 2. Lineament delineation of the research area.

3.2. Land surface temperature calculation

LST calculations are done manually using Microsoft Excel data processing. Steps in using LST calculation formula to get LST value of each Landsat resolution are carried out by following the Split-Window algorithm [5,1].

The results of the calculations are converted onto a map that shows the distribution of surface temperature which can be seen in figure 3. The results of the calculation show that the average surface temperature in the study area is calculated at 16.73 ° Celsius with a temperature range in the study area is 32.89.

It can be seen from the results of the distribution of the LST in the figure above, the western part of the research area has a relatively higher temperature compared to other areas. There are areas with an LST value between 20°-25°C located in the Kawah Putih area. Kawah Putih is one of the craters of Patuha Mountain in the form of a crater lake which has color due to high sulfur content and a certain temperature, which causes the surrounding rock and sand to become white, hence the name [7].

The temperature of the water in the Kawah Putih is quite stable throughout the year between 30°-34°C. This temperature shows a positive result of LST distribution which is proved by successfully identifying relatively higher temperatures area in Kawah Putih.

In the northwestern part of the study area, there is a known hot spring named Ciwalini hot spring that is used by residents to be used as tourist attractions. The hot spring is located in an LST with a value between 20°-25°C. This hot spring is proof that there is a potential geothermal in the northwestern part of the study area.

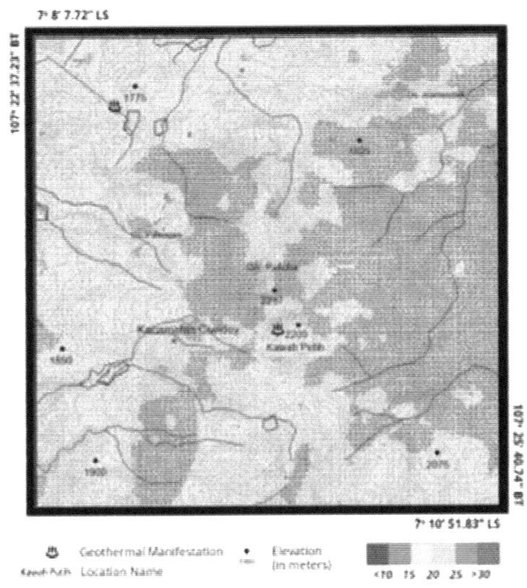

Figure 3. LST distribution in the research area.

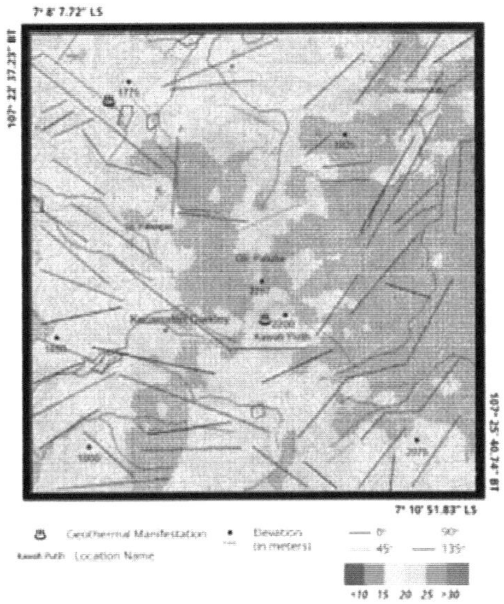

Figure 4. LST distribution map overlaid with lineament delineation result.

The results of lineament delineation and LST distribution are compared to each other. The results indicate that Kawah Putih and Ciwalini hot springs are closely located to delineated lineament. Kawah Putih is shown to be located on the northern side of an E-W oriented lineament, while Ciwalini hot spring is located on the northern side of a lineament with NW-SE orientation. In addition, in the eastern part of Patuha Mountain has a relatively higher lineament density, but there is no manifestations of hot water were found.

The explanation above shows that LST analysis in identifying geothermal potential areas has positive results. Both geothermal manifestations in the research area are located in regions with relatively higher LST values than other regions. The utilization of LST in identifying geothermal potential deserves further investigation for more optimal utilization and can be used in other similar research.

4. Conclusion

After all the methods are carried out in this research, i.e. lineament delineation and calculation of LST distribution in the research area with geothermal manifestations, it can be concluded that the application of LST analysis to determine geothermal potential areas shows positive results. This conclusion is drawn based on the result that shows all the compared elements show a relationship with each other. Further studies need to be carried out to optimize the utilization of LST and also to find out all factors that need to be considered in further utilization.

References

[1] Rajeshwari A and Mani N D *International Journal of Research in Engineering and Technolog* **3**(5) 122–6

[2] Rifani A, Saputro E A, Invanni I and Maru R 2017 *Proceeding of 9th International Graduate Students and Scholars' Conference in Indonesia (IGSSCI)* (Yogyakarta: Universitas Gadjah Mada) 179–190

[3] Sentosa R A, Syabi H F and Haryanto I 2017 *Proceedings of the 5th Indonesia International Geothermal Convention & Exhibition (IIGCE) 2017* (Jakarta: The Indonesian Geothermal Association)

[4] US Geological Survey 2016 *Landsat — Earth Observation Satellites Landsat Missions : Imaging the Earth Since 1972, Supersedes USGS Fact Sheets 2012–3072 and 2013–3060,* 1–4

[5] Gentana D, Sukiyah E, Sulaksana N, Yuningsih E T and Balia L 2017 *FIG Working Week 2017, Surveying the world of tomorrow - From digitalisation to augmented reality*
[6] Skokovic D, Sobrino J A, Jiménez Muñoz J C, Soria G, Julien, Y, Mattar C and Cristóbal J 2014 *land product validation and evolution* (Madrid: European Space Agency Document)
[7] Nurinsiyah D N, Gunawan B, Husodo T, Uchiyama T 2015 *Journal of Rural Problems* **51**(3) 197–202

The analysis of ignition and combustion properties of the burning briquettes made from mixed biomass of rice husk and corn cob

S Suryaningsih[1], O Nurhilal[1], R A Widyarini[1], N Suhendi[2] and Rukiah[3]

[1]Department of Physics
[2]Department of Electrical Engineering
[3]Department of Chemistry,
Faculty of Mathematics and Natural Sciences, Universitas Padjadjaran, Jatinangor Indonesia

Email: sri@phys.unpad.ac.id

Abstract. This research aims to determine the characteristics of briquettes made from charcoal mixed with rice husk and corn cobs. The test that has been done is to determine the ignition and combustion properties of the burning briquettes that have been produced. The test of the time duration of briquette ignition is divided into two, the first is the time duration of burning briquette to boil one liter of water and the second is the time needed for the burning briquettes which is started when the briquette is burned until it has come intothe ash. Correlation between the time duration of the boiling water in the furnace and the briquettes combustion temperature was also observed. The next step is the analysis of carbon monoxide (CO) emissions from its combustion and the quality of the carbonization furnace, and also calculates the thermal efficiency value of the carbonization furnace. The resulting briquettes can be implemented and can be used as an alternative fuel for economic and environmentally friendly in the community.

1. Introduction

Biomass is a renewable source of fuel derived from biological materials and modified anthropogenic material such as agricultural and industrial waste and municipal waste [1]. Burning using biomass does not damage the atmosphere. Many authors emphasize carbon neutrality, where the emissions of carbon from the combustion of these materials are balanced by capturing carbon in the next planting period [2].

Biomass energy from agricultural waste, especially rice husks have long been known to the public, but its limited to the household and small-scale industries. The use for various sectors in the modern application is still quite a bit compared to the existing potential source such the coal, due the rice husks are considered to have a low value in efficiency.

Rice husk is the outer part of the rice grain which is a byproduct of the milling process. About 20-30% of the weight of the rice grain is the rice husk, according to the Ministry of Agriculture 1 kg of rice husk has a heating value of 3300 Cal/g [3]. There are several ways to increase the heating value such as mixing the rice husk with corn cobs which have higher heating value, the densification process, and the torrefaction process. The biomass densification process generally consists of drying and reducing particle size. Whereas to increase the heat value, the torrefaction process is one of the right choices. The torrefaction of the rice husk can hold more than 70% carbon in rice husks, even though oxygen in rice husks is delowficient [4]. The combination of the densification and torrefaction processes will produce high-quality biomass solid fuels in the form of briquettes.

This study aims to find out the best alternative of solid fuel in the form of charcoal briquettes, such as bio-briquettes that can be made from mixed agricultural wastes. By using the appropriate percentage of the composition ratio and chosen the right particle size will produce better quality briquettes, and also expected to have the low in CO emissions.

2. Materials and methods

2.1. The process of making charcoal material

The material used in this study is dried rice husk as the base material for making briquettes which are added with teak sawdust and coffee bark powder. Water and tapioca flour are used as adhesive materials for making briquettes. The first step is the carbonization process, the dry test materials are placed in an electric furnace equipped with a temperature controller and carbonized at a certain temperature. The speed of the furnace temperature rise is 10 °C/min from room temperature (29 °C) until the desired temperature is reached in the range of 100 °C - 250 °C. The duration of carbonization is varied, 1, 2, and 3 hours depending on the moisture content of each material. The second stage is drying the charcoal using a temperature of 50 °C in an oven for 1 hour. The third stage is refining charcoal particle size using the sieve of 40 and 100 mesh.

2.2. The bio-briquette making process

The stages of making biobriquette consist of making adhesive material from a mixture of water and tapioca flour. A comparison of the concentration of rice husk base material with mixing material (100:0, 70:25, 50:50). The sample name of the many briquettes composition selected in this experiment is shown in table 1 below.

Table 1. The composition of the rice husk and corn cob.

Name of sample	Rice Husk (%)	Corn Cob (%)	Sieve Size (Mesh)
A1	100	0	40
A2	75	25	40
A3	50	50	40
B1	100	0	100
B2	75	25	100
B3	50	50	100

The cylindrical briquette printing with a height of 7 cm and a diameter of 4 cm in the middle hollow cylinder. Finally, pressing briquette using a 45 kg weight, and drying the briquette in an oven for 4.5 hours at 50 ° C.

2.3. Characterization of bio-briquettes

Characterization of bio-briquette materials carried out using thermal test, water heating value using briquette stove, and carbon monoxide (CO) gas emission test:

2.3.1. Thermal test. The thermal test is an essential part of qualification and characterization programs for virtually all heat-sensitive products and materials. In this case, the thermal test is done by determining the combustion rate of each briquette. The weight of the initial mass of the briquette has to be measured first. After the bottom of the briquette is burned, then insert the briquette into the combustion container and let the briquette burn out. During the combustion of the briquettes, the temperature of the combustion was observed using an infrared thermometer. The change was observed for every 2 minutes until the briquettes tested were burned out. The calculation of the combustion rate of briquettes done by using these equations (1) and equation (2).

$$Combustion\ rate = Burning\ briquette\ mass\ (g)\ /Burning\ time\ (minutes) \quad (1)$$

$$Burning\ briquette\ mass\ (g) = Initial\ briquettes\ mass - Remaining\ briquette\ mass \quad (2)$$

2.3.2. Briquette test for boiling 1 liter of water. The procedure for determining the calorific value (Q) can be known by boiling 1 liter of water using briquette fuel as follows:

- weighing the mass of briquettes,
- prepare 1 liter of water,
- put the burnt briquettes on the briquette stove,
- d.put a pan containing 1 liter of water on the briquette stove,
- e.record the temperature of water and coals every 2 minutes until the water boiled and the maximum temperature is reached,
- f.continue to record until the briquette burns out,
- g.measure the ash mass and the water mass after the water boiled process.

Each substance has a different heat quantity to increase the temperature of a certain mass. The calculation of heat given to substances as shown in equation (3).

$$Q = m.c.(t2 - t1) \quad (3)$$

where: Q is calories needed (J), m is mass (kg), c is specific heat capacity (J/kg•C) and (t2-t1) is temperature difference (°C).

2.3.3. Carbon monoxide (CO) emission test. Testing of carbon monoxide gas emissions from briquettes that have been made is done by burning them. First, weighing the initial mass of briquettes to be tested is ± 80 grams. The bottom briquette is burned first, then put the briquette into the combustion container and let the briquette burn out. During the combustion of carbon monoxide gas emissions, it was observed with the carbon monoxide gas sensor for every 2 minutes change.

3. Results and discussion

3.1. Thermal test results of the briquettes

In this test, the initial mass of briquettes and the mass after being ash is measured, the difference between these two is the material that has been burned and the burning time can be observed as seen in table 2.

Table 2. The mass of briquettes tested versus burning time.

Parameter	Samples					
	A1	A2	A3	B1	B2	B3
Initial briquette Mass (g)	80	79	78	82	82	76
Remaining briquette Mass (g)	46	34	27	37	26	18
Burning briquette mass (g)	34	45	51	45	56	58
Burning Time (s)	3960	5760	6600	5040	6240	6000

From the results, the combustion of briquettes obtained the data of temperature and combustion time. The results of the measurement of the combustion temperature of briquettes are taken for every 2 minutes until the briquette changes to ash. The observations of the burning duration are shown in figures 1 and 2.

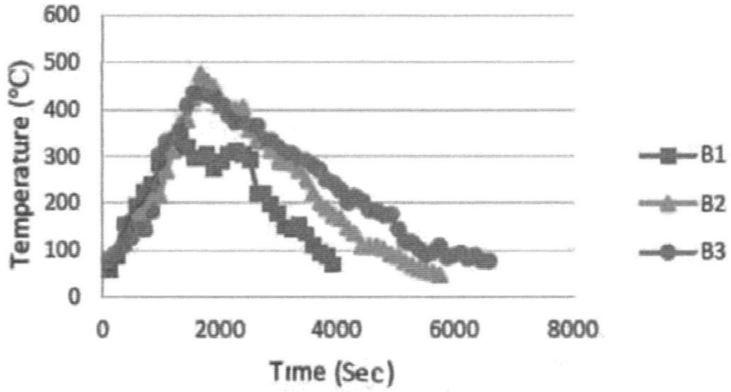

Figure 1. Graph of temperature versus briquettes burning time at 40 mesh particle.

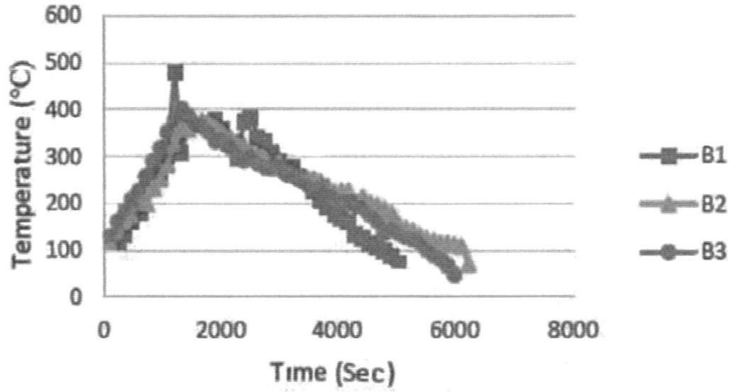

Figure 2. Graph of temperature versus briquettes burning time at 100 mesh particle.

Based on figures 1 and figure 2 it can be seen that the combustion produces temperatures that are getting higher and higher, each briquette has a maximum temperature. After reaching the maximum temperature, the combustion temperature will decrease, which indicates that the combustion starts to stop. The briquette with a sample name A2 reaches a maximum temperature of 475.17 °C. The briquette with sample name B3 reaching the maximum temperature at 401.07 °C. These data show that the particles size can affect the combustion temperature, as seen in the sample A3 & B3, whereas the temperature of A3 is higher than B3 but need more time to achieve.

3.2. Water boiling test

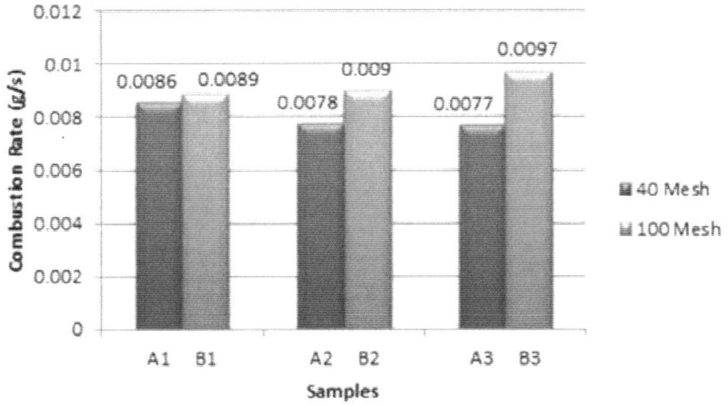

Figure 3. Combustion rate against particles size.

From the six samples in figure 3, the briquettes that have large particles at 40 mesh (B1, B2, B3) have a higher average combustion rate. The highest combustion rate is obtained on sample B3 with a composition of 50:50 and particles of 100 mesh equal to 0.0097g /s.

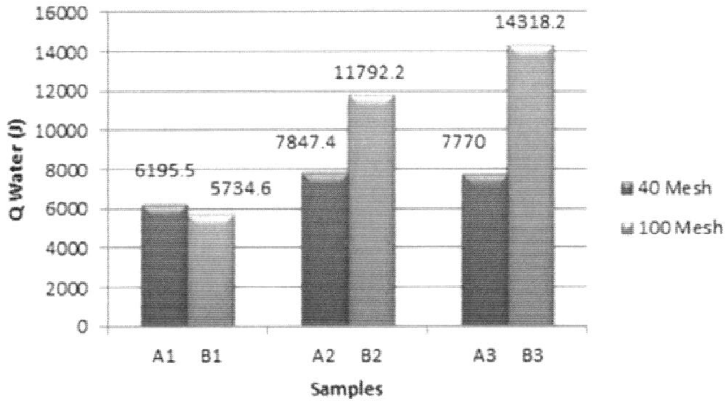

Figure 4. Heat received by water against briquettes particles size.

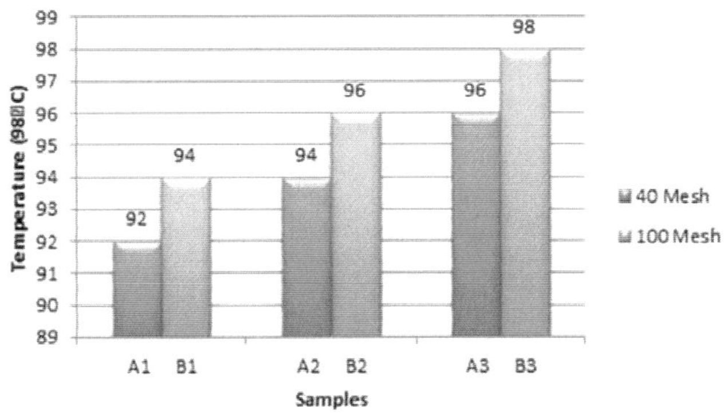

Figure 5. Temperature received by water against briquette particles size.

In figure 4 and 5, the maximum temperature (98°C) and the maximum amount of heat received by water (14,318.18 J) are at composition B3. From both of these figures, the differences in composition ratio are no significant effect on maximum temperature & the amount of heat (Q) received by water.

3.3. Carbon monoxide emission

In table 3 shown, the carbon monoxide was 783 ppm at particles size of 100 mesh but slightly reduced to 632 ppm at particle size 40 mesh. The reduction of ppm value is because smaller particles (100 mesh) have more porosity to hold more oxygen efficiently; hence the combustion perfectly happen. Even though there are differences in composition ratios, the effect was insignificant.

Table 3. CO emission and particles size in correlation with the composition ratio.

Compositions	Sample	Particle Size (Mesh)	CO Emission (ppm)
Rice husk:Corn cob (75:25)	A2	40	632
	B2	100	783
Rice husk:Corn cob (50:50)	A3	40	621
	B3	100	765

4. Conclusions

The advantages of using smaller particles size (in this case: 100 mesh), will increase the combustion rate, increase the temperature achievement in Water Boiling Test & increase the amount of heat (Q) received by water, but also increased the CO emission.

The CO emission could be attributed to the fact that the amount of carbon monoxide is a function of the availability of oxygen during briquettes burning process. Therefore, it cannot be reduced by densification and torrefaction. Even though there are differences in composition ratios, the increasing or decreasing of CO emission that affected by the differences were insignificant.

The briquettes sample B3 that has a composition of 50:50 of rice husk & corn cob, and by using particles of 100 mesh is the best sample in results and ready to apply in a household community and the industrial sector.

References

[1] McKendry P 2002 Energy production from biomass (part 1): Overview of biomass. J. Bioresource Technology **83**(1) 37–46
[2] Strezov V and Evans T J 2014 Biomass Processing Technologies (CRC Press. New York)
[3] Kementrian Pertanian 2000 Sekam Padi Sebagai Sumber Energi Alternatif dalam Rumah Tangga Petani. Badan Litbang Pertanian
[4] Chen et al. 2014 Upgrading of Rice Husk by Torrefaction and its Influence on Fuel Properties, J Bio Resources 9(4) 5893-905

Acknowledgments

We would like to thank the Directorate of Research, Community Engagement and Innovation (DRPMI) – Universitas Padjadjaran for providing funds and facilities through Fundamental Grant Research, Universitas Padjadjaran, the fiscal year 2018.

Integrated model of utilization of organic waste into bio briquettes with community empowerment in West Java

C Mulyana and S Suryaningsih

Departement of Physics, Universitas Padjadjaran, Jl. Bandung-Sumedang Km 21, Jatinangor Sumedang, West Java, Indonesia

E-mail: c.mulyana55@gmail.com

Abstract. Scarcity of fuel gas causes an obstacle of small and medium enterprises (UKM) acceleration. West Java has abundant biomass energy resources, but not yet utilized. Bio-briquette is a biomass product that has high energy content and can substitute LPG. In this research, converting biomass into bio briquette from upstream to downstream, a community empowerment model is engineered. The program begins with collecting potential biomass data in several districts in West Java. Furthermore, encouraging the public to participate in the collection and manufacture of biomass into bio-briquettes. Bio-briquettes making technique was adopted from the research that has been done in various universities. Similarly, prototype chopping-machines, briquette molding, and simple oven have been made in university. The market is households and UKM that have been using LPG. In terms of economic, value of briquette selling more profitable than LPG. The problem is, burning bio briquettes traditional wasteful energy. To overcome this, cooperation with the manufacture of stoves, burners, steamer, and kettle is done and it's more energy efficient. In each stage, mentoring is done. The plan of this program will be completed within two years. The hope, this model can be adopted in other areas, given the biomass resources spread throughout Indonesia.

1. Introduction

Elpiji is a life necessity of many people, both households or small and medium enterprises (SMEs). The use of LPG in Indonesia increases with time. This is related to the success of the government's policy of replacing the use of kerosene into LPG. However from 2007 until 2014 the price of LPG significantly increase, from Rp. 4,250/kg to Rp. 13,000/kg [1].Economic growth accompanied by the increase of LPG demand, resulted in the imbalance between the LPG need and its availability. This is due to the scarcity of oil and gas. In line with presidential rules no. 5 of 2006 concerning national energy policy to develop alternative energy sources as substitution of oil and gas, by diversifying energy the supply and utilization of various energy sources. In addition, a framework for developing energy availability was made by optimizing affordable prices for the community. This policy opens up opportunities for the use of renewable energy sources such as geothermal energy sources, biomass, solar energy as a substitute for oil and gas. Since 1994, coal briquettes have been developed in Indonesia [2]. Coal briquettes have succeeded in substituting the use of oil and gas. This is because coal briquettes heat content is high enough above 5000 calories/gram. But the use of coal briquettes has the disadvantage such as high carbon emissions that pollute the air. This coal briquette so far replaces some of the roles of oil and gas as fuel used in the industry. Indonesia has abundant agricultural commodities such as rice tree (padi), corn tree, palm oil tree, rubber, cocoa coffee and forestry. That biomass wasted are leading commodities for acceleration bio briquette product. Coal briquettes have a high level of heat with low levels of air pollution. Therefore biomass has great potential to be an alternative energy source that is utilized in the form of briquettes.

Some countries have applied the utilization of biomass briquettes as an energy source [3]. Indonesia itself has started to use bio briquettes as an alternative fuel source other than LPG. However, the use of bio briquettes in Indonesia still has weaknesses. Utilization of biomass in the form of briquettes in Indonesia is still not well developed because the management system is not good. Management and distribution of biomass feedstock supply from source to briquette producers until the user has not been well organized. Then there is no cooperation between the government or institutions with the community. Involvement is good in the supply of raw materials briquettes, making biomass into briquette forms, and provision of equipment both for the manufacture of briquettes and for the utilization of the briquette itself. Another weakness is the equipment that people use to utilize briquettes is still simple. So that the use of briquettes still cannot be effective because it has a high level of heat loss. In addition to high level of heat loss, the use of biomass with simple equipment has a weakness by showing by the unfavorable combustion properties, among others, many smoke, ash, and efficiency is very low. Besides, there is no research that is well integrated and serious about briquettes, both research on briquette quality and research on strategic way of briquette utilization in Indonesia.

The potency of bio briquettes in Indonesia, especially in West Java, is very large, especially from plantation and forestry sources such as Kab. Garut (rice, corn, rubber, hemp, wood, coffee), Kab. Sukabumi (timber, coffee, corn) [4]. This potential has not been fully utilized by the government or the community. Along with the rising prices of gas fuel, there are many entrepreneurs who are turning from the production base of gas production base to bio briquettes base production equipment. But the increase in bio briquettes demand is not balanced with its production. This is because the acceleration of utilization of biomass resources as the raw material of bio briquettes is still slow.

In general, this study provides a solution to the problem of bio briquette utilization. The problem of bio-briquette management starts from the provision of material source, manufacture, and distribution of bio-briquette from producer to consumer, as well as the provision of an effective bio briquette tool. The problem of the study in this study is how the strategy of sustainability of supply of raw materials of various biomass through the empowerment of community groups along with the business scheme, how the strategy and implementation of bio briquette production by the community through community empowerment program, how the business briquette scheme produced by the community group is applied by productive community groups and small and medium enterprises, and how to approach and utilize production systems (machines / equipment) by SMEs by using production-based production machine and equipment bio briquettes.

This research has a general purpose that is to create an integrated production system ranging from briquette production to bio-briquette based SMEs production equipment and its utilization by the community along with briquette based business scheme. The specific objective of this research is to develop the community in the form of SMEs Collectors as a provider of biomass sources of raw materials bio briquettes, community empowerment in the form of SMEs Bio-briquettes production as bio briquette producers that will meet the needs of users. The next step develops of an efficient bio briquette system through the development of bio briquettes production machines ranging from spraying machine, printer, author and packaging, approach to productive society to use bio briquettes and stove as alternative energy, development of bio briquettes based production system through development of burner, boiler and kettle, approach to SME producers to implement new briquette-based production system is the production of a community of its own, and establishes a briquette-based community business management system. This research can increase community productivity and energy independence so that ultimately improve the welfare of local communities to SMEs.

2. Literature review

2.1. Biomass briquettes

Biomass is an organic material produced by photosynthesis in the form of products and waste, which then these organic materials are used as a fuel source. With the process of photosynthesis, plants store solar energy and storage of solar energy is used as energy. Biomass consists of several components: moisture content, volatile matter, fixed carbon, and ash. The biomass burning mechanism consists of three stages: drying, devolatilization, and charcoal combustion. The drying process will remove the water content, devolatilization which is the pyrolysis stage will release volatile, and charcoal burning which is the reaction stage between carbon and oxygen will release the heat. The rate of charcoal burning depends on the reaction rate between the carbon and oxygen at the surface and the rate of oxygen diffusion in the boundary and inner layers of the charcoal. The surface reaction mainly forms CO. Outside the particles, the CO will react further to form CO_2. The burning process will leave the material ash. The carbon contained in the charcoal reacts with oxygen on the surface to form carbon monoxide according to the following reaction [5]:

$$C + \tfrac{1}{2} O_2 \rightarrow CO \quad (1)$$

The carbon surface also reacts with carbon dioxide and water vapor with the reduction reaction as follows:

$$C + CO_2 \rightarrow 2CO \quad (2)$$
$$C + H_2O \rightarrow CO + H_2 \quad (3)$$

During the carbonization process, flammable gases such as CO, CH_4, H_2, formaldehyde, methane, formic acid, and acetic acid and non-combustible gases such as CO_2, H_2O and liquid tar are released. The gases released in this process have a calorific value that can be used to meet the heat requirements of the carbonization process [5]. Various biomass can be used as bio-briquette through the process of briquettes such as hemp tree [6], corn stalks and cobs [7], wood processing waste, coconut shell, coconut husk, stem and peanut skin, straw, rice husk, durian skin, banana peel [8]. Each biomass source has a diverse energy content. The coconut shell content of about 5,780 cal/gram is the highest non-coalic biomass (Table 1). While the highest content of heat is coal that has been carbonized for 6,158 cal/gram.

Table 1. Biomass content [9].

No	Type	Heat Values (cal/g)
1	Coconut shell	5780
2	Sawdust teak wood	5479
3	Rice husk	3073
4	Coal is carbonized	6158
5	Non-carbonized coal	6058
6	Corncob	5351
7	Wood charcoal	3583

Some parameters determine the quality of bio-briquette ie moisture content, ash content, volatile substances, carbon-bound, heating value, density, and compressive strength. The briquetting pressure can control the combustion rate of the bio briquette. The greater busting pressure slows the rate of combustion and reduces the rate of COB of corncob bio briquette [7]. As a glue, many researchers use starchy starch. Maryono [10] reported 1-2% starch content producing bio briquettewith ash content below 8% according to SNI standard. Mixing of variations in biomass materials determines the quality of the bio briquette. Cow manure mixed with biomass in the form of husk, straw, and shell (in the ratio of 1:3) can increase the heat content to near the SNI standard [11]. Bio briquette made from mixed biomass between durian and banana peels is reportedly still needing an effort to increase its calorific value [8]. In fact, biomass water hyacinth mixed with coal briquettes and carbonized at temperatures of 400 - 600 ° C can increase the content of briquette heat [12]. The biomass type also determines the combustion characteristics, such as the ignition time (to ash), the amount of smoke and initial startup time. Jamilatun [13] reported that it takes a long time to ignite 116.1 minutes for a coconut shell weighing 220.51 grams, longer than sawdust teak, rice husk and coal. However, coconut shells produce more smoke and require a longer initial start-up time.

2.2. Bio briquette stove

The performance of combustion is determined by various parameters of the biomass briquette production process and the characteristics of the bio briquettes itself. The result of this briquette can be used for the actual combustion process on the stove but to obtain efficient combustion it needs a furnace in accordance with the characteristics of the bio briquettes. An efficient furnace can provide the right combustion air supply so that briquette burning is perfect. With an appropriate furnace, the resulting briquettes can be used for real applications. Saputro [14] performed stove testing by measuring combustion temperature, mass reduction rate during combustion and combustion rate; and reported that the fastest combustion rate and the highest temperature of the bio briquettes (corncob with a starch adhesive) were achieved at an airflow velocity of 0.6 m/s. Therefore, the combustion temperature and length of briquette combustion in operation can be controlled by air flow rate.

2.3. Social engineering

The effort to handle a social problem depends on how we view or the glasses used to observe it. Let's look at the general view of the authorities/government against the community. Some time ago, the view of the community still revolves around that the community needs to be helped, the community has many needs that need to be met, people spend the State's money because it needs to be given many public facilities and others. This is a problem-based or needs-based perspective. The implication is that community issues must be solved and or the needs of the community must be met. This view assumes that the community is not able to solve the problem and can't fulfil its own needs so it must always be assisted, and fulfilled its needs. As long as the approach used is still like this, so long as that society considers itself a party to be met needs, so that would never think to do something for themselves and for the closest social environment.

Gary Paul Green [15] reminds us of a new worldview (though practiced quite a bit in Indonesia), which he named Asset Based Community Development. This is a mindset that assumes that almost every community group has assets or resources that have not been maximally utilized, rather than having needs or problems to be met / solved. This is a view that believes that every member of society actually has the power to overcome its own problems and empowers its potential instead of having weaknesses and shortcomings that result in them always shackled in various difficulties.

So, if the first thing to try to see or map is the weaknesses, problems and needs that exist in the community, then the development of the asset-based community would like to focus on the potential, assets or resources of any existing in the community that can be developed. As for assets that try to be described further in this case include community capital such as human capital, physical capital, social capital and including environmental capital.

This approach will encourage us to map the community in different ways. For example, instead of describing the number of poor people, we are trying to identify potential opportunities for skilled-based economic development and work experience of people in an area. Or, instead of identifying the damage to natural resources, we can describe what natural potentials have not yet been maximized and can still be cultivated for value enhancement.

In general, Asset Based Community Development classes, will encourage us to think about all the potential that exists in society. By using this asset-based approach, we will not see the residents are those who are unemployed, some have narcotic abuse, some are vulnerable, some dropout and others. But we will view them as a group of individuals who have different skills, varying work experiences, different network ownership etc., so it only needs to be synergized to produce a great 'power', including to cultivate their own garbage.

Furthermore, this view will encourage us (government, private or other individual societies) who want to help, can choose the role appropriately. The government, no longer necessarily acts as a 'sanner' who distributes aid, but may act as (1) facilitators who facilitate their own independent forces, or (2) as mediators, who can mediate works and initiatives has been generated by the community to add value, or (3) as advocates, who are reminded of their rights and their potential and resources.

3. Methods

This program will be conducted with several steps, namely:

3.1. Mapping biomass potential to become bio briquettes in West Java

Mapping biomass potential into biobriquette has been carried out by Cukup Mulyana [4]. West Java has the potential to produce biobriquettes because it has a variety of plants beside it has forests whose waste can be converted into bio briquettes. From rice husk energy will be equal to 68,921 1011 kcal while from corn cobs the energy is equivalent to 11.75 1010 kcal, and from the coffee bark it is equivalent to 5,752 x 1012 kcal.

3.2. Sorting raw materials of biomass

In this stage, biomass is selected according to its quality. Biomass that is suitable for making biobriquettes is separated and then molded into biobriquette. In West Java, the potential for bio-briquettes is almost found in every district. The process of biomass sorting is carried out by the community. There are two types of biomass, first biomass from agricultural waste such as stem and head of maize, rice husk, and coffee skin; the second is biomass which is intentionally planted as industrial plants, such as from kaliandra trees, rami trees.

3.3. Designing social engineering

Social engineering is needed to involve the community in the processes, it's start with collecting biomass, making and marketing of bio-briquettes, and changing the habits of using fossil fuel to renewable energy bio briquette. The process is done in the form"Appreciative Inquiry Application for the development of potential community ". In general, activities will be conducted in four stages:

1. Asset Social Assessment: with the aim to uncover the potential of social assets that exist in the community
2. Socialization of assessment results: conducted to sensitize the public related potential
3. Appreciative Inquiry: done to generate a positive atmosphere among community members, share dreams and also design things that can be done in the future
4. Follow-up with plan and group formation

The expected outcomes from these social activities are:

1. There is a change in people's mindset about the benefits and potential use of briquettes in the future
2. There is a desire to maximize the use of briquettes, both for personal and business needs
3. A jointly agreed action plan is formed, namely groups, and if possible a management system of collectors producer and user can be created

3.4. The supply system of the bio briquette production machine

The supply system of bio briquette machines should involve capital owners as funders, society and government as determinants of policies. Parties from universities, research institutions and polytechnics and contribute to the design of making bio briquette prototypes, machines/tools to produce small-scale bio-briquettes and stoves, as well as industrial scale. Some examples of products can be found in bio briquette manufacturers, but the numbers are still very few with poor quality. Manufacturing is needed that is able to prepare machines such as crushing machines, drying machines with good quality but the price is still affordable. To meet these needs, capital suppliers are needed, community involvement and positive support from the local government.

3.5. Manufacture of biomass products

In the process of making biomass products involved the role of universities (PT), outside agencies, local governments, and communities that must be mutually beneficial to each other. The procedure for making bio briquettes is done with the following steps:

1. Milling
   Milling is done by feeding the raw materials alternately into the grinding machine to form the powder. Milling is done again using a blender to form a finer powder.
2. Preparation of adhesive material
   Adhesives that are prepared to bind the powder to make it easy to print. Adhesive of tapioca added hot water then stirred until the dough becomes sticky.
3. Mixing of raw materials and adhesives
   Before mixing the material first weigh the powder, the powder of the composition variant and the adhesive material corresponds to a certain concentration for different variants.
4. Molding process
   The molding process is done by inserting the mixture into the briquette pressing device with the pressure and temperature varied.

3.6. Provision of bio briquette production equipment and machinery

The supply of machine and equipment to produces bio-briquettes is done by prototypes that have been designed by researchers in higher education and researchers then collaborate with small-scale machine-making industries that usually work with SMEs. The equipment is in the form of a chopper, drying machine, molding machine, bio briquette stove for use in the household, a burner for use in SMEs and a kettle for processing products (food).

3.7. The design of distribution channels

Distribution channels for the manufacture of bio-briquettes from biomass include channels for the supply of biomass that are ready to be made bio-briquettes. For this channel to be formed, cooperation between bio-briquette producers and the community is needed as the owner of biomass waste. While the channel for the manufacture of bio briquettes is formed by bringing together the owners of capital, the community and the local government so that the mechanism for making bio-briquettes is cheap, involving the community but still of economic value. While marketing channels are formed by producers of bio-briquette makers and consumer consumers and consumers of SMEs, which are quite large in West Java.

4. Results and discussion

The well-integrated briquette management system is depicted in Figure 1. The management consists of three major activities. It's begin with the provision of raw material briquettes and then bio briquette production and finally selling activities with the market targets community group such as *pesantren*, household need and small + medium industry (SME's) such as tofu Industry etc- the box with dot line. In developing activity of biomass collector need social engineering program because it's involves communities, need an expert for guiding, funding and government policy.

IOP Conf. Series: Materials Science and Engineering **550** (2019) 012007 doi:10.1088/1757-899X/550/1/012007

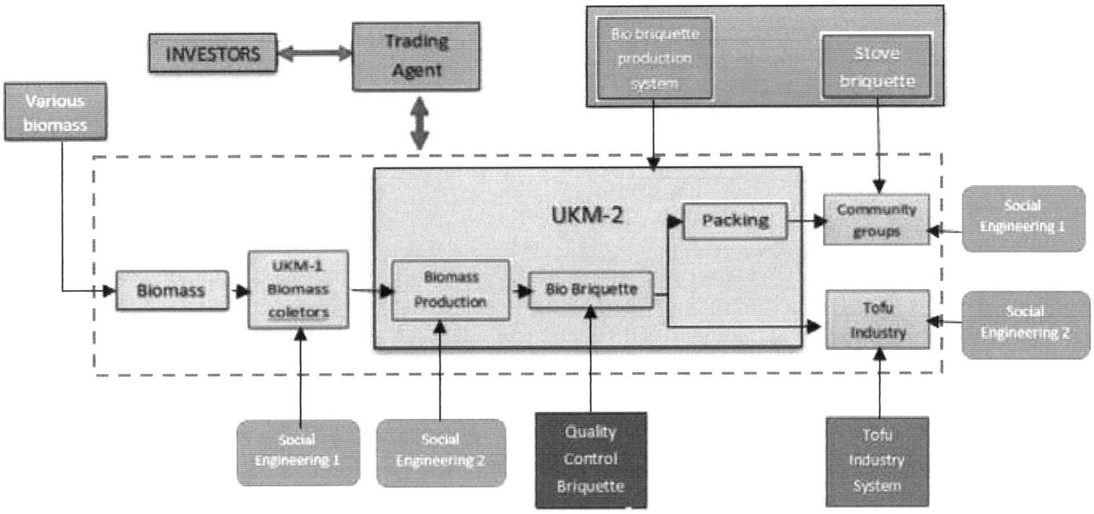

Figure1. Flowchart of biomass management.

4.1. Provision of biomass raw materials
The type of biomass chosen to be bio briquette depends on the availability in each district. From the results of previous studies the types of biomass that will be made into bio briquettes for each. For each district, the type has been determined. The process of collecting biomass includes transportation from the field to the location of collecting, sorting and winding. This activity involves people who were previously unfamiliar with industrial processes. Therefore social engineering is needed to change attitudes and paradigms. This program was designed and prepared by a team involving plant waste experts, social science experts, communications and teams from psychology. In addition, there are economists to calculate the price of biomass to remain economically valuable when it has been converted to bio briquettes.

4.2. Bio briquette production
After the biomass is dry, the biomass is cut to a certain size and then dried again under the sun. Then there are carbonized, mashed, filtered, mixed with adhesive, then molded and dried to reduce the water content. For making bio briquettes, a chopper machine is needed, molded machines, and carbonized machine. All necessary equipment and machinery are prepared from the results of research and development by workshops that have been tested for their functional abilities and costs. The process of making bio briquettes requires standardization of the heat content, density, compressive strength, ash content, and gas content that appears when burned. Also in the burning time test. Standardization and quality control continue to be carried out and accompanied by researchers and authorized institutions.

4.3. Bio briquette support equipment
Bio-briquette support equipmentis bio-briquette stove, burner, steamer, and kettle. The bio briquette stove is different from a gas stove. To get a stove that works efficiently requires good design and engineering. Likewise, for other components, the Bio Briquette Stove has been designed and has been made a prototype. Its performance has been measured but it hasn't been made on an industrial scale. Whereas for steamer,burner and kettle there are already manufactured and have started to be marketed. The market demand is small industries, such as tofu factories, crackers or various food products produced by SMEs not only those that are located around West Java and even across islands in Indonesia.

4.4. Investor and trading agent
To run a biobriquette program from biomass, the employers need funding. Even though its potential is wide open, a supportive business atmosphere is needed, government policies that provide opportunities supporting biobriquette programs as a substitute for LPG fuel and provide economical benefits. A good synergy from various parties are needed in order to success the integrated model of bio briquette management from upstream to downstream

5. Conclusion
A well-integrated bio briquettes management system is done by involving the role of the community in it. Thus, there is good cooperation between communities, universities, institutions external, and governments. The community is empowered from starting the gathering of biomass raw materials and then making the briquettes to the distribution system. Utilization of bio briquettes is maximized by designing equipment that suits the needs of consumers. The designed equipment is briquette stove, burner, boiler, and kettle. Indonesia has an abundant source of biomass that has not been utilized so far. This program can be duplicated in other provinces in Indonesia, so if this program is realized it will increase the energy independence of each province and become a new source of income for the community.

References
[1] Ahniar and Nur Farida 2014 "Sejak 2007, LPG 12 Kg di bawah Harga keekonomian", (https://katadata.co.id/berita/2014/01/08/sejak-2007-lpg-12-kg-di-bawah-harga-keekonomian, accessed on 1 Januari 2019)
[2] Kementerian Energi dan Sumber Daya Mineral 2015 *Perkembangan Penyediaan dan Pemanfaatan Migas Batubara Energi Baru Terbarukan dan Listrik* (Jakarta : Pusat Data dan Teknologi Informasi Energi dan Sumber Daya Mineral Kementerian Energi dan Sumber Daya Mineral
[3] Gladstone S, Tersigni V, Kennedy J and Haldemana J A 2014 Targeting briquetting as an alternative fuel source in Tanzania *Procedia Engineering* **78** 287 – 29
[4] Mulyana C, N I Fitriani, A Saad and Y Yuliah 2016 *IOP Conf. Ser.: Earth Environ. Sci.* 65 136
[5] Francescato V, Antonin E, Bergomi L Z 2008 Wood fuels handbook, Ed. House (Legnaro: AIEL-Italian Agroforestry Energy Association)
[6] Wulandari A P, Anne N and Taslim 2012 *Bioteknologi Pemanfaatan Limbah Hasil Pengolahan Serat Rami (Boehmeria nivea) untuk Pengembangan Eko-teknologi Industri Rami (ETIR) di Indonesia* Report on MP3EI No 213/SP2H/PL/Dit.Litabmas/V/2013
[7] Surono B U 2010 *Jurnal Rekayasa Proses* **4**(1)
[8] Paisal and Karyani M S 2014 *Proceedings Seminar Nasional Teknik Mesin Universitas Trisakti*. (Jakarta: Fakultas Teknik Mesin Trisakti)
[9] Jamilatun S 2008 *Jurnal Rekayasa Proses* **2**(2) 37–40
[10] Maryono, Sudding and Rahmawati 2013 *Jurnal Chemica* **14**(1) 74 – 83
[11] Anugrah S P 2010 *Studi Variasi Komposisi Bahan Penyusun Briket dari Kotoran Sapi dan Limbah Pertanian* Master Thesis (Padang: Faculty of Agriculture Technology Universitas Andalas)
[12] Fachry, Rasyidi A, Sari TI, Dipura AY and Najamudin J 2010 *Seminar Nasional Perkembangan Riset dan Teknologi ke 16* (Yogyakarta: Pusat Studi Ilmu Teknik UGM) 52–58
[13] Jamilatun, Siti 2011 *Prosiding Seminar Nasional Teknik Kimia "Kejuangan*
[14] Saputro D D 2009 *Jurnal Kompetensi Teknik* (**1**)1 15–19
[15] Green, Gary P and Haines A 2002 *Asset building & community development* (New York: Sage Publications)

Acknowledgement
The authors would like to thank to Universitas Padjadjaran for the partially financial support under Academic Leadership Grant (ALG) 2018.

IOP Conf. Series: Materials Science and Engineering **550** (2019) 012008 doi:10.1088/1757-899X/550/1/012008

The SCADA system using PLC and HMI to improve the effectiveness and efficiency of production processes

A Setiawan, Sugeng, K I Koesoema, S Bakhri and J Aditya

Teknik Elektro, Universitas Pamulang, Kampus Viktor,
Jl. Raya Puspiptek, Setu, Kota Tangerang Selatan, Banten 15314, Indonesia

E-mail: dosen00935@unpam.ac.id

Abstract. One of the problems in the industry is waste. There are 7 wastes, namely excessive production, inventory, defect/damage, transfer/transportation, movement, waiting time and excessive processes are things that absorb and waste resources, expenses, and additional time but don't provide any added value in the activity, and therefore this research is fundamental to create an effective and efficient production process. The purpose of this research is to find out how much effectiveness and efficiency can be achieved by a SCADA system using PLC and HMI on industrial machines. This research uses a comparative method between a conventional control system with a SCADA system using PLC and HMI to get cost and time savings results. The results obtained in this study are SCADA systems using PLC and HMI can reduce production from 1.1% to 0.3%, increase productivity 6, 1%, saving production time from 396 minutes to 352 minutes and reduce costs from 6, 71%. The conclusion obtained in this research SCADA system using PLC and HMI can improve the effectiveness and efficiency in the production process in industries.

1. Introduction

There are so many problems in the industrial world, one of them is the waste. It's called 7 waste because it consists of 7 categories of waste [1], including excessive production of products that are ready to sell or products that are still in the process stage; inventory waste, occurs due to accumulation of finished products, semi-finished products and raw materials that require storage and additional employees to control and to document; defects or damage waste, occurs due to poor quality of production resultsso that needs extra repair costs and labor costs; transportation waste due to poor layout of production areas and machines that requires activities to move goods from one process to the next process; movement waste, due to movements of machines or employees that don't provide added value; waiting time waste, due tothe machines or workers don't do work because of waiting for status, orders, engine damage or lack of materials, so that the production process is hampered; the process wastes, because to produce a product requires reprocessing or repairs such as a repeat inspection process.

Therefore this research is v ital to create an effective and efficient production process. The purpose of this research is to find out how much effectiveness and efficiency can be achieved by the SCADA system using PLC and HMI on industrial machines. This research uses a comparative method between a conventional manual control system based on physical buttons and relays with the SCADA system using PLC and HMI.

IOP Conf. Series: Materials Science and Engineering **550** (2019) 012008 doi:10.1088/1757-899X/550/1/012008

2. System presentations

The SCADA system in this research shown in figure 1. SCADA (Supervisory Control and Data Acquisition) is a computer-based monitoring and control system that is used for data controlling, monitoring and data acquisition [2-5] of a process such as manufacturing production processes, electricity generation and so on. The SCADA system has four functions: data acquisition, network communication, data presentation, data control and monitoring. The SCADA has four main supporting components, namely:

1. Input and output devices are connected directly to the PLC which is used as a data acquisition tool to find out the current conditions on the running system, such as temperature sensors, air pressure sensors, count sensors, etc.
2. RTUs (Remote Telemetry Units) are computers, in this case is a PLC (programmable logic controller) which functions to collect data from sensors and send it to a SCADA master or HMI. RTUs convert analog/digital signals provided by sensors into signals that can be sent to the SCADA Master according to the protocol used.
3. SCADA Master / Human Machine Interface (HMI), that connects users directly with the system to be able to carry out data controlling, monitoring and data acquisition.
4. Communication network, which connects SCADA Masters with RTUs either through RS232 network.

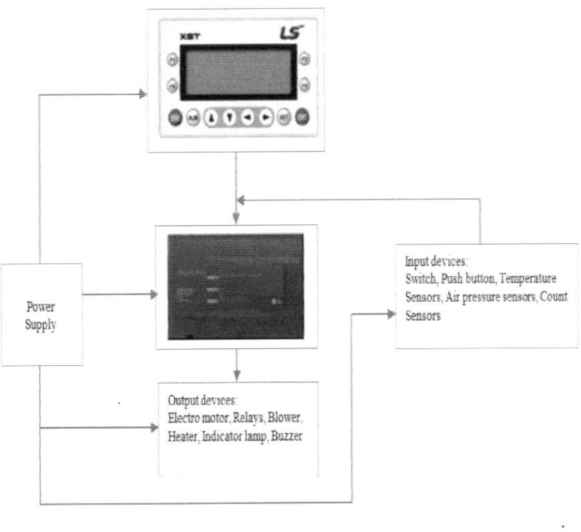

Figure 1. Block diagram of SCADA systems.

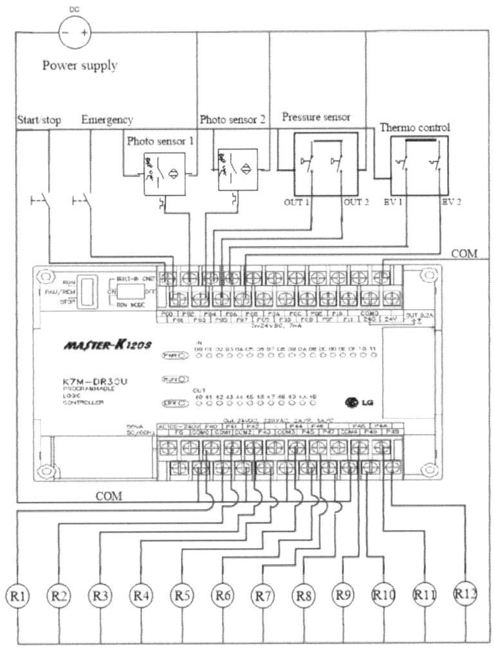

Figure 2. Wiring diagram of input-output PLC.

Figure 2 shows a PLC (programmable logic controller) LS Master K120s with wiring input-output. PLC (Programmable Logic Controller) is an electronic computer that is easy to use (user-friendly) which has control function for various types and levels of difficulty. PLC (Programmable Logic Controller) is a computer specifically designed to control process or machine. This controlled process can be in the form of variable regulation continuously as in servo systems or only involves two-state control (On/Off) but is done repeatedly as we commonly find in a drilling machine, conveyor system and so on.

IOP Conf. Series: Materials Science and Engineering **550** (2019) 012008 doi:10.1088/1757-899X/550/1/012008

Definition of PLC (Programmable Logic Controller) is an electronic system that is operated digitally and designed for use in an industrial environment, this system uses programmable memory to internally store instructions that implement specific functions such as logic, sequence, timing, enumeration and arithmetic operations to control machines or processes through digital I/O (Input / Output) modules or analog [6-7].

Based on the name, the PLC concept is:

1. Programmable; shows the ability in terms of memory to store programs that have been made that are easily changed in function or usefulness.
2. Logic; shows the ability to process inputs in arithmetic and logic that is to do operations such as comparing, summing, multiplying, dividing, reducing, negation, AND, OR and so on.
3. Controller; shows the ability to control and manage the process to produce the desired output.

This PLC is designed to replace a sequential relay circuit in a control system. Besides being programmable, this tool can also be controlled and operated by people who don't have knowledge in the field of computer operation specifically. PLC has a programming language that is easy to understand and can be operated if the program that has been created using software that is in accordance with the type of PLC used has been entered.

This tool works based on existing inputs and depends on circumstances at a particular time which will then ON or OFF the outputs. (1) indicates that the expected situation is fulfilled while (0) means that the expected condition is not fulfilled. PLC can also be applied to control systems that have a lot of output. Table 1 shows a list of input-output PLC.

Table 1. List of input-output PLC

Input		Output	
Address	Description	Address	Description
P0000	Switch start/stop	P40	Relay 1: conveyor A
P0001	Emergency switch	P41	Relay 2: conveyor B
P0002	Photoelectric sensor 1	P42	Relay 3: circulating fan
P0003	Photoelectric sensor 2	P43	Relay 4: blower
P0004	Pressure sensor OUT 1	P44	Relay 5: heater
P0005	Pressure sensor OUT 2	P45	Relay 6: spray disc A
P0006	Thermocontrol EV 1	P46	Relay 7: spray disc B
P0007	Thermocontrol EV 2	P47	Relay 8: start lamp
		P48	Relay 9: indicator lamp (low temperature) and buzzer
		P49	Relay 10: indicator lamp Operate temperature
		P4A	Relay 11: indicator lamp low Air
		P4B	Relay 12: indicator lamp normal Air

Figure 3 shows the PLC program created for this SCADA system.

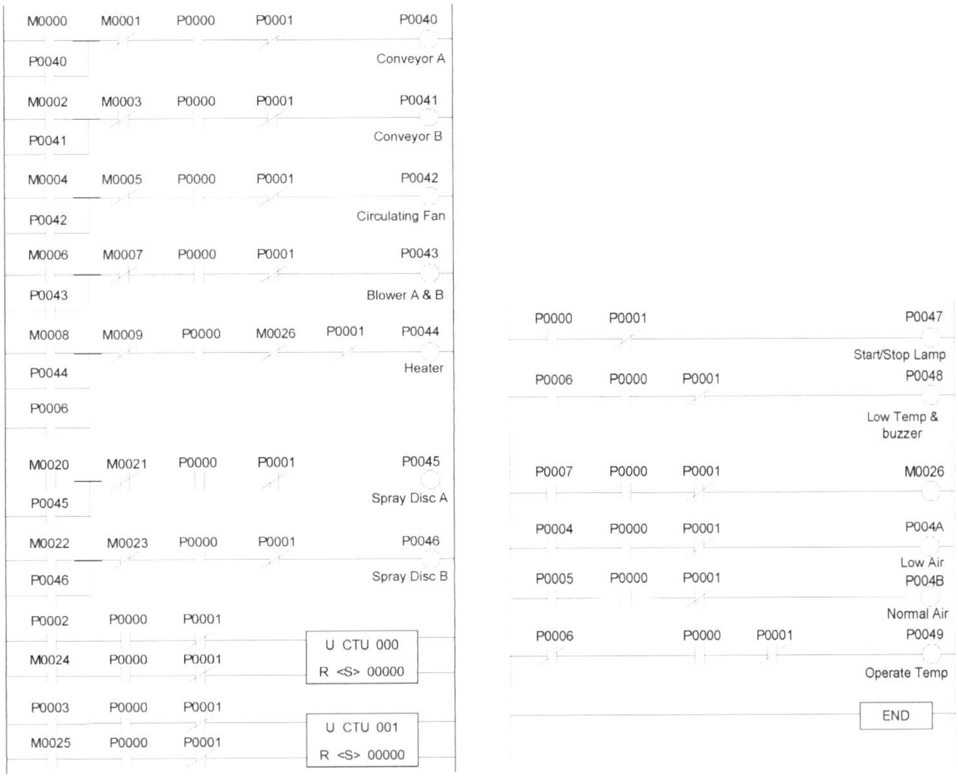

Figure 3. Program of PLC.

Figure 4 shows the power wiring of SCADA systems, the power circuit works by activating the MCB to the ON position so that the voltage goes into the circuit through a fuse. Turn the power switch to ON so that the zero relays (R0) will be active and R0 contacts change from NO (Normally Open) to NC (Normally Close) so that the voltage enters and activates PLC (Programmable Logic Controller), PSU (Power Supply Unit), TC (Thermocontrol), and power indicator lights. Step to turn off the circuit by turning the power switch to OFF and deactivate the MCB to OFF so that all equipment returns to OFF.

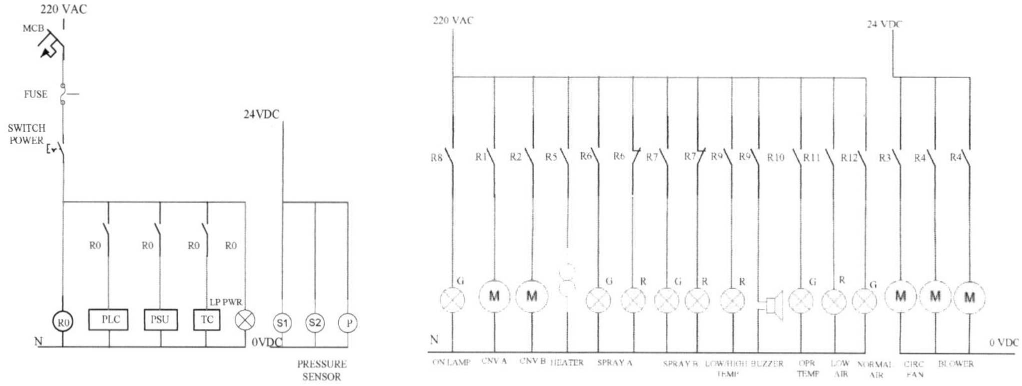

Figure 4. Power wiring of SCADA systems and devices.

Table 2. List equipment of power wiring.

Equipment	Description
PLC	Programmable logic controller
PSU	Power supply unit
TC	Thermocontrol
LP	Indicator Lamp
S1	Count sensor 1
S2	Count sensor 2
P	Pressure sensor
CNV (M)	Motor conveyor
	Heater
	Buzzer
	Circulation fan
	Blower

The HMI (Human Machine Interface) is to make visualization of technology or system in real terms so that the HMI design can be adjusted to facilitate physical work [8-9]. The HMI aims to increase the interaction between the machine and the operator through the computer screen, touch screen and HMI micropanel display and meet user needs for system information. HMI in the manufacturing industry is a Graphic User Interface (GUI) display on a computer screen that will be faced by machine operators and users who need machine working data. In the HMI, there are various kinds of visualizations for monitoring and machine data that are connected online and in real time. HMI will provide the description of machine condition in the form of a production machine map that can be seen which part of the machine is working. In HMI there's also a machine controller visualization in the form of buttons, sliders and so on that can be used to control the machine as it should. Furthermore, the HMI also displays an alarm in case of danger in the system. HMI also displays machine work summary data including graphically.

By using online and real-time systems on the HMI can be obtained a system that can be controlled and monitored as soon as events occur. With the system integrated HMI, all events can be observed from the monitor screen and can control the system immediately through a computer monitor. So that this system can save time and employee to monitor and control each production workstation. Changes online can include adding tags to HMIs, controllers, editing logic on the program and controlling the OS process. An online and real-time system in HMI requires the connection between computer, control system and production machine. Under these conditions, simultaneous information can be transferred from sensors and actuators to the controller and from the controller to the computer, namely to the HMI and database. Likewise in the opposite direction. The real-time system is a system that provides information on a situation, where the information displayed on the receiving side matches the side observed in both times in circumstances. Figure 5 and 6 shows HMI displays.

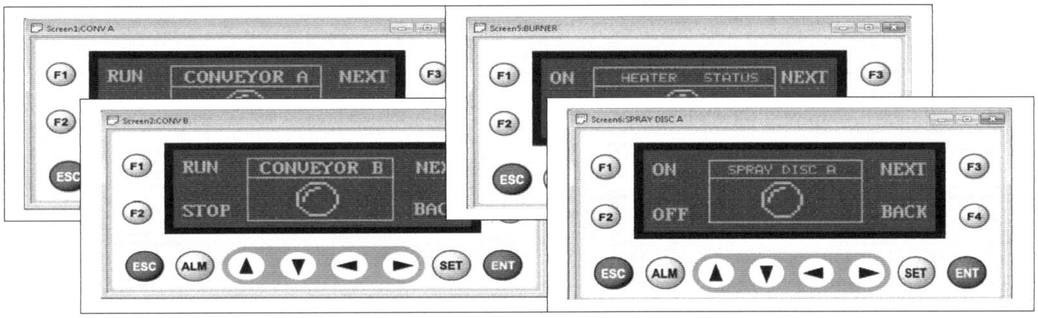

Figure 5. HMI display page 2-5.

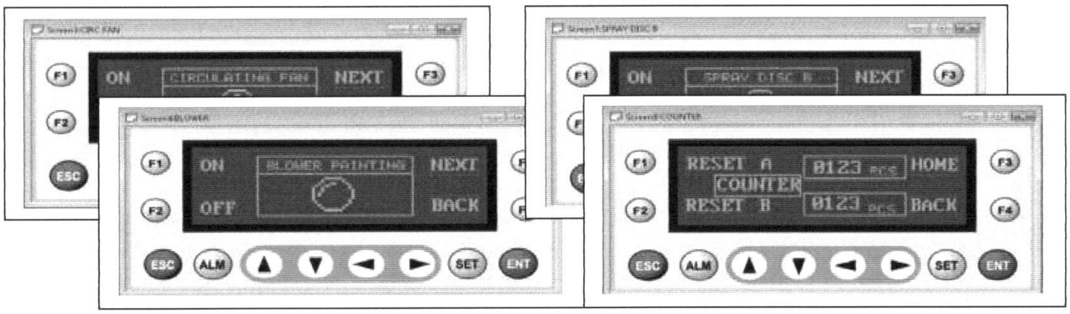

Figure 6. HMI display page 6-9.

The work process of SCADA system using PLC and HMI starting from the control section provides input to the PLC process section that will process it according to the program that has been made and then forwarded to the output module to activate the relay junction. If the relay is active, the relay contact used to supply voltage of 220 VAC toward the load that requires a voltage of 220 VAC and if the relay is inactive, it must be checked on the PLC input and output modules. Line painting model consists of various equipment as a display of the system. The readings result of the thermocouple and photoelectric sensors provide feedback to the process as part of the control.

The following is the input and output path of each switch and keypad that is processed by the PLC

- The start switch will send the input signal to the PLC which is then processed according to the program that has been entered into the PLC. If the PLC gives outputs, the R8 will be active so that the start indicator light turns on and if the light doesn't turn on, it must be checked to start from the PLC input-output section of the relay and its indicator light.
- The F1 keypad screen 2, 3, 4, 5, 6, 7 and 8 gives the input signal to the PLC which is then processed according to the program that has been entered into the PLC. If the PLC gives outputs, the relay will be active so that the loads will be active and if the loads are inactive, it must be checked to start from the PLC input-output section of the relay and the load part.
- The F2 keypad screen 2, 3, 4, 5, 6, 7 and 8 gives the input signal to the PLC which is then processed according to the program that has been entered into the PLC which means that if the keypad is pressed, it will stop work rather than those loads.
- The F1 and F2 Keypad screen 9 only to reset the counter count. The status of the load and counter is displayed digitally by the HMI.

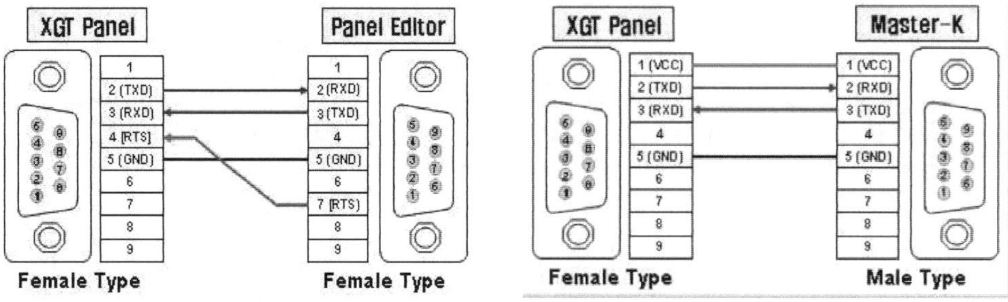

Figure 7. Communication network.

Figure 7 shows a communication network. The XGT panel that contains the program to be able to communicate with the PLC can be connected by using the RS232C serial port and cable and programs that have been created in the editor panel can be uploaded or downloaded by using the RS232C serial port and cable, the configuration of the cable arrangement has adjusted to figure 5.

3. Results and discussion

Based on the system representation in the previous section, the circuit models is realized and tested with the previous arrangement given in figure 1. Procedures of SCADA systems is Data Acquisition (Sensor): Obtain data from the equipment in the field to find out the real-time condition of the running processes, for example: the condition of the device on/off, the value of motor speed, temperature, pressure, counter, etc. Network Communication (Remote Telemetry Units): Data communication between sensors with RTU/PLC and RTU/PLC with SCADA master/HMI. In this case, RTU has an important role to bridge between sensors and SCADA network, RTU converts the sensor/relay input signal into the corresponding protocol format and sends it to the SCADA master. Data Presentation (HMI): The SCADA system can present the data needed by the user in the desired format. Control System (HMI): Control of the ongoing process can be carried in full through HMI, automatically or manually. Table 3 shows the comparison between a conventional control system with the SCADA system.

Table 3. Comparison between a conventional control system with the SCADA system.

Process	Conventional system	SCADA system
Data acquisition	To obtain data have to look directly at the position where the measuring instruments	
Data presentation	Data collection and Data input is done manually on the computer, data presentation take a long time	All can be done in one screen (HMI) easily and quickly
Control and Monitoring	Control and monitoring is done directly on the machine	
Troubleshooting	Problem analysis is done by checking one by one on the equipment	With an alarm system
	Count sensor 1	

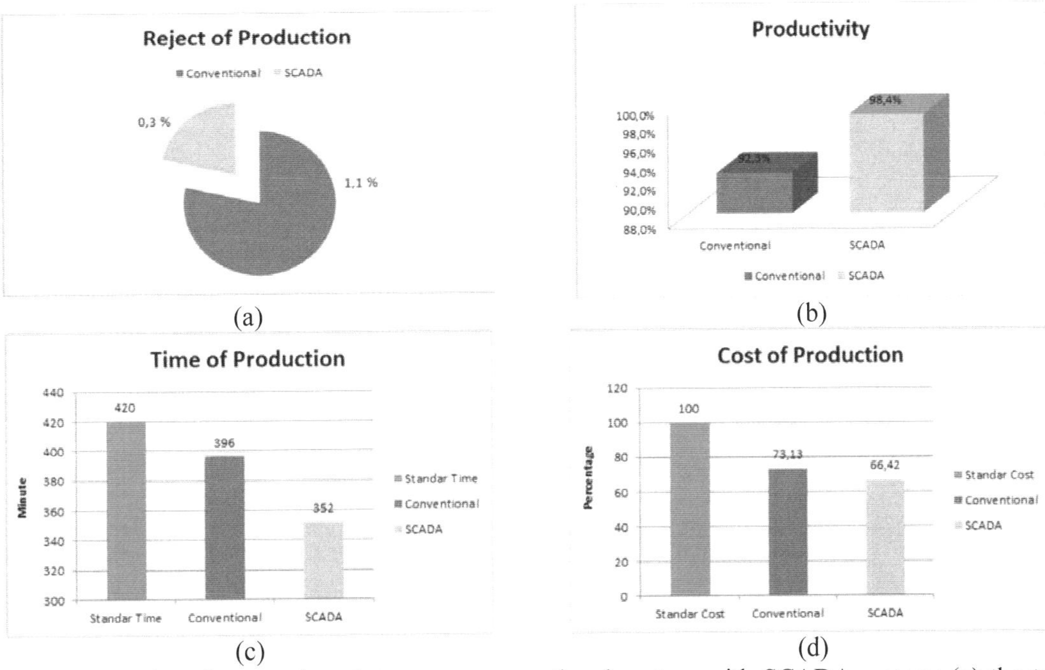

Figure 8. Results of comparison between a conventional system with SCADA system; (a) the total value of the rejected product; (b) productivity of conventional and SCADA system; (c) the length of time of production; (d) the cost of producing conventional systems and SCADA systems.

Figure 8 shows the results of comparison between a conventional system and a SCADA system in the pie chart (a) shows the total value of the rejected product that is produced from the production process for a conventional system of total rejection 1.1% and the SCADA system producing a better total rejection of 0.3%. In the bar graph (b) the SCADA system hasa higher productivity than the conventional system which is a productivity increase of 6.1%. In the bar graph (c) shows the length of time the SCADA system is able to do work faster than conventional systems. In the bar graph (d) shows the cost of producing conventional systems and SCADA systems compared to the predetermined standard costs SCADA system costs more economically 6.71% compared to conventional systems.

4. Conclusion

The process of designing and manufacturing SCADA systems based on PLC (Programmable Logic Controller) and HMI (Human Machine Interface) consists of two designs, namely hardware and software, for hardware consisting of PLC input/output design, power and power circuits for software comprising of a ladder diagram PLC and make an appearance on the HMI. After testing the system, the whole system can work properly, operators can more easily control, monitor and retrieve the data needed, easier to maintain and solve problems that occur, the use of HMI (human machine interface) makes installation wiring more simple, that it can be concluded that the results obtained in this research are two benefits, the first SCADA system can improve the effectiveness of the production process with operational ease. The second SCADA systems can increase efficiency by saving the cost and time used in the production process.

References

[1] Jakfar A, Setiawan Wahyu E and Masudin I 2014 Waste reduction using lean manufacturing approach Scientific *Journal of Industrial Engineering* **13** 01
[2] Panchal P, Mahesuria G, Panchal R, Patel R, Sonara D, Pitroda D, Nimavat H, Tanna V and Pradhan S 2016 Up-gradation in SCADA and PLC of existing LN control system for SST-1

Fusion Eng. Des. **112** 883–7
[3] Sangeetha Lakshmi A, Naveenkumar B, Ganesh Balaji A and Bharathi N 2012 Experimental validation of PID based cascade control system through SCADA–PLC–OPC and internet architectures *Measurement* **45** 643–9
[4] Morsi I and El-Din Mohy L 2014 SCADA system for oil refinery control *Measurement* **47** 5–13
[5] Figueiredo J and Costa José S 2012 A SCADA system for energy management in intelligent buildings *Energy and Buildings* **49** 85–98
[6] Priyanka E B, Maheswari C and Thangavel S 2018 Online Monitoring and Control of Flow rate in Oil Pipelines Transportation System by using PLC based Fuzzy-PID Controller *Flow. Meas. Instrum.* **62** 144–51
[7] Cristani M, Demrozi F and Tomazzoli C 2018 ONTO-PLC: An ontology-driven methodology for converting PLC industrial plants to IoT *Procedia Comput. Sci.* **126** 527–36
[8] Ravaie K and Valizadeh A-h 2002 Selecting human machine interface package in integratinga process automation system *ISA Transactions* **41** 115–26
[9] Priyadharson A S M, Ganesan R and Surarapu P K 2015 PLC – HMI Automation based Cascaded Fuzzy PID for Efficient Energy Management and Storage in Real Time Performance of a Hydro Electric Pumped Storage Power Plant *Proc. Technol.* **21** 248–55

Analysis harmonic distortion of side outgoing transformer 20 kv/ 1.2 kv in traction substation of Tanjung Barat KRL Station

A Triyanto[1], A Sofwan[2] and A Sunardi[1]

[1]Electrical Engineering, Pamulang University, Viktor Campus, Jl. Raya Puspiptek, Setu, South Tangerang City, Banten 15314
[2]ISTN Campus, Jl. Duren Tiga No. 38, Pancoran, South Jakarta. 12760

E-mail: arifintriyanto87@gmail.com

Abstract. Operational Electrical Railway of PT. KAI Commuter Jabodetabek is supplied by overhead contact system 1500 VDC. Traction substation of Tanjung Barat station uses 6 pulse rectifiers to change alternating current (AC) become direct current (DC) with silicon rectifier type. The effect of the rectification process is harmonic current that causes overheating even fire in transformator. This paper aims to resolve the problem on the rectification process by a simulation using an ETAP 12.6.o application by introducing a single tuned passive filter of the 5th order which is customized to the converter circuit load system. The simulation results showed that filter installation caused reduction of average total harmonic distortion (THD) value where 20% load reduced from 20.19% to 1.02%; 40% load reduced from 20.19% to 3.93%, in 60% load reduced from 20.19% to 1.70% and 80% load reduced from 20.20% to 4.91%. It was highlighted that the obtained total harmonic distortion (THD) reduction below 5 % that meet the standards of IEEE 519-1992.

1. Introduction

Transportation in the modern world today is a component that cannot be separated from daily human activities. Despite various types of transportation available, electric train (KRL) isan alternative transportation that has advantages such as reducing congestion and more time efficiency [1], KRL is a type of transportation driven by DC electric motors supplied from a converter in a traction substation. The KRL traction substation requires an electricity supply to meet the operational needs of the KRL. A quality electricity supply system is needed to expedite the operation of transportation [2], The supply of electrical energy in its distribution is inseparable from various disturbances that can disruptof KRL operations. Disorders that often occur in the supply of electrical energy, one of which is harmonic distortion. The power supply of the KRL traction substation uses AC voltage on the incoming side of the transformer with 6 silicon rectifier system resulting in harmonics. The KRL traction substation consists of transformers as a distribution of AC PLN sources, from transformers distributed to silicon rectifiers (SR). The SR is the producer of DC sources needed by top-flow electricity (LAA) as a supply of DC motors and utilities in the KRL system. In the KRL traction substation system, a 20 kV PLN AC voltage is lowered on the incoming transformer side 1.2 kV on the outgoing transformer side. A 1.2 kV AC source as a supply of DC 1500 V source generating converter for LAA requirements. This paper aims to resolve the problem on the rectification process by a simulation using an ETAP 12.6.o

IOP Conf. Series: Materials Science and Engineering **550** (2019) 012009 doi:10.1088/1757-899X/550/1/012009

application by introducing a single tuned passive filter of the 5th order which is customized to the converter circuit load system [3-5].

2. System presentations

The West Tanjung KRL traction substation has an installed power of 4,330 kVA PLN and an output power of 3919 KW. There are 6 pulses on SR (Silicon Rectifier) that it affects harmonic distortion. The more KRL load, the more current needed, caused anwill increase in the harmonic distortion value. The following figure 1 explained the substation capacity of the Tanjung Barat station.h

KERETA API

PT.KERETA API INDONESIA (PESERO)

DATA KAPASITAS GARDU LISTRIK

Resor	: 1.11 Pasar Minggu
Gardu	:Tanjung Barat
Tanggal	: 31-10-2016
Lampiran	: 00

LOKASI GARDU LISTRIK	DAYA PLN (kVA)	KAP. SR (kW)	KAP. TRAFO (kVA)	TEG. PLN (kV)
Tanjung Barat	5.540	4.000	4.530	20

Figure 1. The capacity of the West Tanjung electrical substation.

KRL crossing from Depok Station crosses Tanjung Barat Station which goes to Pasar Minggu Station until Manggarai Station is called Hilir, while the KRL from Manggarai Station to Duren Kalibata Station crosses Tanjung Barat Station to Depok Station called Hulu. The following table 1 displays KRL cross -trips on the KRL system Tanjung Barat Station.

Table 1. Cross-trips on the KRL system Tanjung Barat Station.

Direction	Feeder (mm^2)	Messenger (mm^2)	Trolley (mm^2)
TNT - PSM (SP. HILIR)	BC2 x 300 mm^2	St 1 x 90 mm^2	Cu 1 x 110 mm^2
PSM - TNT (SP. HULU)	BC 2 x 300 mm^2	St 1 x 90 mm^2	Cu 1 x 110 mm^2
LNA - TNT (SP. HILIR)	BC 2 x 300 mm^2	St 1 x 90 mm^2	Cu 1 x 110 mm^2
TNT - LNA (SP. HULU)	BC 2 x 300 mm^2	St 1 x 90 mm^2	Cu 1 x 110 mm^2

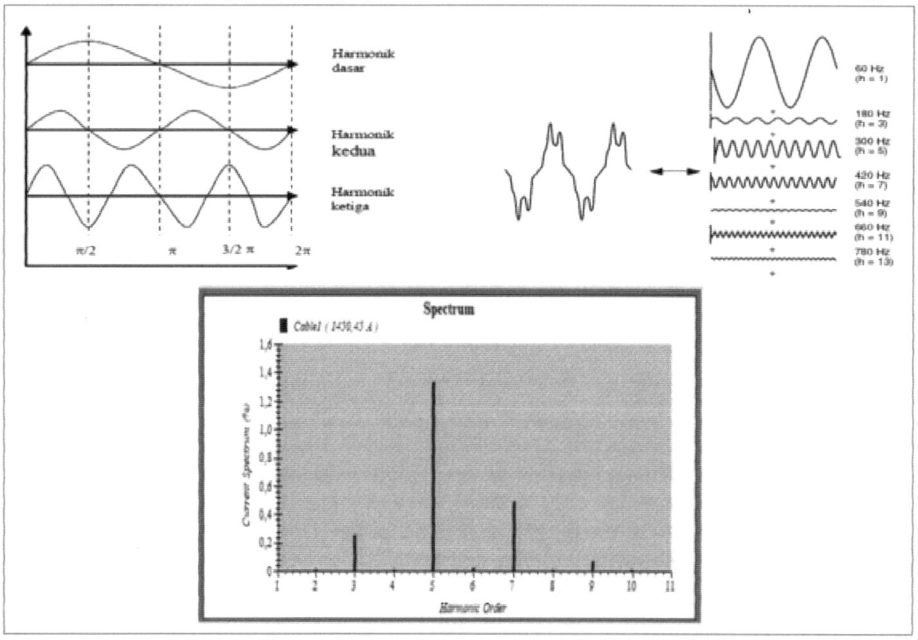

Figure 2. The illustration of fundamental harmonics, harmonic waves, and harmonic order charts.

IOP Conf. Series: Materials Science and Engineering **550** (2019) 012009 doi:10.1088/1757-899X/550/1/012009

Generally, voltage and current waves are transmitted and distributed from source to load in the form of pure sinusoidal waves. However, in the process of transmission and distribution, there are various kinds of interference so that the waveform is no longer sinusoidal pure. One of the deviation phenomena of sinusoidal waveforms is harmonic distortion. If a superposition occurs between the fundamental frequency waves and the harmonic frequency waves, a distorted wave will be formed so that the waveform is no longer sinusoidal. Harmonics caused by infected non-linear loads in the power system. Harmonics usually occur in odd order with increasing harmonic order, and the magnitude tends to decrease.

The standard limitation of harmonic distortion values for voltage and current must conform to international standards IEEE-519-1992 so that the equipment used remains in a safe and durable condition. Standard harmonic voltages and currents are shown in table 2.

Tabel 2. Standardization of voltage harmonics.

No	Power Rail Voltage PCC	Individual Voltage Distortion (IHDv)	Total harmonics Voltage Distortion (THDv)
1	≤ 69 kV	3.0	5.0
2	69	V < V ≤ 161 kV	1.5
3	> 161 kV	1.0	1.5

To determine the current harmonic limit according to IEE-519-1992 standard according to the value of the Short Circuit Ratio (SCR). Where SCR is the ratio between short circuit current and nominal load current as in table 3:

Table 3. Standardization of current harmonics.

No	Order of Harmonics (in %)						Total Demen Distortion (TDD)
	Isc/I_L	< 11	11-17	17-23	23-35	> 35	
1	< 20	4.0	2.0	1.5	0.6	0.3	0.5
2	20 < 50	7.0	3.5	2.5	1.0	0.5	8.0
3	50 < 100	10.0	4.5	4.0	1.5	0.7	12.0
4	100 < 1000	12.0	5.5	5.0	2.0	1.0	15.0
5	> 1000	15.0	7.0	6.0	2.5	1.4	20.0

Silicon rectifiers are one of the main components of the traction substation which serves to rectify 1200 VAC voltage to 1500 VDC voltage. At the electrical substation of PT. KAI there are two types of silicon rectifiers, namely SR 6 pulses and SR 12 pulses. At present, the silicon rectifier technology used for the traction substation system is a 12 pulses rectifier. The semiconductor component used by the rectifier for rectifier is the press pack version diode. Silicon Rectifier (SR) 6 pulses are SR which rectify 6 AC wave pulses, the number of pulses rectified depends on the type of connection transformer [6-8].

IOP Conf. Series: Materials Science and Engineering **550** (2019) 012009 doi:10.1088/1757-899X/550/1/012009

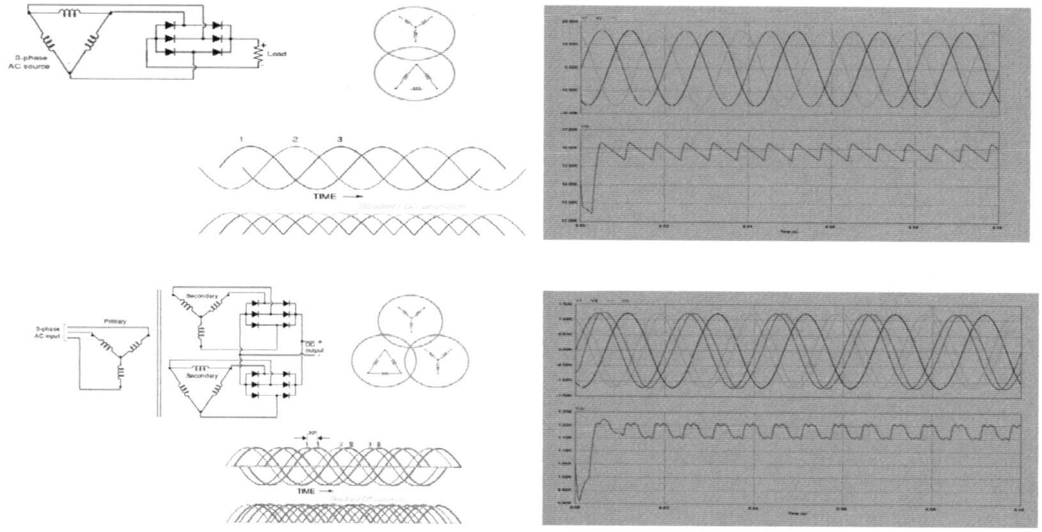

Figure 3. A transformer circuit with SR 6, 12 pulses and AC wave rectifier becomes DC.

ETAP (Electric Transient and Analysis Program) PowerStation 12.6 is application software that is used to simulate an electric power system. ETAP caton work in offline conditions for simulating electric power. The features contained in ETAP vary, among others, features used to analyze power plants, transition systems and electric power distribution. The advantages of using ETAP software include. Figure 4 shows display of ETAB software. From the above settings, the ETAP software is obtained in accordance with the specified amount of load. This explanation can be seen in figure 5.

Data sources in this study were obtained using calculations and assisted by using ETAP 12.6.0 software adapted to the specifications of the Tanjung Barat KRL traction substation. The calculation is done on the transformer bus at a voltage of 1.2 kV before being distributed to SR. In addition to using calculations also use ETAP simulations to determine the THDi value. Voltage and harmonic currents from these calculations, a comparison is made using harmonic standards. So that the results of voltage and current are close to harmonic standards so that the equipment in the traction substation system remains safe when operating. In ETAP simulations, we know the harmonic distortion values based on calculations with minimum and maximum loads. So that it can be planned to install passive filters using ETAP simulations to reduce harmonic distortion values. The calculation and simulation above aim to determine the value of harmonic distortion, plan the installation of filters and to reduce the value of harmonic distortion. By observing the harmonic distortion value and adjusting the harmonic distortion value according to harmonic standards, the equipment in the traction substation used remains in a safe condition and lasts longer [9].

IOP Conf. Series: Materials Science and Engineering **550** (2019) 012009 doi:10.1088/1757-899X/550/1/012009

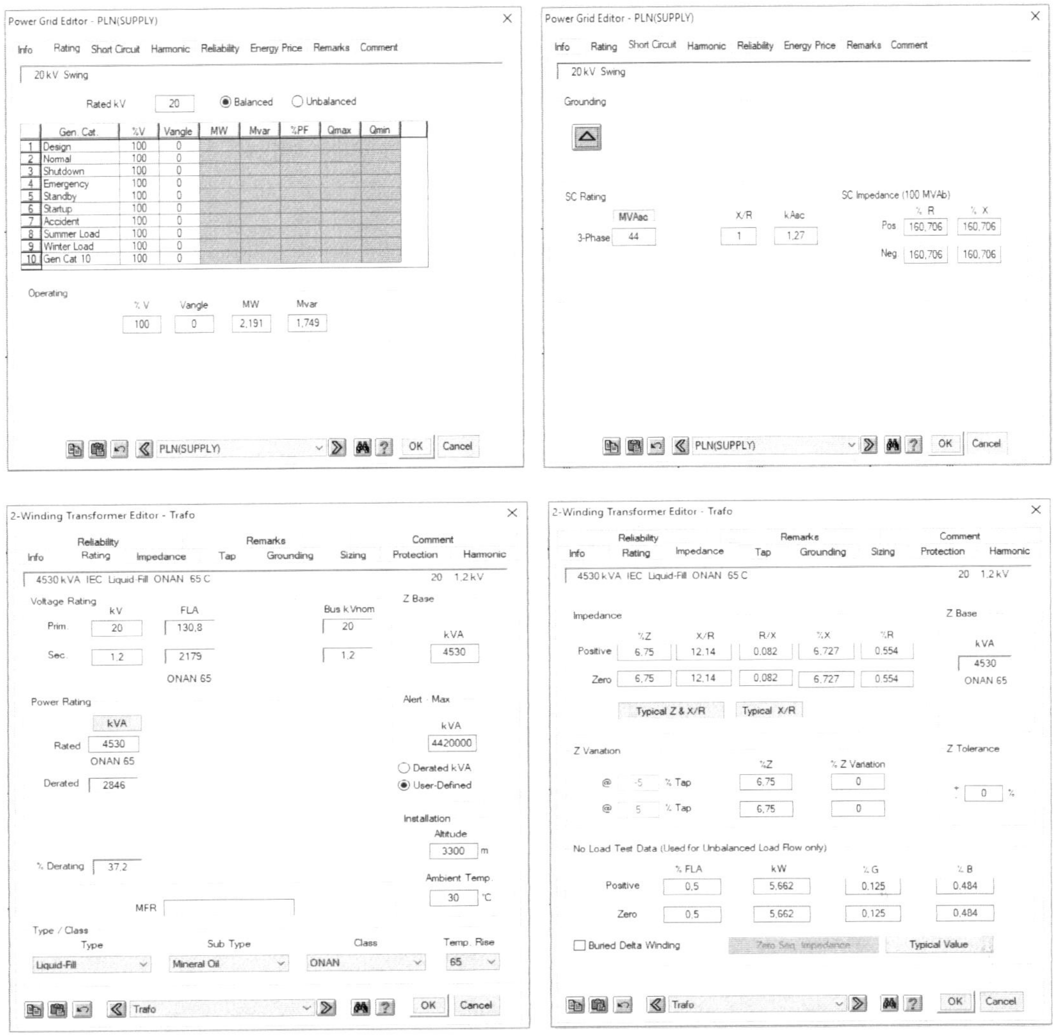

Figure 4. Display transformer voltage and input settings in ETAP software.

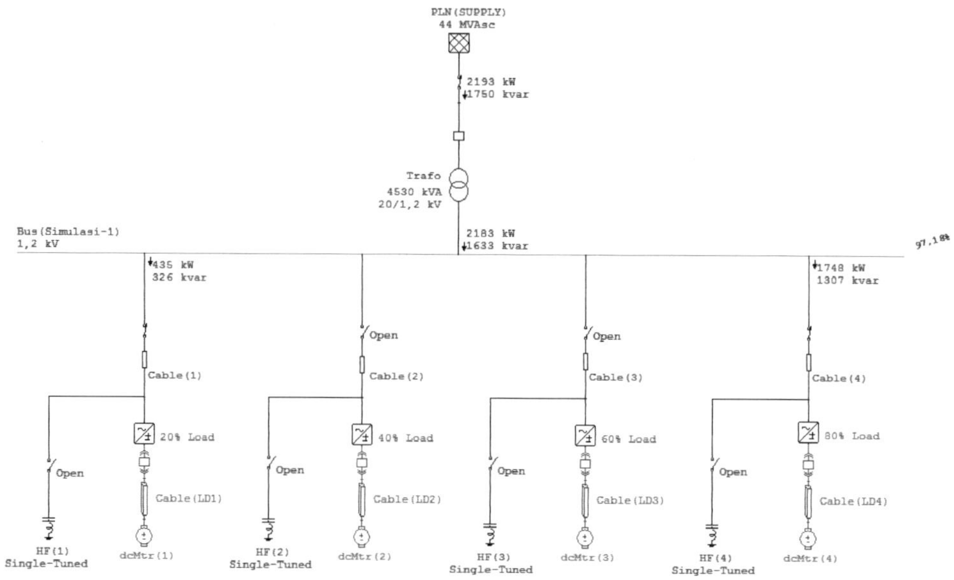

Figure 5. Display of load flow analysis software ETAP.

3. Results and discussion

Data processing and data analysis in this study used two methods. The research method of harmonic distortion in the KRL traction substation is analyzing the data from the simulation using SR load. After the simulation is done, analyze the data to plan passive filters that aim to reduce the value of harmonic distortion. The data that has been obtained from the simulation is compared with the data obtained from calculations with the theory of harmony to adjust to harmonic standards.

3.1. Power flow simulation

This power flow simulation is needed to get apparent power (kVA), real power (kW), receptor power (kVAR) and power factor (PF) needed to do passive harmonic filter design to reduce harmonics that occur in the system. In this power flow analysis is carried out by varying loading from the maximum 80% transformer capacity of 3625 kW, so that a measurement variation can be achieved thatcan represent the actual traction substation loading system. The loading variations include:

1. When the load is 20% of the transformer capacity of (725 kW)
2. When the load is 40% of the transformer capacity of (1445 kW)
3. When the load is 60% of the transformer capacity of (2175 kW)
4. When the load is 80% of the transformer capacity of (2899 kW)

After loading variations are obtained, then can simulate the system power flow with two simulations, simulating the load of 20%, 80%, 40% and 60%:

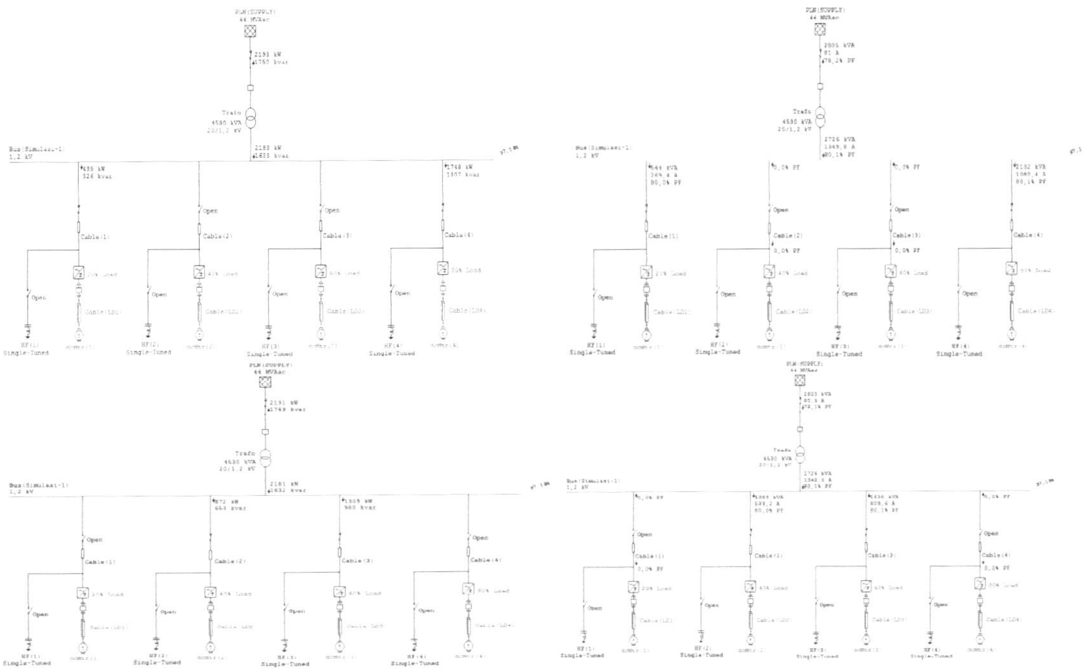

Figure 6. Display power simulation images at loads of 20%, 40%, 80% and 60%.

From the simulation results the power flow in the load is obtained by variable real power (kW), reactive power (kVAR), apparent power (kVA), current (A), and power factor (PF).

Table 4. Results of simulation of system power flow.

Simulation result	Load variation				Total
	20%	40%	60%	80%	
Real power (kW)	435	872	1309	1748	4364
Reactive power (kVAR)	326	653	980	1307	3266
Apparent power (kVA)	544	1089	1635	2182	5450
Current (A)	269.4	539.2	809.6	1080.4	2698.6
Voltage (V)	1200	1200	1200	1200	1200
Power factorDaya (%)	80	80	80	81	80.25

Before simulating harmonics, the parameters of the simulation program are filled out, so that the resulting order, waveform, and spectrum harmonics are produced. Based on equation (2.3), at the 1st order angle 0° is fundamental at a frequency of 50 Hz, then the 5th order at a frequency of 250 Hz as a discussion in the highest value research.

IOP Conf. Series: Materials Science and Engineering **550** (2019) 012009 doi:10.1088/1757-899X/550/1/012009

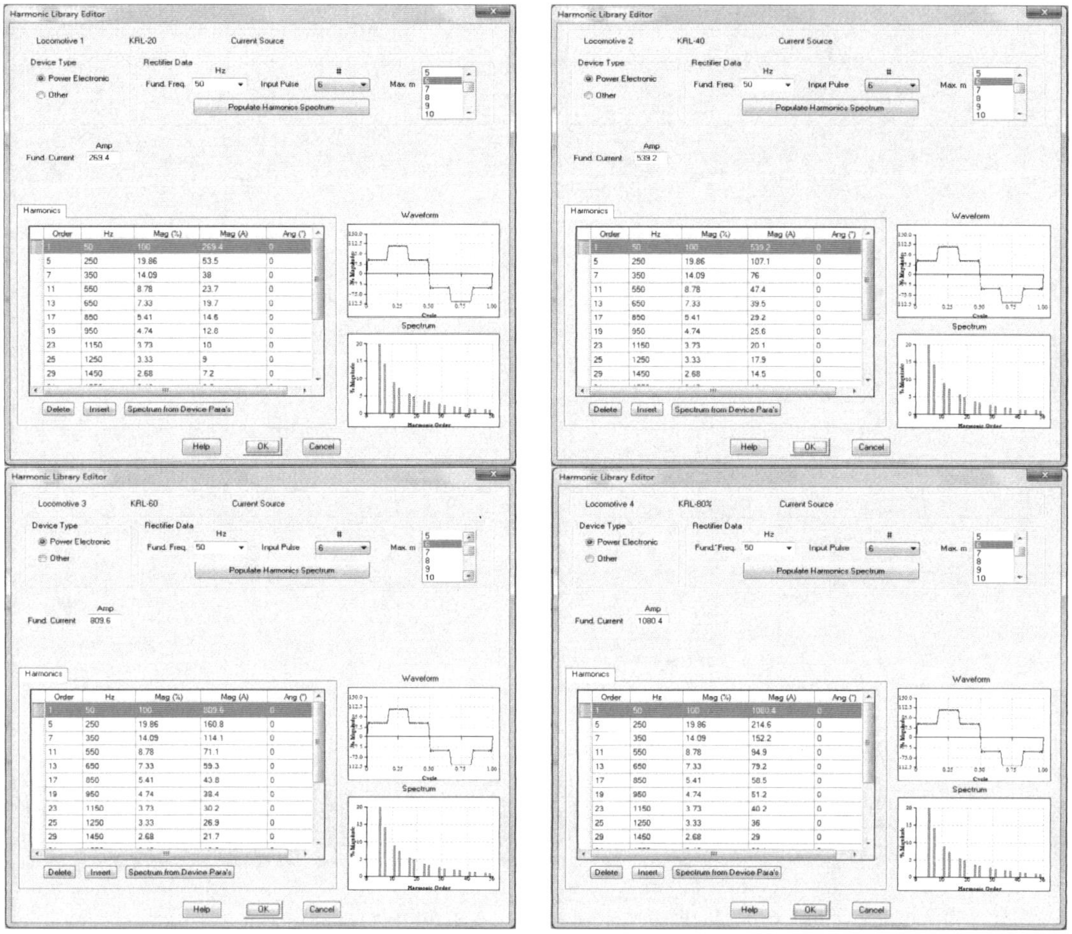

Figure 7. Display library harmonic loading of 20%, 40%, 60% and 80%.

Then a harmonic simulation can be done with the aim to get the value of the SR load on the outgoing transformer side of the Tanjung Barat traction substation. The effect of KRL loading on SR and the form of harmonics prior to filter installation is shown in figure 8 below:

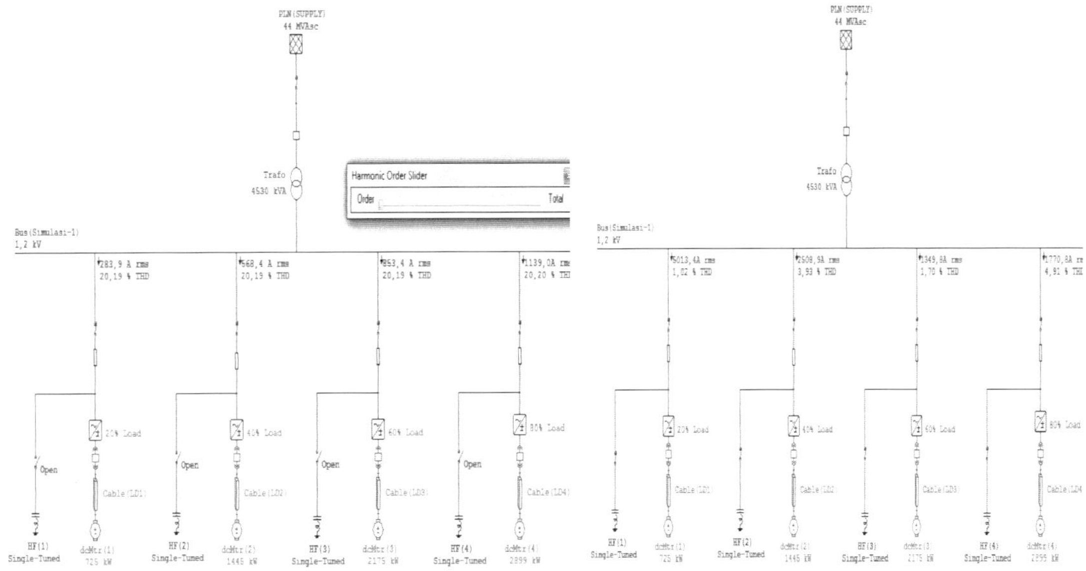

Figure 8. Harmonics simulation before and after filter installation.

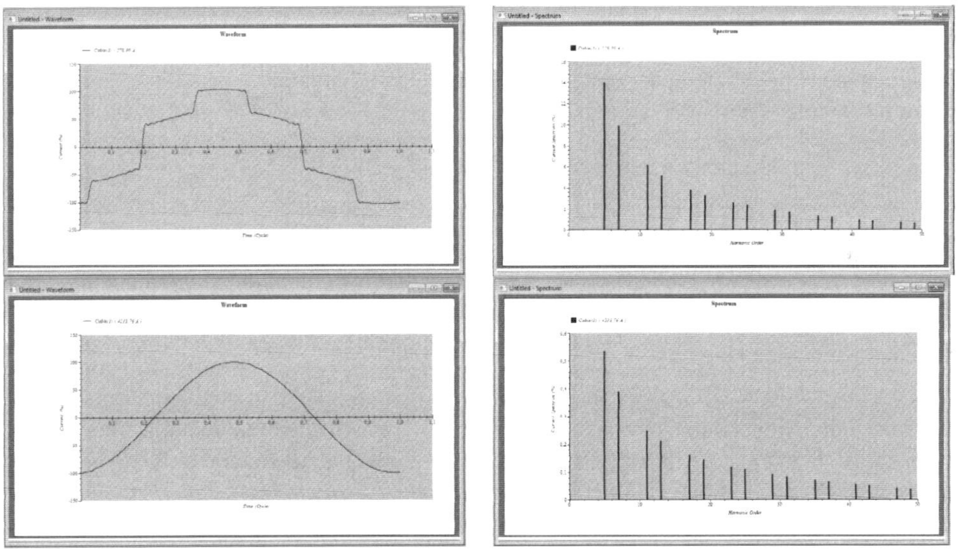

Figure 9. Harmonic waves 20% before and after the filter.

3.2. Filter design

To improve the power factor of the transformer unit, the following calculations can be done. The following is an example of 20% loading:

$$I_L = 269.4\ A$$

$$S = 544\ kVA$$

Reactive power needed to improve the power factor can be calculated as follows:

$$Q_1 = 326\ kVAR$$

If the power factor (power factor) is increased to 0.9, then the amount of reactive power:

$$PF = \theta_2 = 0.9$$

$$Q_2 = 544 \times \sin[\cos^{-1}(0.9)] = 544 \times \sin[25.84] = 237\ kVAR$$

Therefore $C = 326 + 237 = 563\ kVAR$ and the reactive required passive harmonic filter to compensate according:

$$Q_C = Q_1 - Q_2$$

$$Q_C = 326 - 237 = 89\ kVAR$$

Then to determine the reactance of the filter (X_filter):

$$X_{filter} = \frac{(V_{LL})^2}{Q_C}$$

$$X_{filter} = \frac{(1.2\ \text{kV})^2 \times (1000)}{89\ kVAR} = 16.2\ \Omega$$

While each inductive reactance value and capacitive reactance, at a tuning frequency of 250 Hz (5th order harmonics):

$$X_c = \left[\frac{h^2}{h^2 - 1}\right] \times X_{filter}$$

$$X_c = \left[\frac{4.5^2}{4.5^2 - 1}\right] \times 16{,}2 = \left[\frac{20{,}25}{20{,}25 - 1}\right] \times 16.2 = 17\Omega$$

Where the value h:

$$h = n - (n \times 10\%) = 5 - (5 \times 10\%) = 5 - (0.5) = 4.5$$

Determine the value of the capacitor:

$$= \frac{1}{2\pi f X_C}$$

$$C = \frac{1}{2 \times 3.14 \times 50 \times 17} = 1.87\ \times 10^{-4}\ \mu\text{F}$$

While the value for the inductor:

$$X_L = \frac{X_C}{h^2}$$

$$X_L = \frac{17}{4{,}5^2} = \frac{17}{20{,}25} = 0.84\ \Omega$$

Determine the value of the inductor:

$$= \frac{X_L}{2\pi f}$$

$$L = \frac{0{,}84}{2 \times 3.14 \times 50} = 0.0268\ H$$

Passive harmonic filter circuit consists of R, L, and C which have values and obtained from the calculations above, then the determination of the R value, with Q is between 20-100, Q = 50 is adjusted according to the data in the ETAP simulation, then obtained:

$$Q = \frac{n \times X_L}{R}$$

$$50 = \frac{5 \times 0.84}{R}$$

$$R = \frac{5 \times 0.84}{50} = 0.084\ \Omega$$

The effect of passive filter installation is that the sub-harmonic distortion value is permitted by the IEEE-519-1992 standard that harmonic distortion must be below 5%. Then the KRL system is in a safe condition to operate and the equipment is protected from damage caused by harmonic distortion.

Table 5. Comparison of harmonic simulation results.

Load variation	Harmonic simulation	
	Before applying filter (THD, %)	After applying filter (THD, %)
20%	20.19	1.02
40%	20.19	3.93
60%	20.19	1.70
80%	20.20	4.91

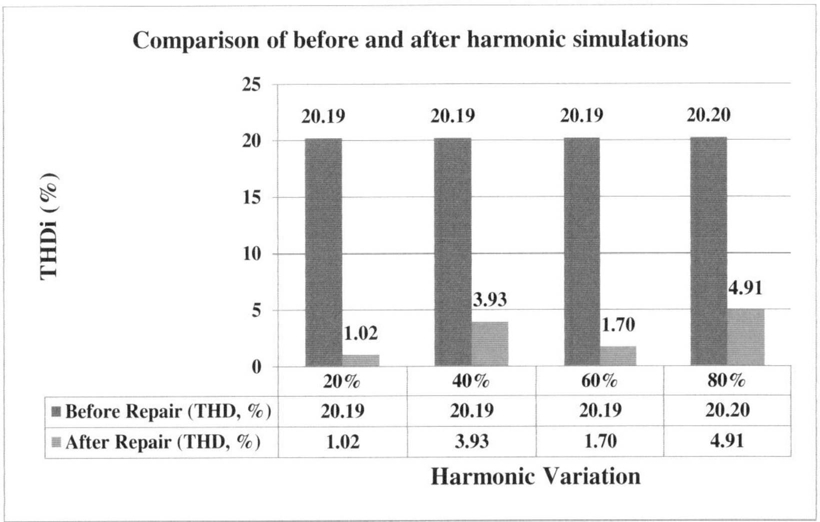

Figure 10. Load comparison chart after and before using the filter.

4. Conclusion

From the results of the study obtained the total harmonic distortion current value of 20.19%. This shows the harmonic distortion value above 5% and exceeds the standard IEEE-519-1992 harmonics limit, it is necessary to improve the value of the harmonic distortion.

Power flow analysis simulations on the 1.2 kV outgoing transformer side using varying loads adjusted for to obtain apparent power values (kVA), real power (kW), receptor power (kVAR) and power factor (PF).

Passive filter design results obtained parameter values: for 20% load are: C = 653 kVAR, X_L = 0, 84Ω, and R = 0,084 Ω; for 40% load are: C = 712 kVAR, X_L = 0.419Ω, and R = 0.0419 Ω; for 60% load are: C = 950 kVAR, X_L = 0.08Ω and R = 0.008 Ω; for 80% load are: C = 1188 kVAR, X_L = 0.079Ω and R = 0.0079 Ω;

The reduction of the total harmonic distortion current in the load (20%) dropped from 20.19% to 1.02%; expenses (40%) decreased from 20.19% to 3.93%; expenses (60%) decreased from 20.19% to 1.70%; and expenses (80%) decreased from 20.20% to 4.91%. So that the total harmonic distortion reduction results are below 5%. Based on IEE 519-1992 standards that the harmonic distortion value must not exceed or a maximum of 5%.

IOP Conf. Series: Materials Science and Engineering **550** (2019) 012009 doi:10.1088/1757-899X/550/1/012009

References

[1] Sabri Y 2012 *Bandung : STEI ITB Journal* pp. 1-7
[2] PT KAI 2016 *Capacity of Top Flow Electricity* (Jakarta: PT. KAI)
[3] Irshadi I 2014 *Technical Journal* pp. 1-10
[4] Fauzan M R 2015 *ELECTRICIAN-journal of Electrical Engineering and Technology* **9**(3) pp. 192-202
[5] Sunanda W and Rahman Y A 2011 *Foristek Journal* (1)1 pp. 16-21
[6] Sugiarto H 2012 *Vokasi Journal* **8**(2) 80-89
[7] Rinas I, Wayan 2011 *Teknologi Elektro Journal* **10**(1) pp. 20-26
[8] Prihatama E 2013 *Journal of Engineering* **2**(2) B-228-B233
[9] Multa L and Aridani R P 2013 *ETAP Training Module* (Yogyakarta: UGM) pp. 1-58

Development of DC smart grid model for voltage stability disturbance

D W Maulana[1], Fery Faizal[1,2], F F Florena[3], F Alfandi[2], C Panatarani [1,2], I M Joni[1,2] and B M Wibawa[4]

[1] Nanotechnology Research Center, DRPMI, Universitas Padjadjaran, Jl. Raya Bandung-Sumedang KM 21, Jatinangor, Sumedang, West Java, 45363, Indonesia.
[2]Department of Physics, Faculty of Mathematics and Natural Sciences, Universitas Padjadjaran, Jl. Raya Bandung-Sumedang KM 21, Jatinangor, Sumedang, West Java, 45363, Indonesia.
[3] Department of Physics, Faculty of Mathematics and Natural Sciences, Bandung Institute of Technology, Jl. Ganesha No 10, Bandung, Indonesia.
[4]Department of Electrical Engineering, Faculty of Mathematics and Natural Sciences, Universitas Padjadjaran, Jl. Raya Bandung-Sumedang KM 21, Jatinangor, Sumedang, West Java, 45363, Indonesia.

E-mail: imadejoni@phys.unpad.ac.id

Abstract. DC smart grid and distributed generation have evolved rapidly as the integration of renewable energy. The robustness parameter of DC smart grid can be defined by the ability of the system to maintain stability under normal or perturbed conditions and provide fast restoration after faults. This research aims to design DC smart grid to simulate the voltage stability again disturbance. A model of DC smart grid was tested by key parameters in power station such as voltage stability. Voltage stability is the ability of power system to maintain steady voltage within permissible range at all buses in reasonable condition and after having been subjected a perturbation, such as an incremental change in load. The obtained duration time for a system to recovered voltage stability problem was approximately two seconds until three seconds. It was concluded that designed DC smart grid could simulate the voltage stability disturbance.

1. Introduction

A smart grid has many renewable energy sources and connects into the power transmission. A smart grid also composed of a distribution system as an integral component [1]. The smart grid concept provides the possibility to detect, analyze and respond to various voltage stability disturbance. A smart grid used for improving efficiency, sustainability, flexibility, reliability, and security of the electrical system [2].

Since the 1920s, the stability of the power system against operational security has become a severe problem. The diversity of disturbances and instability in the power system can cause damage to the generator which causes the system to die (black-outs), so that stability analysis on the smart grid becomes very important in planning, designing and operating. The toughness parameter of a DC smart grid is defined by the system's ability to recover quickly when a fault occurs [3]. Research on the DC smart grid has been carried out in the form of an analysis of voltage stability without interruption and with the presence of interference by Amty Ma'rufah [4]. In that study, a relay has been used as a power supplier actuator and caused recovery time around 4th to 6th seconds.

In this research, a smart DC grid model was developed to simulate the voltage stability again disturbance. Voltage stability is the ability of power system to maintain steady voltage within permissible range at all buses in normal condition and after having been subjected to a severe system perturbation, such as a simulation of an increase of load or create an overloaded condition. MOSFET was used as actuators for controlling their power allowed to work at high frequencies. In addition, the DNP3 protocols were utilized as protocols of the communication.

2. DC smart grid model

The general smart DC grid system in Figure 1 consists of a smart-grid master (SG-M), smart-grid 1 (SG-1) and smart-grid 2 (SG-2). SG-M have a function as the main grid with stable resources using 12V 10A batteries. The SG-M can supply power to another grid if there is an increase in load and loss of resources power on a distributed grid. The distributed grid consists of SG-1 and SG-2. The SG-1 using 12V 5A battery whereas the SG-2 using the 12V 3.5A battery. Rechargeable batteries (Accu) are considered as storage from the renewable energy source.

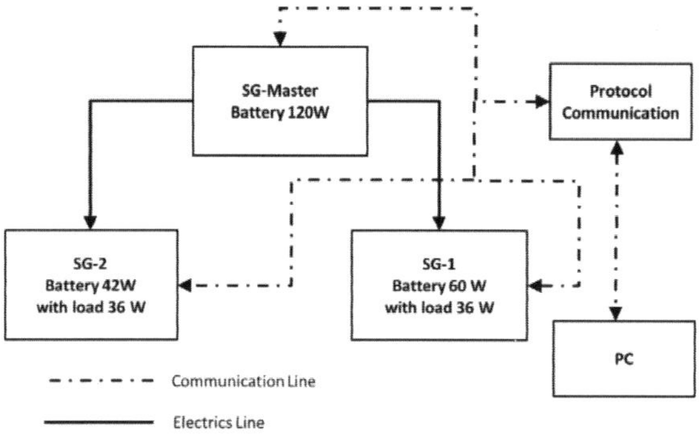

Figure 1. The general DC smart grid system.

The developed DC smart grid system equipped with two-way communication using RS-485 protocol allows operating data management systems in a Personal Computer (PC). Based on the information from the distributed grid, if the distributed grid needs more power, PC will launch command to SG-M to give power to the distributed grid. PC displayed a table data and online measurement and monitoring graph from each smart grid. Thus the status of the system was under control. The DNP3 protocol was employed for the communication between subsystems as shown in Figure 2. The DNP3 protocol is a data communication standard for the DC smart grid as outlined in IEEE P1815-2010.

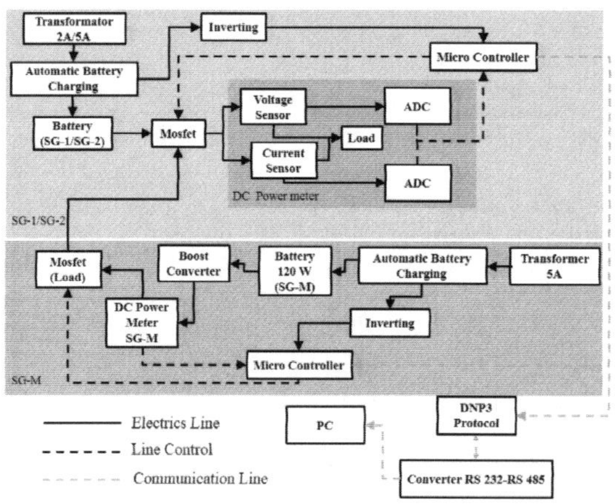

Figure 2. The DC smart grid subsystem

IOP Conf. Series: Materials Science and Engineering **550** (2019) 012010 doi:10.1088/1757-899X/550/1/012010

The DC smart grid subsystem are comprises of the battery as stable power source, an ATmega 8535 controller, a personal computer (PC), a Mosfet as a power supply actuator, an automatic battery charger circuit that keeps the power of battery, DC power meter, boost converter, max 485 circuit and max 232 as a serial communication line. We are used boost converter only in SG-M wherein for maintained possibility reverse current flow to SG-M. In the previous research [4], analysis stability was evaluated from voltage response against a disturbance. They used a simple relay as actuator power distribution. In this developed-DC smart grid, we employed a Mosfet, which fast response was expected well to recover the system against disturbances than a simple relay.

This system was applied to simulate disturbance by using 2 DC lamp (12V/35watt) with 2 scenarios. The first scenario was SG-M without a used boost converter and the second scenario was SG-M with a used boost converter. Both scenarios were simulated on SG-1 and SG-2. Each scenario had 3 steps. The first step, the SG-1 or SG-2 would not be loaded from the lamp. The second step, the SG-1 or SG-2 would be loaded with a reasonable load from the lamp (12V/35watt). Then the third step, the SG-1 or SG-2 received an extra load from 1 more lamp (12V/35watt). In the third step, the distributed grid received burdened status and requested assistance from SG-M. The SG-M received the information from PC to open Mosfet gate and give the power. The SG-M checked own battery state of charge (SOC) status first, before SG-M open Mosfet gate. When SG-M battery was under 50% SOC, the SG-M give no power to the distributed grid. During the simulation, the distributed grid cannot help each other.

3. Result and discussion

The integrated smart DC grid system was tested their performance by sending data to the PC via RS-485 and RS-232 using the DNP3 protocol.

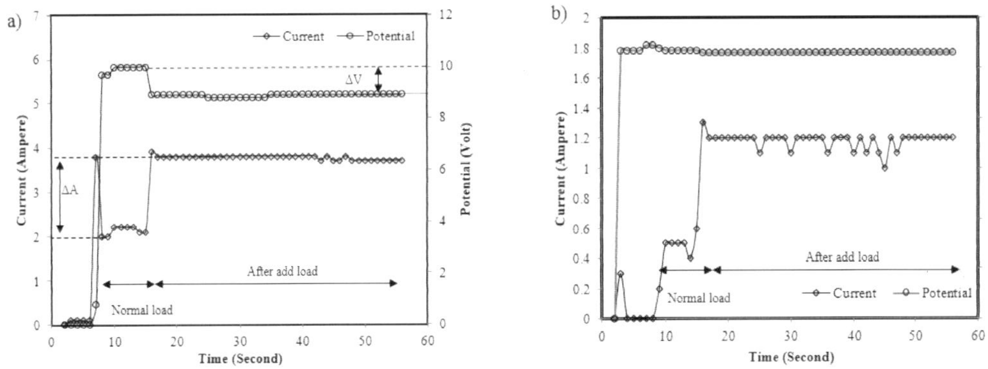

Figure 3. Current and potential profile of a) SG-2 with assisted of SG-M and b) SG-M without using a boost converter.

Figure 3a showed the voltage stability under disturbed load without using the boost converter for the grid SG-2. Initially, SG-2 was operated with a normal load (lights with a power of 12V/35W (3A)) which was noted as normal in the figure. The voltage was stable at 10 Volt and current raised to 2.2 A. Then in 16th seconds, the SG-2 load was purposely added with 35W power lamp in parallel to simulate load to the system indicated by the rise of the current to 4A. In this situation, SG-2 delivered an aiding signal to SG-M because the battery specifications on SG-2 could only drain a maximum current of 3.5A. SG-M responded to the received signal by providing power supply aid (Figure 3b). Utilized the power aid from SG-M, the current flow to a maximum value of 3.8 A, however, the voltage did not return to stable values i.e., 10 Volt. The system could not resolve the voltage on SG-2 to initial stable condition (10 Volt) even though the power at SG-M was greater than SG-2. Furthermore, additional effort need to be applied to resolve this problem.

Furthermore, in the second test, the scenario was carried out with similar treatment as a previous test (SG-M using boost converter). In this scenario, the distributed grid under test was SG-1. Initially, SG-1 was operated without load and then the system added a normal load of a lamp with a power of 12V / 35W (3A).). When the load was added to the SG-1 with 35W power lamp in parallel, the current reached a maximum current of 4.5A. As shown in Figure 4a, before the SG-1 got assisted from SG-M, the voltage drop to 8.2 volts. Whereas when the SG-1 got assisted from SG-M, the voltage back to 9.4 volts.

By applying a smart system when the voltage dropped, SG-1 automatically send a signal to the PC that SG-1 required additional power. The PC delivered a signal to SG-M then transmitted additional power to SG-1. Figure 4b showed the current increased on SG-M after SG-1 had received additional power supply. The SG-M checked the SOC of SG-M battery first before sending additional power. If SOC of SG-M battery under 50% so the SG-M would not open the Mosfet. The recovery time of this system was around 2 seconds. Utilization of boost converter at SG-M produced the voltage of SG-M became 12 V and helped SG-1 voltage returned to initial voltage. It was highlighted, the SG-M have supplied to a maximum current to the SG-1, and maintained voltage over 11 V.

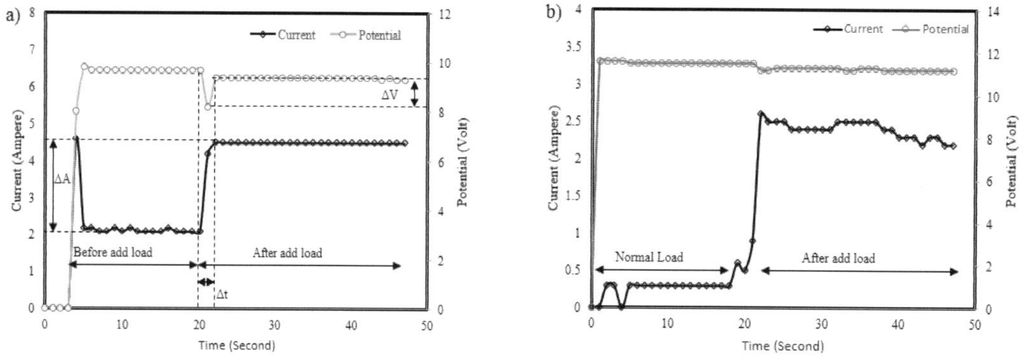

Figure 4. Current and potential profile of a) SG-1 with assisted of SG-M, b) SG-M using boost converter.

Similarly occurred on SG-2 when this grid was added to to simulate overloaded this grid was unable to reach maximum current (Figure 5). It was highlighted that when a smart system via a PC automatically controls by delivering additional power from SG-M to SG-2. It was clearly showed in Figure 5a the voltage raised soon after getting the power supply aid from SG-M and yielding a recovery time of 3 seconds.

The conditions that occurred in the SG-M before and after being loaded are shown in Figure 5b. Smaller resources on SG-2 (12V 3.5A) cause the power transmitted to SG-2 was greater than the power transmitted to SG-1. This clearly in turned that the magnitude of the current flowing to SG-2 (3.5A) which was greater than the current flowing to SG-1 (2A).

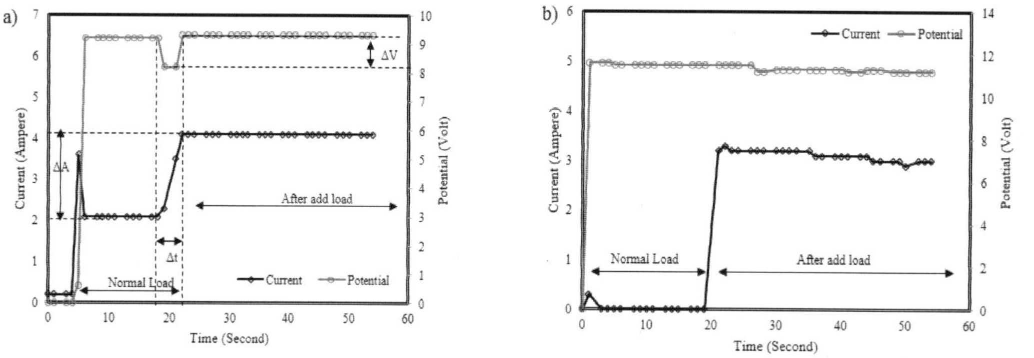

Figure 5. Current and potential profile of a) SG-2 with assisted of SG-M., b) SG-M using a boost converter.

IOP Conf. Series: Materials Science and Engineering **550** (2019) 012010 doi:10.1088/1757-899X/550/1/012010

4. Conclusion

The developed smart DC grid model successfully simulated the performance of a smart DC grid when the disturbance occurred and the system responded by controlling the distribution of DC power to achieve voltage stability. It was highlighted that the response time to resolve the voltage stability was approximately two seconds until three seconds

References

[1] Wang W, Xu Y and Khanna M 2011 *computer network* **55** 3604–3629.
[2] Colak I, Sagiroglu S, Fulli G, Yesilbudak M and Covrig C F 2016 *Renewable and Sustainable Energy Reviews* **54** 396–405.
[3] Hidayatullah N A, Paracha Z J and Kalam A 2011 *Smart Grid and Renewable Energy* **2** 99-109.
[4] Dalimunthe A M A, Mindara J Y, Panatarani C and Joni I M 2016 *AIP Conference Proceedings* **1719** 030055

Acknowledgment

The author acknowledges the financial support from Riset Hilirisasi Produk Unggulan, Universitas Padjadjaran, Bandung, Indonesia.

IOP Conf. Series: Materials Science and Engineering **550** (2019) 012011 doi:10.1088/1757-899X/550/1/012011

Online air velocity control, temperature and humidity monitoring system for clean room using Raspberry Pi

A Sunardi, D Agus, H Aditya and W Susila

Electrical Engineering Department, Pamulang University, Kampus Viktor, Jl. Raya Puspiptek, Setu, Kota Tangerang Selatan, Banten 15314

E-mail: dosen00332@unpam.ac.id

Abstract. The requirement of a clean room in the pharmaceutical industry is to maintain the airflow between 0.36-0.54 m/sec and record the temperature and humidity parameters. The laminar airflow machine is used to maintain the air velocity. The air velocity settings in laminar airflow machine are usually done manually by adjusting the speed control knop. In this study, we have designed a prototype of automatic air velocity control, temperature, and humidity monitoring system using Arduino nano and Raspberry Pi. The airflow control is done by the PID method. From the test results, the prototype has worked as expected. The accuracy of the measurement is ± 5% for air velocity, ± 2 °C for temperature and ± 5% for moisture. The system can control the air velocity between 0.36 m/s to 0.54 m/s. The measured parameters then send to the server and utilized for online monitoring the air velocity, temperature, and humidity in the clean room.

1. Introduction

The requirement for a clean room on a laminar air flow machine in the pharmaceutical industry which is regulated in the operational manual applies the guidelines for how to make a good drug [1], which maintains air velocity between 0.36-0.54 m/s and records temperature and humidity [2]. The laminar air flow machine is a machine that produces a laminar air flow direction because it consists of a blower motor that blows air regulated by speed control through several filters [3]. In the process of setting the speed set manually by rotating the potentiometer on speed control, this is done when the engine installation is first and every six months an operational check is carried out on the amount of airspace to ensure the airspeed is in accordance with the conditions. In these conditions, records are also carried out to monitor the results of airspeed, temperature and humidity manually into the operational qualification form as one of the controlled written reports to the drug and food regulatory agency [4]. Airspeed, temperature and humidity are not monitored so one to find out these results is to use an airspeed, temperature and humidity monitoring system through a web server so that it can be monitored online [5]. Still the manual controls on the laminar air flow engine in the airspeed setting that is required to encourage recording of airspeed data online.

2. Research methods

The process flow diagram is intended to describe the operational processes of the system that will later be made, and so that if later there are problems it can help to solve the problem. figure 1. Process flowcharts can be seen.

IOP Conf. Series: Materials Science and Engineering **550** (2019) 012011 doi:10.1088/1757-899X/550/1/012011

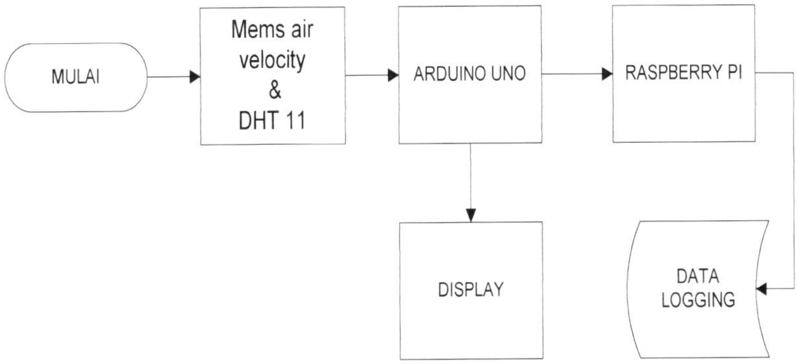

Figure 1. Process flow chart.

Data sender block diagram to website contains the power supply, raspberry pi, and modem. The working principle is all data from Arduino is sent from Raspberry Pi which contains the results of airspeed, temperature and humidity to be sent to the website that is connected to the modem WiFi SSID as shown in figure 2.

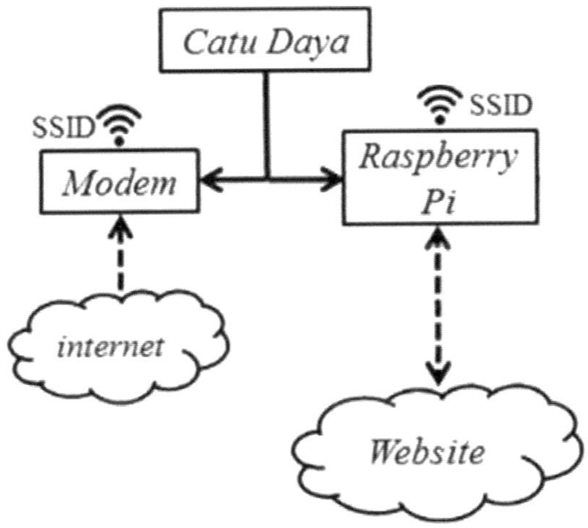

Figure 2. Block diagram receiving response from website.

The overall diagram contains block phenomena, input blocks, process blocks and output blocks. Block phenomenon contains the phenomenon of temperature, humidity, and air velocity which will later be captured or felt by the sensor. The input block contains the airspeed mems air sensor, DHT11, Arduino Uno, the principle works is airspeed as an input phenomenon, which will be captured by the sensor mems air velocity into an analog signal and converted into digital data by Arduino to be sent via serial USB connection. The process block contains the website and raspberry pi, the principle works is when data is sent to the website, then the website will manage the data to become information. The output block contains a blower motor which is controlled by a motor drive that uses a PWM system, and a display, the working principle is as an output. For logging data on temperature, humidity and velocity regulated by raspberries. Blower motor serves as an airspeed control system if the air is less or more than 0.36 - 0.54 m / sec which is regulated by Arduino. While the display serves to display information on temperature, humidity, and airspeed or from the website. Figure 3 system block diagram can be seen below.

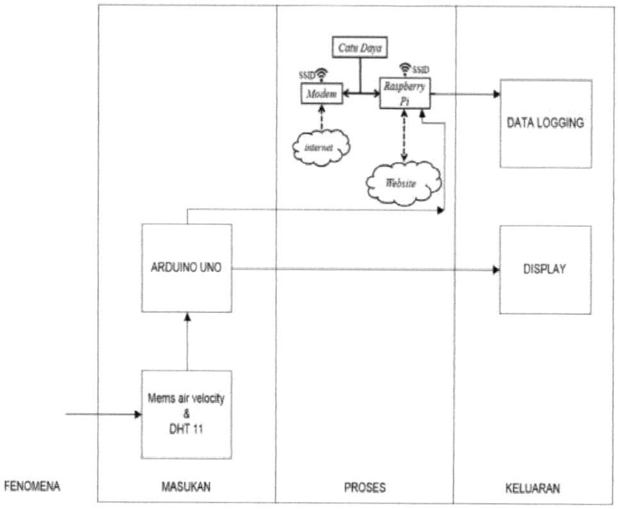

Figure 3. Block diagram of the overall system.

Hardware designing is to combine hardware design for sensor data senders to websites and hardware design for recipients of responses from websites, for the overall design of this system can be seen in figure 4 below.

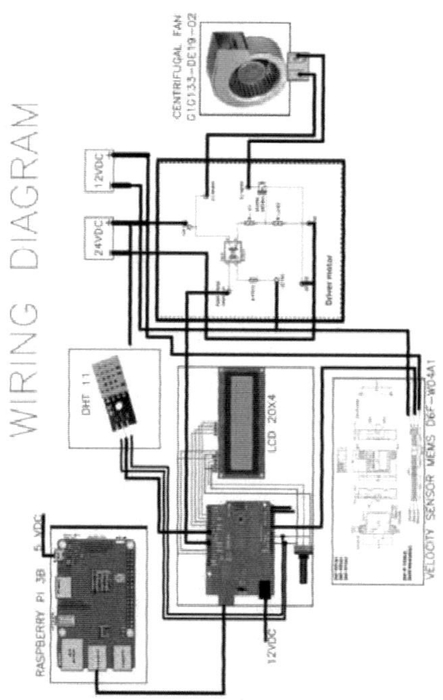

Figure 4. Hard system design.

Website design is intended to receive data from the DHT1 sensor and mems the air velocity through Arduino sent to Raspberry Pi via USB serial communication and displayed on the website.

The design of the Arduino Uno Program is intended to receive data from the DHT 11 sensor, mems air velocity, and the PID control system via Arduino Uno using the Arduino IDE program.

The design of the Raspberry Pi program is intended after getting a trigger from the website to enable the Raspberry Pi GPIO pin to be used. In this case what is connected to the Raspberry Pi GPIO pin is the servo motor on the GPIO 7 pin and the buzzer on the GPIO 0 pin. Raspberry Pi GPIO pins are by default programmed using Python language, where the Python language runs on an operating system (OS) installed on raspberry pi, for this final project I use Debian Linux OS. Before programming on raspberry pi, it is necessary to update and upgrade the system, and install the Python library for GPIO pins first through the raspberry pi command terminal on the Debian Linux OS, so that when the program starts it does not experience problems.

Database design is intended to receive data from the DHT sensor 11 and mems air velocity through Arduino that is sent to Raspberry Pi via USB serial communication and stored in the database via XAMPP.The overall design of this system is made in the form of a prototype as a system modeling, the shape of the current prototype seen in figure 5.

Figure 5. Tool prototype.

3. Results and discussion

Testing and analysis of the tool aim to test and determine the extent of the performance of the design results discussed in Chapter III and determine the level of success of each specification that has been submitted. Tests carried out include part testing and the whole system.

3.1. Airspeed detection verification Results with a calibrated tool.

This test is conducted to find the value of the air velocity captured by mems air velocity based on the datasheet (14) that will be compared against the velocity of the calibrated instrument:

a) Airspeed verification in PWM settings 0.
   This test is carried out 10 times the result of the mean air velocity of the tool 0.21 m / sec, while the average value of the instrument air velocity is 0.114 m / sec, so the average accuracy is 0.84%.
b) Airspeed verification in PWM settings 11.
   This test is carried out 10x the result of the mean air velocity of the instrument is 0.413 m / sec, while the average value of the instrument air velocity calibrated is 0.463 m / sec, so the average accuracy is -0.108%.
c) Airspeed verification on PWM settings 64.
   This test is carried out 10 times the result of the tool air velocity average value of 0.46 m / sec, while the average value of the calibrated instrument air velocity is 0.544 m / sec, so the average accuracy is -0.15%.

d) Airspeed verification at PWM settings 127.
This test is carried out 10 times the result of the mean air velocity of the tool 0.483 m / sec, while the average value of the calibrated instrument air velocity is 0.635 m / sec, so the average accuracy is -0.24%.

e) Airspeed verification at PWM settings 191.
This test is carried out 10 times the result of the mean air velocity of the tool is 0.513 m / sec, while the average value of the instrument air velocity is calibrated 0.681 m / sec, so the average accuracy is -0.25%.

f) Airspeed verification on PWM settings 255.
This test is carried out 10x the result of the mean air velocity of the tool is 0.622 m / sec, while the average value of the instrument air velocity is calibrated 0.737 m / sec, so the average accuracy is -0.25%.

3.2. Motor RPM results use a calibrated tool

a) PWM motor RPM results 64 settings.
This test is carried out 10x standard 500 meter RPM, while the average value of Rpm meter instrument is calibrated 556.8, so the average accuracy is = 0.1%.

b) PWM Motor RPM Results Settings 127.
This test is carried out 10x standard 1000 meter RPM, while the average value of Rpm meter instrument is calibrated 1258.2, so the average accuracy is = 0.205%.

c) PWM Motor RPM Results Settings 191
This test is carried out 10x standard 1500 meter RPM, while the average value of Rpm meter instrument is calibrated at 1480.5, so the average accuracy is = -0.013%.

d) PWM Motor RPM Results Settings 255
This test is carried out 10x standard 2000 meter RPM, while the average value of Rpm meter instrument is calibrated 2140.7, so the average accuracy = -0.066%.

Airflow testing was carried out 10 times to test the airflow of a prototype laminar air flow machine. Table 1. shows the results of testing airflow. The procedures in this test are as follows:

1. Turn on the Engine Prototype.
2. Prepare a smoke test.
3. Place the smoke test inside the engine chamber
4. See the results whether laminar air flow or turbulence

Table 1. Airflow test results.

No	Result
1	Laminar
2	Laminar
3	Laminar
4	Laminar
5	Laminar
6	Laminar
7	Laminar
8	Laminar
9	Laminar
10	Laminar

IOP Conf. Series: Materials Science and Engineering **550** (2019) 012011 doi:10.1088/1757-899X/550/1/012011

Test results for the temperature of the DHT 11 Sensor with a Calibrated Tool. Tests that have been done on the results of the average value of DHT 11 temperature are 29 ° C, while the average value of the instrument is calibrated 29.6 ° C, so the average accuracy = -0.020%.

Test Results for DHT 11 Sensor rH with a Calibrated Tool. Tests that have been done result in the average value of DHT 11 rH is 47%, while the average value of the instrument is calibrated 44.7%, so the average accuracy = 0.051%.

4. Conclusion

Based on the design, implementation and testing, conclusions can be drawn as follows: This tool works to provide information on temperature, RH, and airspeed that can be displayed via the web. This tool produces Laminar Airflow. Average airspeed accuracy = 0.84% in PWM setting 0 Average airspeed accuracy = -0.108%, average airspeed accuracy = -0.15% in PWM setting 25. Average air speed accuracy = -0.24% in PWM setting 50. Average accuracy of DHT temperature 11 = -0.020% and Average accuracy - rH = 0.051%.

References

[1] Taylor J R 1999 *An Introduction to Error Analysis: The Study of Uncertainties in Physical Measurements.* (Sausalito: University Science Books) p 128–9

[2] Sayuti A 2015 *Essay. Designing a Temperature Monitoring System Using Web and Android Raspberry Pi in the Server Room of Darma Persada University.* (Jakarta : Informatics Engineering Faculty of Engineering Darma Persada)

[3] Nugroho Y A 2011 Application of autocopler sensors on wind speed gauges based on the Atmega8535 AVR microcontroller. *lib.unnes.ac.id.* 7969/1(2011)/8556

[4] Giant F R, Darjat and Sudjadi 2015 *Transmisi Journal* 17(2) 71

[5] Kristiningsih 2015 *Design of PID Control in Water Discharge Plant with Ziegler-Nichols and Cohen-Coon Methods Using Matlab and Arduino* (Bandung : Politeknik Negeri Bandung)

Rail information system prototype with web-based magnetic detection method

A Sunardi, F I Su'ud and Widiyantoro

Electrical Engineering Department, Pamulang University, Viktor Building, Jl. Raya Puspiptek, Setu, Kota Tangerang Selatan, Banten 15314

E-mail: dosen00332@unpam.ac.id

Abstract. Trains are the most popular means of transportation by some people because it is cheap, fast, safe and comfortable. Because of carrying a sense of security, the construction of supporting infrastructure must be made safe, ranging from signs to the implementation of the door guard system in the railway crossing. Still often found in some railway crossing areas there is no guard door crossing system, so there are many accidents that resulted in the absence of a railway door guard system. The prototype was built to develop a railway control system by applying microcontroller technology, minicomputer and Hall effect sensor. The system will automatically detect the train that will pass by sending railway traffic safety information via the internet near the railway crossing. So that road users can find out early if there is a train that will pass by seeing the information monitor.

1. Introduction

Trains are the most preferred means of transportation for some people because they are cheap, fast, safe and comfortable. Because it carries a sense of security, the construction of its supporting infrastructure must be made safe, starting from its signs to the implementation of a crossing guard system for railroad crossings [1]. Frequently, there is no system for crossing the railroad guard found in some railroad crossing areas. From 2015 to August 3, 2017 it was noted that the highest number of train accidents in non-official crossings reached 274. In a study conducted in Switzerland and Russia, magnetic fields that occurred on the railroad tracks were produced by boogie-driven traction motors, the results of the study explained the magnetic fields on the railroad fluctuating [2].

The strength of the magnetic field enlarges as the locomotive speeds up, brakes, climbs or pulls more and more loaded carriages. In the aspect of vibration experienced by the rail, if the rail is going to be traversed by a train, then the rail will experience vibration [3]. The closer the train, the greater the frequency of vibrations experienced by the rail [4].

2. Research methods

In this study, a model of a magnetic field information system on the railroad will be made that can be utilized as early prevention of train accidents using the sensor hall effect. In this case, it measures the sensitivity of the hall effect sensor in capturing the magnitude of the magnetic field [5].

Modeling using block diagrams is divided into two parts, namely the block of sensor data sender to the website and the recipient block of response from the website [6]. The working principle of sending sensor data to a website is a magnetic field as an input phenomenon will be captured by the Hall effect sensor into an analog signal and converted into a signal digital by Arduino to be sent to the website using ESP8266 which is connected to the wifi modem SSID [7]. The working principle of the recipient block response from the website is when the data has been sent to the website by ESP8266, then the website will manage the data so that it becomes an information and an order to the raspberry pi that has a wifi modem SSID to activate the GPIO pin [8], so that the servo motor and buzzer function as a security system when a magnetic field is detected. Whereas to display magnetic field information or danger signs from websites using displays. Figure 1 below is an overall block diagram.

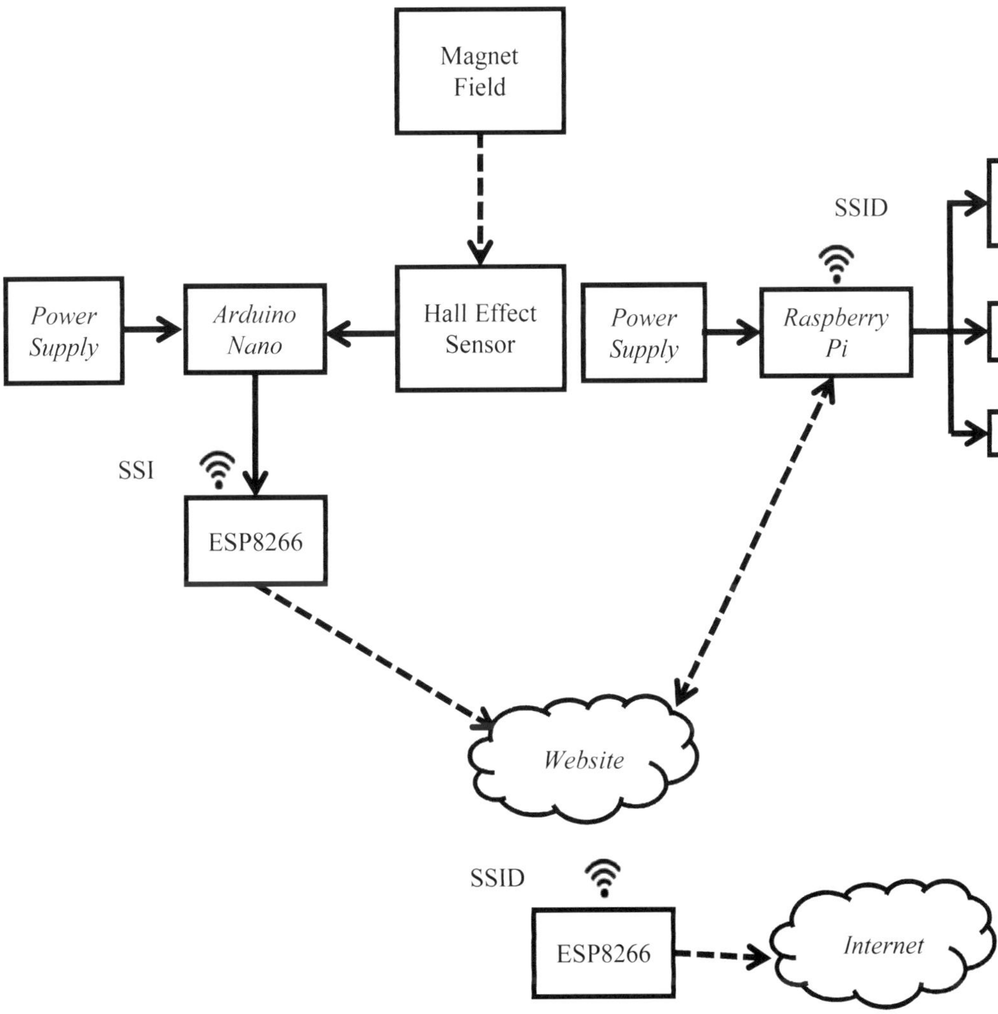

Figure 1. Block diagram of the overall system.

The process flow diagram is intended to describe the operational processes of the system that will later be made, and so that if then there are problems it can help to solve the problem. Figure 2 process flow diagram can be seen below.

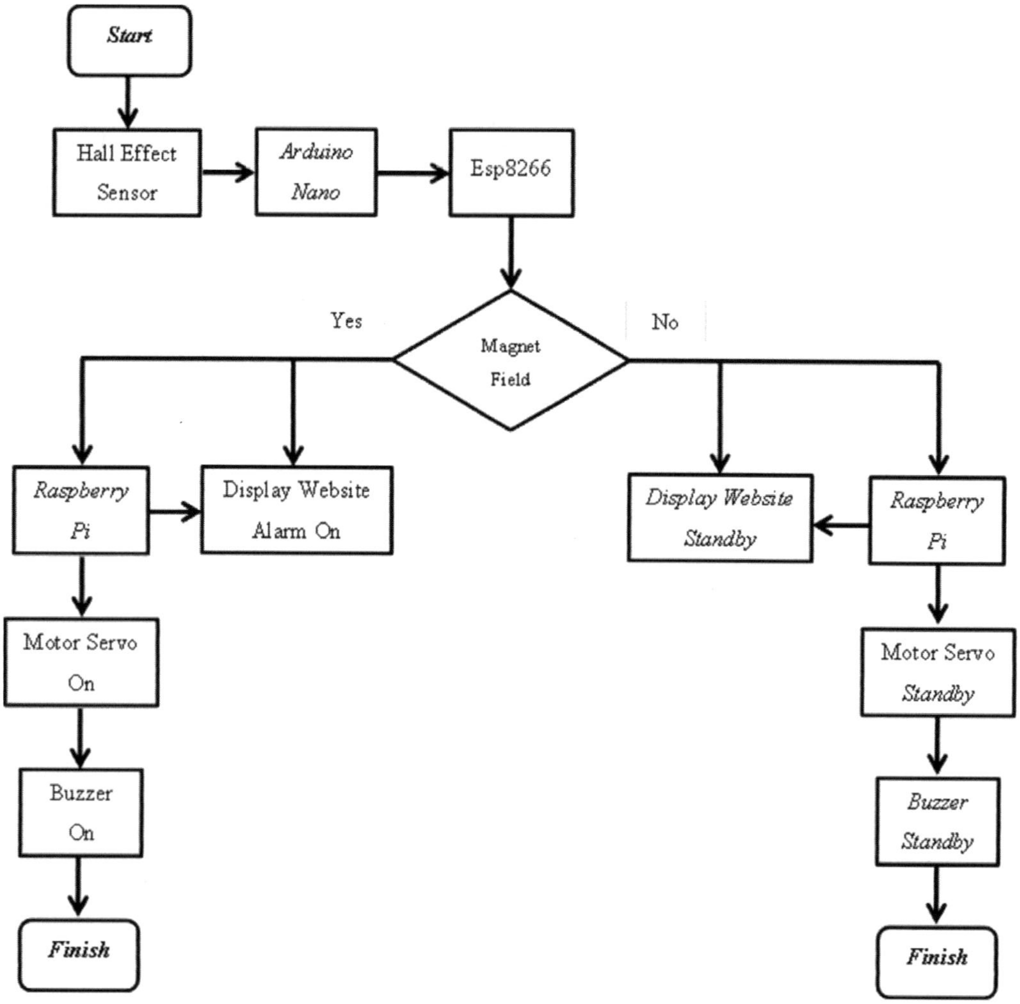

Figure 2. Process flowchart.

The website design is intended to receive data from ESP8266 and make an order to turn on the Raspberry Pi GPIO pin. Besides this website is designed to provide information display to road users when the magnetic field is detected. Display information on the website is divided into two parts, first display with a red column warning where this condition is when the status is not safe/active. The website will provide information that there is a magnetic field as shown in figure 3 below, so that road users cannot pass.

IOP Conf. Series: Materials Science and Engineering **550** (2019) 012012 doi:10.1088/1757-899X/550/1/012012

Figure 3. Website designing is a magnetic field.

Next displays with a blue warning column where this condition is when the status is safe / standby, the website will provide information not detecting magnetic fields as shown in Figure4 below, so that road users can pass.

Figure 4. Website designing there is no magnetic field.

3. Results and discussion

Testing of Magnetic Field Detection is done at a distance of 0-500 mm to find the threshold value of the magnetic field quantity captured by the hall effect sensor, figure 5 shows the test results obtained.

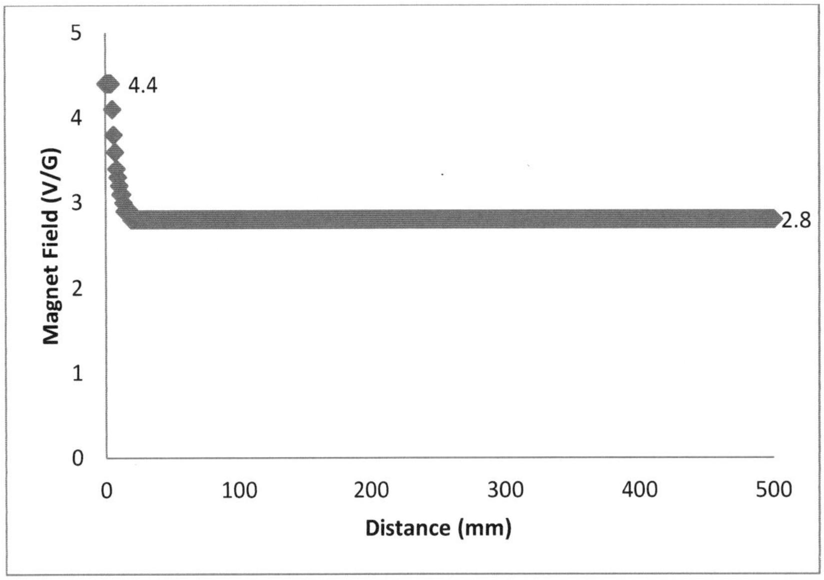

Figure 5. Testing of magnetic field detection.

Based on figure 5 above, the highest value when the magnetic field is detected by the hall effect sensor reaches 4.4 V / G which is at a distance of 0-3 mm and then drops to reach its lowest value of 2.8 V / G at a distance of 19-30 mm. The 2.8 V / G value on the sensor hall effect according to the datasheet is also referred to as the initial value or equal to "zero", because of the Hall effect sensor issues a voltage on the "Out" foot of approximately half of the input voltage. The Hall effect sensor successfully detects a magnetic field at a distance of 0-18 mm, this distance is hereinafter referred to as "there is a magnetic field", while at a distance of 19-500 mm, hereinafter referred to as "no magnetic field".

Testing the response time website is done ten times to test the website changes its appearance or not when receiving incoming data from ESP8266 and knows the time needed when the website responds and after responding, the website can be accessed via the address "www.lab-android.com/kereta". As previously known, the website appearance is designed into two parts, first the condition when the safe/standby status is displayed when there is no magnetic field detected, the second is the condition when the status is not safe / active when a magnetic field is detected. Figure 6 shows the test results obtained.

Based on figure 6, the highest value in testing when there is a magnetic field is the same as the longest time in responding to incoming data from ESP8266 or there is a magnetic field of 6 seconds on the 9th test and the lowest value is the fastest time of 4 seconds on the 5th test. The highest value in testing when there is a magnetic field is the same as the longest time after responding to incoming data from ESP8266 or after there is no magnetic field which 10 seconds on the 6th test and the lowest value equals the fastest time of 3 seconds on the 8th test. In both of these tests, there is an average and after no magnetic field is 5.8 seconds.

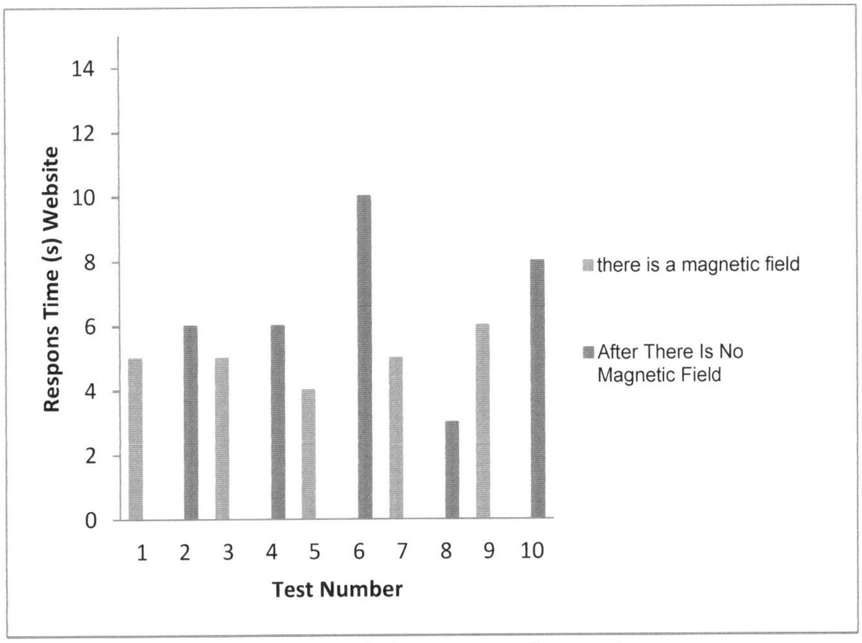

Figure 6. Website response time testing.

Testing the response time of the system when There is a Magnetic Field carried out 10x to test the response time and display of the website when the system detects the magnetic field active status, figure 7 below shows the test results obtained.

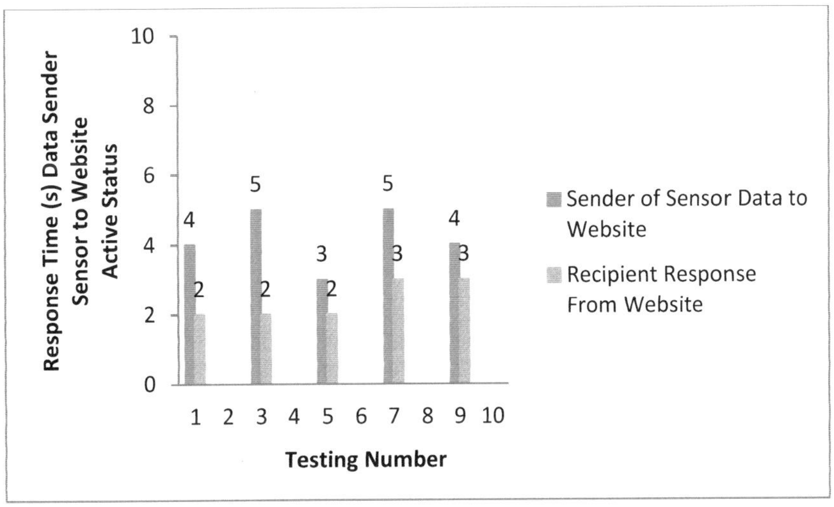

Figure 7. System response time testing detects the magnetic field.

Based on figure 7 above, ESP8266 successfully sends data to the website and the website changes its appearance to be active/alarm so that it can order the output of raspberry pi to be active. The highest value in this test is the same as the longest time, the sender of sensor data to the website is 5 seconds on the third test, and 7, the recipient of the response from the website is 3 seconds in the 7th and 9th tests. The lowest value is the fastest time, the sensor data sender to the website is 3 seconds on the 5th, 7th and 9th tests, the recipient of the response from the website is 2 seconds on the 1st, 3rd and 5th tests.

Testing System Response Time When There Is No Magnetic Field is done 10x to test the response time and website display when the system does not detect the magnetic field passive status, figure 8 below shows the test results obtained.

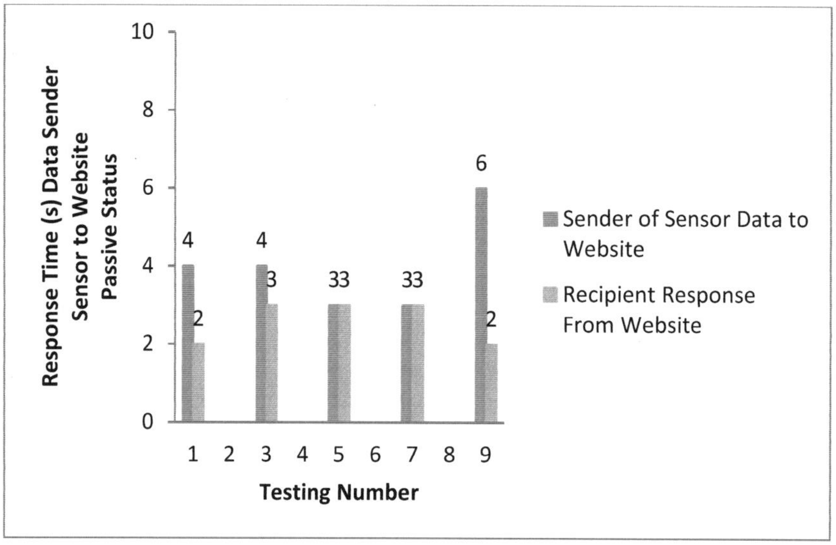

Figure 8. System response time testing detects the magnetic field

Based on figure 8 on the previous page, ESP8266 does not send data to the website and the website changes its appearance to be safe / standby, so it can order the output of raspberry pi to be passive. The highest value in this test is the same as the longest time, the sender of sensor data to the website is 6 seconds on the 10th test, the recipient of the response from the website is 4 seconds on the 2nd and 4th test. The lowest value is the fastest time, the sensor data sender to the website which is 3 seconds on the 6th and 8th tests, the recipient of the response from the website is 2 seconds on the 2nd and 10th tests.

4. Conclusion

Based on the design, implementation and testing, conclusions can be drawn as follows: The hall effect sensor has a sensitivity to detect magnetic fields at a distance of 0-18 mm with an out voltage of 4.4-2.9 (V / G), whereas not detecting a magnetic field has an out voltage of 2.8 (V / G) at a distance of 19-500 mm. The technical response that receives data from the ESP8266 module has an average of 5.8 seconds. The response time of this device when detecting a magnetic field has the fastest time of 3 seconds for the sender of sensor data to the website at the 5th, 7th and 9th test, while the recipient of the response from the website is 2 seconds on the 1st, 3rd and 5th tests The response time of this device when not detecting a magnetic field has the fastest time of 3 seconds for the sender of sensor data to the website in the 6th and 8th tests, the recipient of the response from the website is 2 seconds on the 2nd and 10th test.

References

[1] Ptitsyna N G *et al.* 2003 *Bioelectromagnetics Journal* **24**(8) 546–56
[2] Ram S *et al.* 2016 *International Research Journal of Engineering and Technology* (IRJET) **3**(8) 1666
[3] Gustara M Y 2016 *Essay Detection of Train Magnetic Field for Early Information Based Emergency Stop System Arduino Uno R3 Microcontroller with Solenoid Sensor* (Electrical Engineering Department - Lampung University).
[4] Suresh V and N Arumugam 2016 *Proc. Intl. Conf. on Emerging Engineering Trends and Science*

(ICEETS) (Tamil nadu: Anna University) p 232
[5] Robertson C R 2008 *Fundamental Electrical and Electronic Principles 8'th Edition*. (Slovenia : Charon Tec Ltd., A Macmillan Company).
[6] Giant F R, Darjat and Sudjadi 2015. *Transmisi Journal* **17**(2) p 71
[7] Sujarwata 2013 *Angkasa Journal* **5**(1) p 48-51
[8] Macintyre S A 1999 *Magnetic Field Measurement* (Boca Raton : CRCpress)

IOP Conf. Series: Materials Science and Engineering **550** (2019) 012013 doi:10.1088/1757-899X/550/1/012013

Development of high DC-current measurement system using closed loop Hall Effect configuration

F Faizal[1,2], M Hafidz[1], F F Florena[3], D W Maulana[2], A Abdurochman[1], C Panatarani[1,2] and I M Joni[1,2]

[1]Departement of Physics, Universitas Padjadjaran, Jl. Bandung-Sumedang Km 21, Jatinangor Sumedang, West Java, Indonesia;
[2]Nanotechnology and Graphene Research Center, Universitas Padjadjaran, Jl. Bandung-Sumedang Km 21, Jatinangor Sumedang, West Java, Indonesia;
[3]Department of Physics, Faculty of Mathematics and Natural Sciences, Bandung Institute of Technology, Jl. Ganesha No 10, 40132, West Java, Indonesia;

E-mail: imadejoni@phys.unpad.ac.id

Abstract. Recently, the requirement of high-power electric sources for the vehicle industry has been increased, this trend forces the research of DC power sources and storage to be developed faster. For supporting the development, the requirement of the robust measurement system cannot be avoided. In this paper, non-contact high DC-current measurement system had been developed using Hall effect sensing. By this principle, a wide range of current magnitude spanning from -169 A to 237 A with 0.25 A resolution by design was developed. The design of the instrument was extended from the conventional closed-loop configuration system. The magnetic field obtained from secondary current was used to "purify" offset voltage and by using absolute voltage subtractor, the offset voltage can be removed. By this approach, the error due to permanent magnetic fields of concentrator can be compensated. The sensor and the conditioning signal were coupled with the digital block and the measurement data can be observed from LCD and connected to the PC by interface software.

1. Introduction

Rapid developments of electric cars in order to build "green" low pollution transport mode has reached enthusiasm after the successful product launched by Tesla motor. The grown market proofed that reliable and efficient electric car on the grid is possible to be realized. The fact that it is even difficult to follow by other variants such as fuel cell powered vehicle . As an example, Tesla Roadster can draw electric current around 40 A from allocated 100-200 A of the electric grid at a typical home in USA . Besides the obviously known factor such as size and cost of the vehicle battery , . The efficient management of the electricity system is a critical factor for the electric vehicle. The system which deals with the high DC current should be supported by proper instrument and control system for measurement, monitoring and control of the electrical quantity such as voltage and current.

To deal with high magnitudes of DC current, the instrument system must be equipped with a robust, accurate, safe and reliable sensor. There are various sensors can be used to implement an electrical current instrument. Resistive current sensing (shunt) is a high accuracy method due to direct contact between a sensor and signal conditioning block, which also may cause voltage drop (loading effect) due to the lack of isolation between the blocks. In this case, the non-contact option is a more reliable choice to avoid direct contact between the measured current and the instrument. Several options such as magneto-resistive, current transformer and Hall-effect sensor are available .

Hall effect sensor is a relatively simple sensor that suitable to measure high magnitudes of DC current because the geometrical nature of sensor mechanism which needs a cross-sectional magnetic field to the sensor, as mainly high DC current over a wire generate circular geometry around a wire. The magnetic field can then be concentrated into the sensor by using several concentrator materials such as FeSi, FeNi and 1010 steel with the typical geometrical form of concentrators in the shape of the ring and rectangular metal around the conducting wire (as measurement object) .

A previous study on the electrical current measurement using Hall effect sensor was developed in an open-loop configuration type of signal conditioning circuit . In an open-loop type the detected Hall potential was directly used as the main signal as the representation of the measured electrical current along the conducting wire. Another study, showed that the open loop system relatively susceptible for the deviation, therefore the accuracy of the most open loop configuration of Hall effect sensor is around 1% of full scale, while the open loop configuration accuracy can reach around 0.5 % of full scale. In the same study, the authors also introduced the optimization of the load resistance and the error compensation method for increasing the accuracy of the closed-loop configuration of Hall effect sensor.

In the previous study, the focus mainly deals with the accuracy of the low current measurement. In the current study, an implementation of the method described in the reference , was applied to measure high magnitude DC current, thus the block evaluation was performed and the result will be discussed.

2. Methods

A closed-loop configuration of Hall effect sensor is an optimization of the signal conditioning to enhance the capability of the Hall effect sensor in handling the deviation error appears in the electrical current measurement. A typical close loop Hall sensor consists of Hall effect sensor, coupled with the amplifier to drive the transistor as the switch of the secondary current from power supply to flow into wire turns, around magnetic concentrator. This method will "reset" the magnetization state in the concentrator material due to the primary measured current, thus making the offset of the Hall potential reduced.

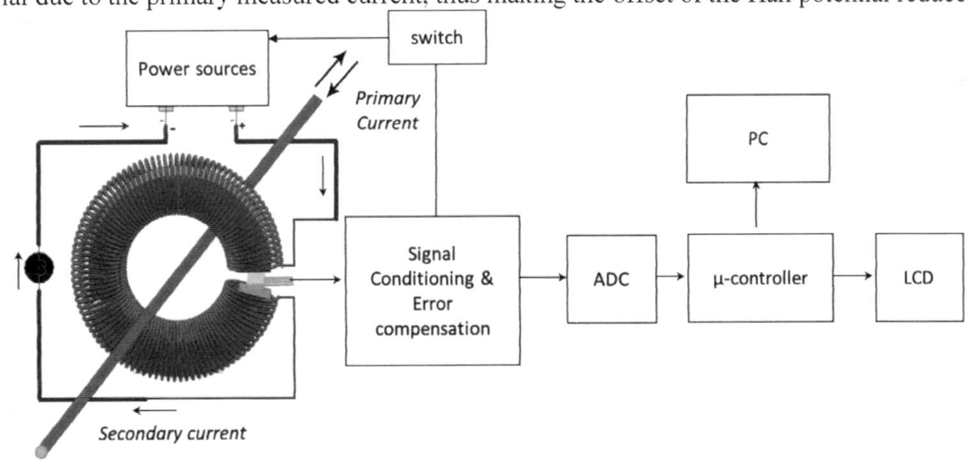

Figure 1. Schematic design of the high-DC current measurement instrument based on closed-loop Hall effect sensor.

The mechanism is described in figure 1 where the schematic design of the designed instrument was also illustrated. The Hall effect sensor used in this study is SS49E (Honeywell Inc.). The error compensation of the Hall-potential adopted the method described in the literature . In the signal conditioning block, the amplifier was used to synchronize the voltage to fit the scale for data conversion, the output signal was connected to 10 bits-ADC (analog to digital converter) integrated into ATmega

328p microcontroller (Atmel Corp.). The microcontroller managed the data to an LCD display and sent the data to a PC via serial port (FTDI1232). In the PC side, the interface software was developed using Visual Basic (Microsoft Corp.)

3. Result and discussion

3.1. Sensor sensitivity and signal conditioning

The sensitivity of the sensor was obtained at 0.016 V/A at 2.835 V/V gain in the forward direction of current, associated with the possible voltage range in our design (1.164 V to 4.956 V) gives the possible current range to be measured is 0 A to 237 A. In the reverse current direction (negative), the voltage range 1.033 V to 4.509 V associated with 0 A to -174 A by the sensitivity of 0.02 V/A at 2.587 gain. The variation in the gain value is possible due to the tolerance value of the different resistor used in the switch channel. It was found that for the increasing secondary current at 0.01 A, the sensor potential was increasing at 0.0068 V and for decreasing secondary current at 0.01 A the sensor potential was decreasing at 0.078 V. From the number of turns in the concentrator it was known that the 0.01 A in the secondary current will be associated with 1 A of the primary measured current.

3.2. Data conversion and interfacing

According to the calculation, 10 bits ADC with a reference voltage of the microcontroller of 5.17 V gives the ADC resolution at 0.005 V/Data. The values were evaluated by connecting the signal conditioning block to the microcontroller and the microcontroller displayed the converted data to LCD.

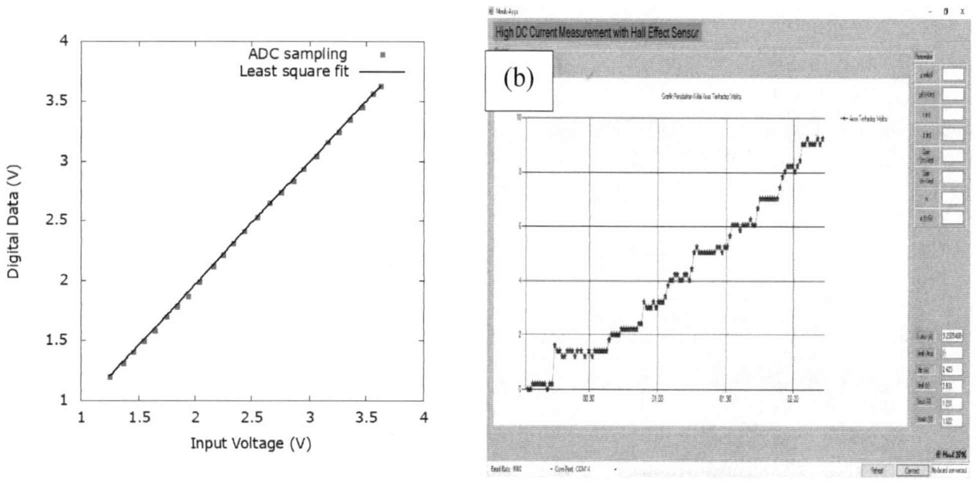

(a) (b)

Figure 2. (a) Properties of ADC evaluated by sampling voltage gives the least square fit at $V_{data}= 1.01\ V_{in}+0.07$ and R=1. (b) screenshot of interface software during the measurement and data acquisition.

The linearity of the data conversion was observed from the plot in figure 2 (a). The sampling voltage data was converted into digital data and gives the fitting line slope at 1.01, offset 0.07 V and correlation coefficient 1.00.

The developed interface software records the real-time values provided by the microcontroller, the screenshot was presented in figure 2 (b) where the increasing values of the DC current were captured vs. time.

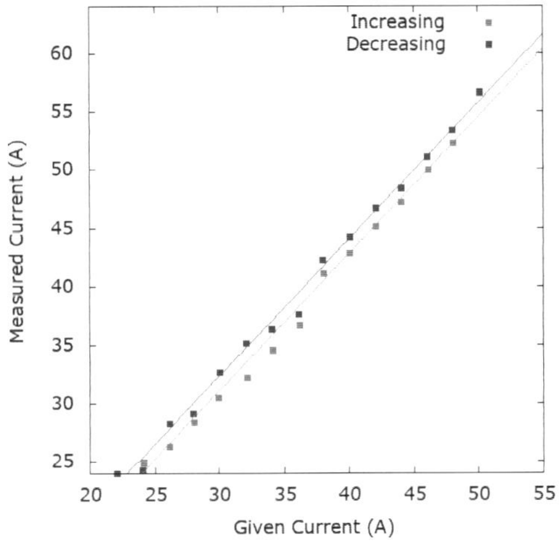

(a)

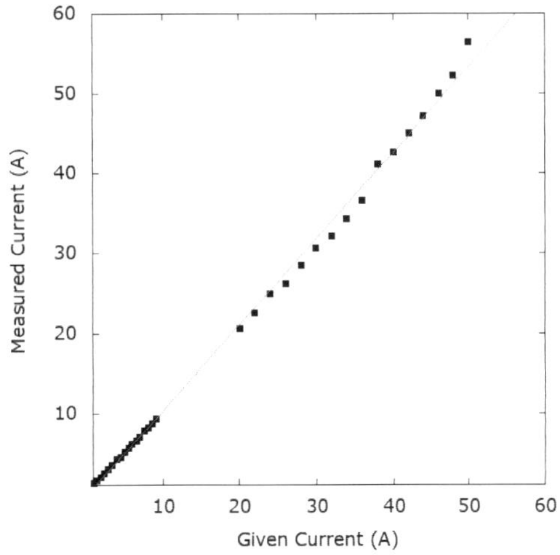

(b)

Figure 3. (a) The plot of measured current and given current at increasing step value and decreasing step value, the relative constant difference was observed. (b) At different current region.

3.3. Integrated system performance

The evaluation to all block in the designed system was implemented by current measurement. The result was presented in the plot of figure 3 (a). While the given current was set in 20 A – 50 A range, it was observed that there are different values between increasing "step" of current, compared to decreasing step values of given current. By using the linear fit in both data, the difference can be analysed and it was observed that the difference is relatively constant at this range. This difference described the "as is" values from the developed circuit, the information of the deviation can be used easily for the adjustment/tuning the instrument in the software side (microcontroller and interface software).

The instrument performance was also evaluated in the lower range of the current signal at 1 A - 10 A, the plot was presented in figure 3 (b). In that plot, the direct comparison with the higher range (20 A- 50 A) was obvious, the highly linear characteristic of accuracy in the lower current range can still be preserved in the higher current range, and the accuracy seems to be in agreement with the previous study . The correlation coefficient for the plot is sufficiently high at 0.99. Although the instrument was designed to work at full range -169 A – 237 A, unfortunately for the current higher than 50 A still cannot be evaluated due to the unavailability of the current source at that range, it will be tested in the future enhancement and integration to control system in the smart DC-grid monitoring and control system development.

IOP Conf. Series: Materials Science and Engineering **550** (2019) 012013 doi:10.1088/1757-899X/550/1/012013

Table 1. The datasheet of developed high DC current instrument.

Characteristic	Biases	
	Forward	Reverse
Output type	Linear	
Range (*designed*)	0 – 237 A	0 – 169 A
	1.164 – 4.956 V	1.033 - 4.529 V
μ_r Concentrator	63.578	84.281
Error	2 – 13 %	
Error	±0.313 A	±0.250 A
Sensor voltage sources	5 V	
Voltage supply, Op-Amp	(+) 12 V & (-) Ground	
Current transducer resolution	0.016 V/A	0.020 V/A
ADC reference voltages	5 V	
Analog to digital reference voltage	0.005 V/Data	
Offset	1.164 V	1.033 V
Calibration	V_{offset}= 2.5 V	V_{offset}= 2.5 V
	V_{reff}= 2.1 V	V_{reff}= 2.9 V
Linearity (*Correlation coefficient*)	0.993	0.997

The summary of the High DC current instrument characteristic was presented in table 1.

4. Conclusion

The measurement system for high DC current has been developed using closed-loop Hall effect schemes. The system was integrated to the digital block for data acquisition. The Integrated system showed the linear profile of measurement data and designed to operate in the current from -169 A to 237 A range. The measurement deviation of the developed instrument was compensated by a closed-loop Hall sensor configuration followed by offset reduction in signal conditioning block and adjustment in the digital data based on the block calibration and evaluation. The implemented system will be ready to use for the monitoring or control system which deals with high magnitude of DC current such as smart DC grid for house electricity and electric vehicle charging system.

References

[1] Hardman S, Shiu E and Steinberger-Wilckens R 2015 Changing the fate of fuel cell vehicles: Can lessons be learnt from Tesla Motors?, *Int. J. Hydrogen Energy* **40**(4) 1625–38

[2] Edson Z 2010 A Snapshot of the Electric Car on the Electric Grid 2010 . Available: https://www.tesla.com/blog/snapshot-electric-car-electric-grid. .

[3] Liu Y, Lin H, Yu W, Luyao L, Sun Q and Wennersten R 2018 Influence of the Electric vehicle battery size and EV penetration rate on the potential capacity of Vehicle-to-grid, *Energy Procedia* **152** 630–5

[4] Nykvist B, Sprei F and Nilsson M 2019 Assessing the progress toward lower priced long range battery electric vehicles *Energy Policy* **124** 144–55

[5] Ramsden E 2006 *Hall-Effect Sensors: Theory and Application. Newnes* (Amsterdam: Elsevier/Newnes) pp 1–2

[6] Gillet C and Friedrich A 2013 Guidelines for Designing a Concentrator for High-Current Sensing Applications with an Allegro Hall-Effect Sensor IC *Allegro MicroSystem* 1–7

[7] Svoboda J 2004 Practical Aspects of Magnetic Methods for Materials Treatment *Magnetic Techniques for the Treatment of Materials* (Netherlands: Springer)p 319–467

[8] Cheng X, Sun Z, Wang X and Liu S 2014 Open-loop linear differential current sensor based on

dual-mode Hall effect, *Meas. J. Int. Meas. Confed* **50** (1) 29–33
[9] Liu J-G, Sanli A, Wang Y and Liu C 2012 Error Compensation of Closed Loop Hall Effect Current Sensors *2012 IEEE International Workshop on Applied Measurements for Power Systems (AMPS)* (IEEE)
[10] Wang Y, Liu J G, Zhao J and Yang Y 2012 Split core closed loop hall effect current sensors and applications *PCIM Europe Conference Proceedings*

Acknowledgments
The author acknowledges the financial support from Penelitian Pengembangan Unggulan Perguruan Tinggi (PPUPT), Universitas Padjadjaran, Bandung, Indonesia.

IOP Conf. Series: Materials Science and Engineering **550** (2019) 012014 doi:10.1088/1757-899X/550/1/012014

Design of 2 dof arm robot control system using ultrasonic sensor

K Sekarsari, D Ikhsan and Marfin

Teknik Elektro, Universitas Pamulang, Kampus Viktor, Jl. Raya Puspiptek, Setu, Kota Tangerang Selatan, Banten 15314

Email: kartika_sekarsari@yahoo.com; daanuikhsan@gmail.com

Abstract. This paper explains about design of 2 DOF hand-shaped manipulator robot prototype with articulated manipulator joints configuration. The manipulator robot joint driven by a servo motor in a closed loop that was controlled using an Arduino uno microcontroller. Movements of joints and end-effector robots were well-designed and allowed real-time observation interfaced by Arduino uno-based computer software. The manipulator robot that is developed can move objects to the right and left. The robotic gripper can grab objects that are less than 5 cm in size and have the ability to move objects with a maximum weight of 1 kg.

1. Introduction

Microchip and microcontroller technologies have produced many new innovations, such as in the case of smart devices and robot technology. The robot technology widely used in industry to help human performance for a routine and sophisticated work. Similarly, in the automotive manufacturing industry and transportation equipment many use this robot technology to lift large enough objects, move them, and also install them. This technology is very helpful for human performance in accelerating the production process in the industry due to its strength and very long working time lead to maximum quality of products. One of the recent important robot technology is the arm robot or usually called a robot manipulator. Arm robots are produced and used in industry for certain functions such as to move, lift, and install objects that have relatively large weights [1][2].

This paper reported the design of a 2 DOF manipulator robot which was applied to the motion control system used to lift and move objects according to the capabilities of the robot. A servo motor was placed in each of the rotation joint and was controlled and reconfigured by the computer through Arduino Uno microcontroller based software. The prototype of DOF 2 manipulator robot was intended as a medium of learning for students, especially in robotics course.

2. Manipulator Robot

Illustration of an industrial robot as a manipulator robot usually has a complete arm shape including: hands, wrists, to the ends of the hands. The entire robot is usually designed with various equipment so that it has the ability to perform the desired task [3]. A manipulator robot designed in the form of a mechanical arm (manipulator) is one of the components of a robotics system which includes: the mechanical arm itself, an external power source, the function of the manipulator robot end arm, such as: gripper, internal and external sensors, computer interface, and the controller computer. Programming software on robots must be considered as a whole of the whole system, because the design of control and programming on the robot will have a major impact on performance, accuracy and repeatability.[1]

The robot manipulator configuration, arms (links) on robotic manipulators connected by joints form kinematic chains. This joint will usually revolute or linear (prismatic). A rotary joint is identified as a hinge that forms a rotary motion between two links. Whereas prismatic joints, will form a linear motion between two links.[3][4] Images of rotary and prismatic joints are represented in figure 1.

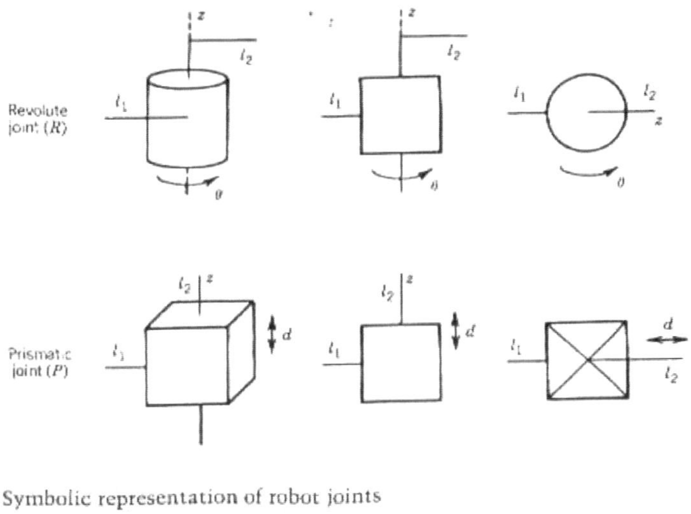

Figure 1. Symbol of revolute (rotary) and prismatic joints [3].

In the motion of robotic manipulators, the design right kinematic control will be able to solve the desired position problem rather than prioritizing dynamics control. Because dynamic modeling is very complicated and if the results are not correct then the kinematic modeling function becomes inaccurate and even not useful in improving the overall quality of control [5][6].

This manipulator robot consists of reverse kinematics and reverse kinematics. Forward Kinematics is a method for determining the orientation and position of the end-effector from the magnitude of the joint angle and the link length of the robot arm. Advanced kinematics equations are obtained based on the DOF number and kinematic chain type of the robot manipulator. Meanwhile, kinematic inverse is the opposite of future kinematics. This method is needed to determine the angular value of the joint needed so that the end-effector can reach the desired position [4].

Kinematic equation of the manipulator robot is explained as follows; figure 2 describes that if two arms are connected with a rotary joint, the end-effector is located at position x_r, y_r can be calculated by knowing the values of θ_1 and θ_2 and arm length-1 (l_1) and arm-2 (l_2) with the following equation [5][7]:

$$x_r = l_1 cos\theta_1 + l_2 \cos(\theta_1 + \theta_2) \quad (1)$$
$$y_r = l_1 sin\theta_1 + l_2 \sin(\theta_1 + \theta_2) \quad (2)$$

The above equation can be obtained using advanced kinematic analysis. By law trigonometric identity [5][7] :

$$\cos a + b = \cos a \cos b - \sin a \sin b \quad (3)$$
$$\sin a + b = \sin a \cos b + \sin b \cos a \quad (4)$$

then obtained,

$$x_r = l_1 cos\theta_1 + l_2 cos\theta_1 \cos\theta_2 - l_2 sin\theta_1 sin\theta_2 \quad (5)$$
$$y_r = l_1 sin\theta_1 + l_2 sin\theta_2 \cos\theta_2 + l_2 cos\theta_1 sin\theta_2 \quad (6)$$

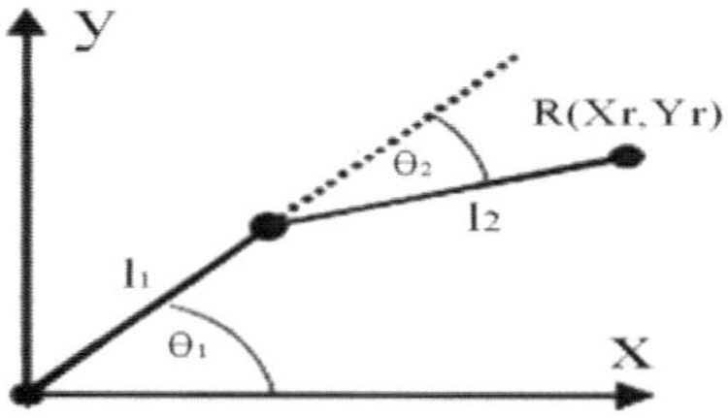

Figure 2. Manipulator 2 joint [5][7].

In inverse kinematics will be obtained [5][7] :

$$\theta_2 = \text{arc cos}\frac{(x_r^2 + y_r^2 - l_1^2 - l_2^2)}{(2\, l_1 l_2\,)}$$

$$\theta_1 = \text{arc tan}\frac{y_r(l_1 + l_2\cos\theta_2) - x_r l_2 \sin\theta_2}{x_r(l_1 + l_2\cos\theta_2) + y_r l_2 \sin\theta_2}$$

3. Stages of making robotic arms

The stages of designing the arm robot are shown in figure 3 . Preparation of tools and materials as well as the mechanical design of the system were the first steps in developing a robot arm. The second step is the installation of a servo motor. The design of the servo motor 1 is different from the other servo because it has a gripper attached to the servo. This gripper is useful for grabbing the desired object. The scale for this design measure is used in units of millimeters (mm). The width of the opening that the gripper can do is 5.6 cm. This makes the Robot only able to grab objects with a size below 5.6 mm. Wrist servo motors (servo 2) have a servo design that is almost the same as servo 1, but it doesn't have additional devices like a gripper. The motor on the wrist plays a role in attaching the gripper. The servo base (servo 3) design is designed to be more robust and large than the other servo to be able to withstand greater loads. The shape of the servo 3 is like a box with the bottom cut off which serves as a place to place the robot . Especially for servo 3, 4 bearings are mounted on the round shaft to reduce the amount of friction when rotating. The next stage, the robot arm that is made requires a sensor to detect objects or objects that will be taken. The sensor used is the SRF04 Ultrasonic sensor so that it can be easily read by the Arduino microcontroller. The Arduino microcontroller has a support circuit that helps work from program operations. Sensor and servo circuits are connected to the microcontroller on certain pins. Testing of microcontroller program on the robot arm movement is the final stage of making a robotic manipulator arm [2].

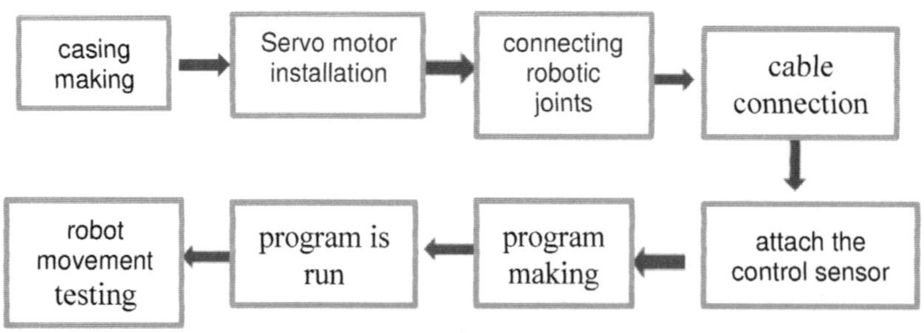

Figure 3. Stages of developing a robot arm.

IOP Conf. Series: Materials Science and Engineering **550** (2019) 012014 doi:10.1088/1757-899X/550/1/012014

3.1. Mechanical design

The manipulator robot that is made has a resemblance to the human arm. This robot has three rotation joints including a gripper. Where the rotation joint on the base becomes the base for the robot's shoulder. This robotic arm has the advantages of being able to take objects of certain size and weight and can perform spontaneous movements [8][9]. Detection of objects to be taken and moved by the robot using the SRF04 ultrasonic sensor.

The placement of the gripper on the end effector in figure 4 is useful for the robot arm to grab the object to be moved. The shape of the gripper is chosen to resemble claws, because this form is easy to design and has synchronous movements between the two parts of the claws. A pair of claws will give a balanced grip on the object being taken. The gripper capit makes it easier for the arm robot to grip the object well in vertical conditions [1][2].

In Figure 4 it is shown that the robotic arm is equipped with an Ultrasonic sensor which functions as a controller of robot movements. The placement of sensors in front of objects in a certain way makes the sensor can detect objects well.

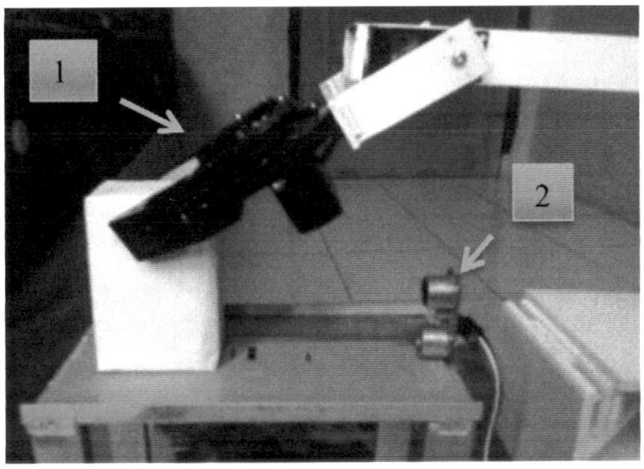

Figure 4. Mechanical of robotic arms show (1). Gripper; (2) SRF04 ultrasonic sensors

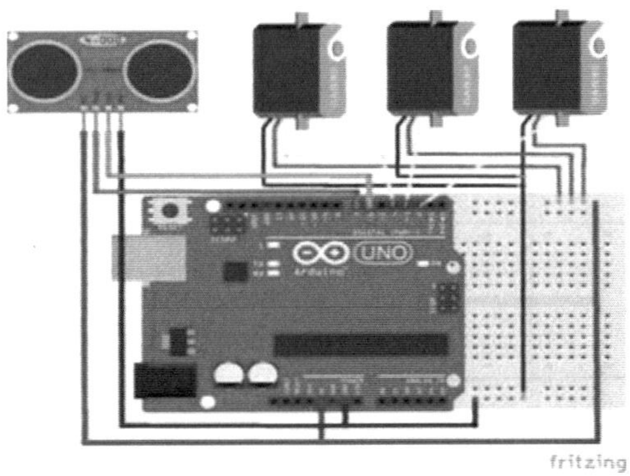

Figure 5. Servo series and robot sensors on Arduino

The Arduino microcontroller has a support circuit that helps work from program operations. The sensor circuit will be connected to the microcontroller at Pin 6 for the triger and pin 7 to echo. while for servo motors there are pins 2,3, and 4. This sequence can be seen in figure 5.

3.2. System flow diagram

The program flow diagram for robot arms is shown in figure 6. The program is made with input in the form of values d1,θ_1, $\theta_2$and output in the form of end-effector position on the robot. Input values d1, θ_1, , θ_2, are limited according to the configuration space of each joint [5]. If the sensor input value exceeds or is smaller than the specified distance, the program generates an error message. This system serves to determine the actual movements that occur in the manipulator and see the accuracy of the motion control system reaching the desired reference point.

IOP Conf. Series: Materials Science and Engineering **550** (2019) 012014 doi:10.1088/1757-899X/550/1/012014

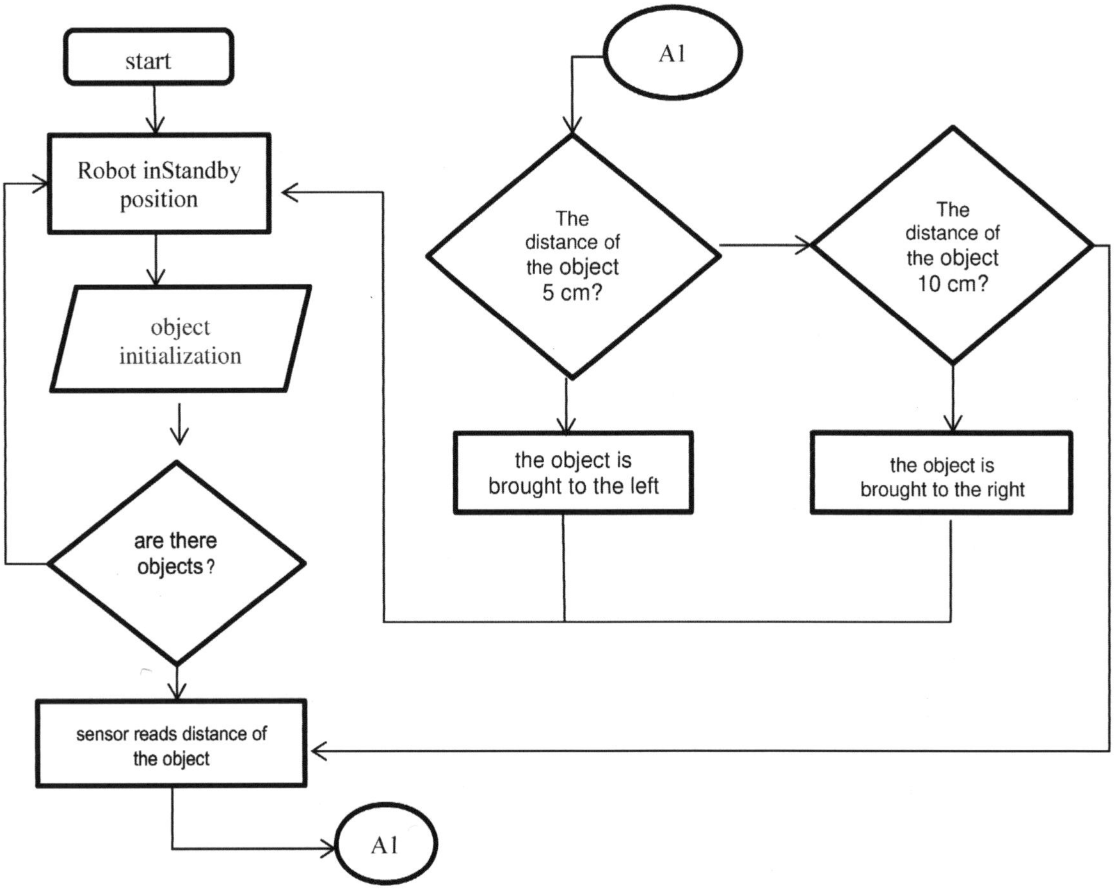

Figure 6. System flow diagram.

4. Results and discussion

Each servo has a pulse width limit to do circular motion with a certain angle. Maximum and minimum pulse widths are obtained from measurements when servo motors are installed. The results of measuring the maximum pulse width and the angle formed are shown in table 1. Table 2 shows the test results data entered in the program to provide the appropriate angle for each servo, so that the robot can perform movements to reach objects that have been specified without errors. The gripper grip is carried out nine times with different object objects. The results of this test are shown in table 3.

Table 1. Results of measurement of pulse width and servo angle

No	Servo type	Max and Min pulse width values (uS)	Angle of rotation on servo (°)
1	Servo *gripper*	600 2400	0° 180°
2	Servo *wrist*	650 2500	0° 180°
3	Servo *base*	550 2600	0° 180°

Table 2 . Angle values for robot movements.

No	Servo type	Angle of rotation on servo (°)	Information
1	Servo *gripper*	20° 80°	Pinch Open
2	Servo *wrist*	45° 90°	Down Up
3	Servo *base*	83.° 130°	Middle position Right position

Table 3. Test results of gripper grip on objects

No	The form of the object being grasped	Object dimension (cm)	Object weight (kg)	Surface conditions of objects	Observation result
1	Acrylic	8.5 x 5	0,0617	Slippery	Good
2	Plastic	10 x 5	0,1263	Rude	Good
3	Batery box	4.,5 x 2.5	0,0845	Slippery	Good
4	Marker	14 x 1	0,006	Slippery	Good
5	Aluminium rod	13 x 2	0,007	Slippery	Good
6	Iron cylinder	13 x 2.5	0,1332	Slippery	Good
7	Big screwdriver	4.5 x 3	0,0609	uneven	Good
8	Medium screwdriver	8,5 x 2	0,0293	uneven	Good
9	Pilers	16.5 x 5	0,0937	uneven	Good
10	Wood	10 x 5	1	Rude	Good

In this test the acquisition of ultrasonic sensor distance measurements with actual distance is used ruler per 10 cm to a distance of 300 cm. Testing based on the object color of the object being transferred, namely: black objects, white objects, glass, and objects that have uneven surfaces.

Table 4. Test results on robot movements.

Distance of objects (cm)	Robot movement
3	Not moving
4	Not moving
5	Not moving
6	Not moving
7	Not moving
8	Move right
9	Not moving
10	Move left
11	Not moving
12	Not moving
13	Not moving
14	Not moving
15	Not moving

The test results in table 4 , show the movement of the robot in accordance with the program made. The robot will move and move the object to the right if the object is at a distance of 8 cm against the sensor. And the Robot will move and move the object to the left if the object is at a distance of 10 cm against the sensor.

5. Conclusion

This paper presents the design of a 2 DOF manipulator robot prototype for the application of lifting and moving light material objects less than 5 cm with the help of a Gripper. The ultrasonic sensor as a detector for the presence of objects will give a signal to the Arduino microcontroller. Furthermore, the Arduino Uno controller is used to control the servo motor in combined motion and robot end-effector. The manipulator robot designed has the ability to move objects with a maximum weight of 1 kg.

References

[1] Microchip.Sept/Okt 2018. The Power Of Two. Micro Solutions. A Microchip Technology Inc. Publication. www.microchip.com
[2] Mohamed Fezari and Ali Al-Dahoud 2014 Easy Way to Learn Robotic Using Microcontrollers. Proceedings of the 2014 International Conference on Circuits, *Systems and Control* 100-105.
[3] Qi A N W, Voon K L, Ismail M A, Mustaffa N and Ismail M H 2015 Design and Development of a Mechanism of Robotic Arm for Lifting Part1. *2nd Integrated Design Project Conference (IDPC)* (Pahang: Faculty of Mechanical Engineering, Universiti Malaysia)
[4] Ismail M H, Qi A N W, Voon K L, Ismail M A and Mustaffa N 2015 Design and Development of a Mechanism of Robotic Arm for Lifting Part 5. *2nd Integrated Design Project Conference (IDPC)* (Pahang: Faculty of Mechanical Engineering, Universiti Malaysia)
[5] Nurhakim H R 2010 The Motion Control System of Continuous Path Tracking Using Cubic Trajectory Planning On 4 DOF Robot Manipulator *Final Project of Electrical Engineering* (Depok: University of Indonesia)
[6] Solichin A 2009 Design Of Robot Arm (Robotic Arm) with Manual Control Final Project (Bogor: Faculty of Fisheries and Marine Sciene, Bogor Agricultural Institute)
[7] Budiharto W 2014 *Modern Robotics, Theory and implementation (revised edition)* (Yogyakarta: ANDI) 42-44
[8] Pitowarno E 2006 *Robotics, Design, Control, and Artificial Inteligence* (Yogyakarta: ANDI)
[9] Iqbal J, Islam R, Khan H 2012 Modeling and Analysis of a 6 DOF Robotic Arm Manipulator, *Canadian Journal on Electrical and Electronics Engineering* **3**(6) 300-306

A Method to Determine Control Limit and to Monitor the Resistance Value of a Multiproduct Calibrator

N F Yayienda, M Azzumar and A Faisal

Metrology Research Center, Indonesian Institute of Sciences
Tangerang Selatan, Indonesia

E-mail: nibrasfitrah@gmail.com; muhammad.azzumar90@gmail.com; agahfaisal@gmail.com

Abstract. Monitoring of the dc resistance value of a multiproduct calibrator had been implemented. This multiproduct calibrator is commonly used as a reference standard to calibrate electrical measurement testers such as insulation tester, earth tester, continuity tester, ground bond resistance tester and many other resistance meters. The monitored value of the multiproduct calibrator can be used to support measurement traceability needs for those electrical measurement testers and meters in a power distribution utility. The monitoring was done by measuring nine data of resistance value within nine respective days to constructa preliminary statistical control limits. The measurement was continued monthly to monitor the dc resistance value that was intended to test the control limits. The monitoring is performed by doing F-test and T-test on the recent measurement data over the all previous measurement data. If a measurement data is found to be outside the control limits, an action should be done to accommodate the information and renew the statistical control limits. The renewal action for the dc resistance value of the observed multiproduct calibrator was performed for several times within a year. It is reported based on the method implementation acquired within a year, that the upper control limit (UCL) and the lower control limit (LCL) for the dc resistance value of 19 kΩ is 18.99964 kΩ and 18.99912 kΩ. Moreover, to verify the performance of the final statistical control limits, the additional measurement is taken and the result lies between the UCL and LCL

1. Introduction

Based on National Standardization Agency of Indonesia (BSN) website, there are already 246 accredited calibration laboratories, 43 labs of them having electrical calibration scope[1]. These accredited laboratories which have a capability on meters and testers calibration such as a digital multimeter, earth tester, insulation tester represent their standard based on a Multiproduct Calibrator (MPC). Research Center for Metrology of LIPI (RCM-LIPI), as National Metrology Institute, provides the calibration services for the laboratories. One of the calibration services is the MPC calibration. During this year there are already 16 MPCs that have been calibrated by RCM LIPI. Some of them have already been calibrated and the rest are not more than two years recalibration period.

On the ISO/IEC 17025, chapter 5.9 said that the laboratory shall have the quality control procedures for monitoring the validity of calibration undertaken. The resulting data shall be recorded in such a way that trends are detectable and statistical techniques shall be applied to the reviewing of the results [2]. However, to acquire the quality assurance of calibration, the laboratory shall do an intermediate check within half calibration period, or make the inter-laboratory comparison [3]. Moreover, another method is recommended by National Accreditation Committee (KAN) to acquire the quality assurance for laboratories which is using statistical control process that is widely used to determine measurement result quality to analyze the resulting data [4, 5].

Practically, sometimes the inter-laboratory comparison becomes a more popular option than any options to acquire the quality assurance of the accredited laboratories. However, the comparison should not only become a pilot test to check the quality assurance. Monitoring of calibration validity is more an early stage than the comparison since it is an internal laboratory activity and not depends on other laboratory activity. The monitoring result can be used as the base estimation to support the laboratory's Calibration and Measurement Capability (CMC).

The quality assurance by the intermediate check mention in [3] does not mean sufficiently enough acquired only one-time monitoring during recalibration period of MPC. The KAN guide as in [3] has already described how to do the monitoring process along a period of time, however, the example to strong for standard mass measurement. By following our study, we propose a method to determine the control limit and to monitor the representing value of MPC. The quality assurance may be set as a recalibration of retained UUC, such a digital multi-meter, that supporting CMC using a standard representation based on MPC acquired more frequent by adapting the KAN guide in term of electrical quantity.

2. Measurement process

The parameter tested in this paper is a dc resistance which is generated by the standard of Multiproduct Calibrator (MPC). It is usually used to calibrate industrial and laboratory ohm meter. In this method, the preliminary data need to be taken to obtain the preliminary control limits, and then the acquired control limits will be tested using a monthly measurement process. To maintain good reproducibility along the process, the personnel, the mean equipment and the method are the same. The environment condition shall be maintained at a temperature of (23 ± 2) °C and relative humidity of less than 75%

The example measurement point used is 19 kΩ, where it is one of a verification point of MPC Fluke 5520A corresponding to the standard resistor of RCM-LIPI. In monitoring calibration value of a DC resistance of MPC can be done through several steps. The steps are determining control limits, taking data of a monthly measurement process, and renewing control limits based on F-test and T-test analysis

2.1. Determining control analysis

Preliminary statistical control limits are constructed from a set of measurement consisting of nine data taken within nine respective days. The measurement is a calibration of retained digital multi-meter which is operated following a direct comparison method as our work instruction. The statistical control limits are determined based on twice standard deviations and the mean of nine preliminary data. The statistical control limits that determined in this paper can be expressed as follow

$$CL = \bar{x} \pm 2\sqrt{\frac{1}{n-1}\sum_{i=1}^{n}(x_i - \bar{x})^2} \quad (1)$$

The statistical control limits consist of both upper and lower limits, for obtaining upper control limit (UCL), add twice standard deviation to the mean of preliminary data and for obtaining lower control limit (LCL), subtract the mean of preliminary data by twice standard deviation.

2.2. Monthly measurement

The measurement is continued monthly to monitor the dc resistance value that was intended to test the control limits. This measurement process is performed to obtain the ten data observed.

2.3. *Monitoring of control limit*

Monitoring of control limits is performed by doing F-test and T-test on the recent measurement data over the all previous measurement data

2.3.1. F-Test. F-test is used to test the equality of variances of two populations. The two populations mentioned are the all previous measurement data and the recent measurement data. F-test is performed by obtaining the test statistic of F_{Test} and the critical region of $F_{Critical}$. F_{Test} and $F_{Critical}$ can be obtained by equation (2) and (3), respectively:

$$F_{test} = \frac{s_1^2}{s_2^2} \tag{2}$$

$$F_{ciritical} = \begin{cases} F_{(1-\alpha/2, N_1-1, N_2-1)} \\ F_{(\alpha/2, N_1-1, N_2-1)} \end{cases} \tag{3}$$

Then indicate the equality of two variances by comparing F_{Test} and $F_{Critical}$ with two conditions. The hypothesis that the two population variances are equal will be rejected if

$$F(\alpha/2, N_1-1, N_2-1) < F_{test} < F(1-\alpha/2, N_1-1, N_2-1) \tag{4}$$

T-Test. T-test is used to know the equality of means of two populations. T-test is performed by obtaining the test statistic of T_{test}. T_{test} can be obtained by Eq. 5 and Eq. 6, where the Eq. 5 is used for unequal in means and the Eq. 6 is used for equal in means.

$$T_{test} = \frac{\overline{Y_1} - \overline{Y_2}}{\sqrt{s_1^2/N_1 + s_2^2/N_2}} \tag{5}$$

$$T_{test} = \frac{\overline{Y_1} - \overline{Y_2}}{S_p\sqrt{1/N_1 + 1/N_2}} \tag{6}$$

Where

$$S_p^2 = \frac{(N_1-1)s_1^2 + (N_2-1)s_2^2}{N_1 + N_2 - 2} \tag{7}$$

The hypothesis that the two population means are equal will be rejected if:

$$T_{test} > T_{1-\alpha/2, v} \tag{8}$$

Where $T_{1-\alpha/2,v}$ is the critical value of the t-distribution at α significant level (we use 0.05) with v degrees of freedom for unequal in mean can be seen in Eq. 9, and equal in mean can be seen in Eq. 10.

$$v = \frac{(s_1^2/N_1 + s_2^2/N_2)}{(s_1^2/N_1)^2/(N_1-1) + (s_{21}^2/N_2)^2/(N_2-1)} \tag{9}$$

$$v = N_1 + N_2 - 2 \tag{10}$$

The results obtained by using F-test and T-test are used to determine the renewing of control limit

3. Results and discussion

The value of preliminary control limits (UCL and LCL) consisting 9 data taken in 9 respective days can be seen in table 1. The values of these control limits can be obtained by using Eq. 1.

Table 1. Value of preliminary control limits.

Nominal	UCL	LCL
19 kΩ	18.99923 kΩ	18.99944 kΩ

These control limits can be renewed based on the monthly measurement process. The values of monthly measurement are analyzed by using F-test and T-test where it can be done by using Eq. 2 and Eq. 5. In addition, the value of UCL and LCL are not more than CMC of F5520A in 19 kΩ, where the CMC value is ± 0.00043 kΩ. The illustrations of UCL, LCL and CMC values can be seen in figure 1

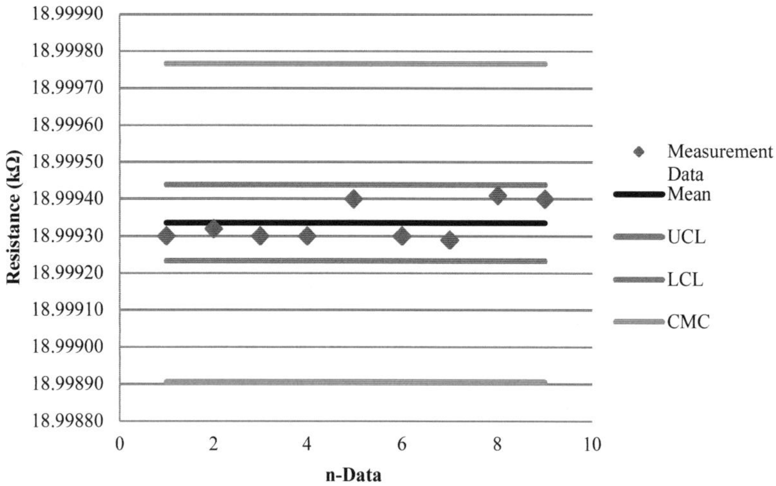

Figure 1. UCL, LCL and CMC.

The values of F_{test}, $F_{critical}$ (1-α/2, N_1-1, N_2-1) and $F_{critical}$ (α/2, N_1-1, N_2-1) on the first month measurement respectively are 2.65 and 0.23 and 4.10. These values show that the dispersion of recent data is equal to the previous data as mentioned in Eq. 4. The values of T_{test} and $T_{critical}$ on the first month measurement respectively are 1.32 and 2.11. These values show that based on T-Test, the mean of the recent data is equal to the previous data because based on the criteria, if $T_{test}<T_{critical}$, the two population means are equal. So the conclusion is the result based on either T-Test or F-Test for first month measurements are lie inside the control limits. So the control limits do not need to be renewed.

For the second month, the standard deviation value for the recent data is 0. It means denominator value for F_{test} is 0, because of this the value of F_{test} is infinity (∞). Meanwhile, the value of T_{test} bigger than $T_{critical}$ because of the standard deviation of recent data is 0. So the conclusion is even though the standard deviation is 0, the control limit still have to be renewed. The renewed control limits can be seen in table 2.

Table 2. Value of control limits in the second month.

Nominal	UCL	LCL
19 kΩ	18.99945 kΩ	18.99925 kΩ

For the third month, the value of F_{test} lies between the value of $F_{critical}$ (1-α/2, N_1-1, N_2-1) and value of $F_{critical}$ (α/2, N_1-1, N_2-1). The value of F_{test} is 1.47, and the value of $F_{critical}$ (1-α/2, N_1-1, N_2-1) and $F_{critical}$ values (α/2, N_1-1, N_2-1) are 0.38 and 3.58. Meanwhile for the T-Test, $T_{Critical}$ is bigger than T_{Test}. The values of $T_{Critical}$ and T_{Test} respectively are 1.72 and 2.03. So the conclusion is the recent data taken lies between the control limits and these control limits do not need to be renewed.

The fourth month has a similar type of data if it is compared with the second month because the standard deviation is 0. So it makes the data lies outside the control limits and needs to renewing the control limits. The new control limits can be seen in table 3.

Table 3. Value of control limits in the fourth month.

Nominal	UCL	LCL
19 kΩ	18.99946 kΩ	18.99927 kΩ

For the fifth month and sixth month, the recent data are out of control. So the control limits need to be renewed. The new control limit for the fifth month and sixth month can be seen in table 4.

Table 4. Value of control limits in the fifth and sixth month.

Nominal	UCL	LCL
19 kΩ (5th)	18.99952 kΩ	18.99917 kΩ
19 kΩ (6th)	18.99956 kΩ	18.99918 kΩ

For the seventh month, the value of Ftest lies outside the Fcritical values it means that the dispersion of previous and recent data are unequal. Even though the F-Test shows that the dispersion of two populations is different, the T-Test shows that the mean of previous and recent data are equal as the value of Ttest is 2.25 and the value of Tcritical is 1.99. Although the dispersion is different, the control limits do not need to be renewed because the mean is equal.

For the tenth and eleventh month, the result of F-Test shows that the dispersions of previous and recent data are equal. But based on the T-Test, the mean of previous and recent data are unequal. Although the dispersion two populations are equal, the control limits needed to be renewed because of the T-Test result. If the result of T-Test is unequal, it means the value of recent data is out of control. The new value of tenth and eleventh control limits can be seen in table 5.

Table 5. Value of control limits in the tenth and eleventh month.

Nominal	UCL	LCL
19 kΩ (10th)	18.99955 kΩ	18.99917 kΩ
19 kΩ (11th)	18.99954 kΩ	18.99917 kΩ

For twelve months, the dispersion and the mean of recent and previous data are equal. So the control limits do not need to be renewed. Based on the F-Test and T-Test, the dispersions of previous and recent data are equal but the mean of two data are unequal. So the control limits need to be renewed. The last control limit can be seen in table 6.

Table 6. Value of control limits in last month.

Nominal	UCL	LCL
19 kΩ	18.99964 kΩ	18.99912 kΩ

The last value of UCL and LCL verified for the following year. The verification is done by measuring the nominal in the following year and analyzing based on F-Test and T-Test. The values of F_{test}, $F_{critical}$ $(1\text{-}\alpha/2, N_1\text{-}1, N_2\text{-}1)$ and $F_{critical}$ $(\alpha/2, N_1\text{-}1, N_2\text{-}1)$ of verification respectively are 1.87, 0.45 and 3.39. The values of T_{test} and $T_{critical}$ respectively are 0.31 and 1.98. Based on the Eq. 4 and Eq. 8, the dispersion and the mean are equal to the previous data. So the control limits do not need to be renewed and it means that the last control limits (based on one year observation) verified and can be used for the following year. In addition, the value of verified UCL and LCL are bigger than the preliminary UCL and LCL but still not more than CMC of F5520A. The illustration of UCL, LCL, uncertainty and all measurement data can be seen in figure 2.

During the monitoring period, there are several measurement data that lie outside the CMC, this data can be caused by several factors such as measurement configuration and environmental condition. Since the measurement configuration for this paper using 4 terminal dc resistance measurement is the same from the first month until now based on the work instruction, the remainder is assumed that the measurement configuration is not the main cause for the several data that lie outside the CMC. Otherwise, the environmental condition such as drastic changes in humidity, temperature and atmospheric pressure can affect the resistor and make the dc resistance value fluctuate although it is not more than its floor value [6]. The possibility data that lie outside the CMC for the following time is still in observation.

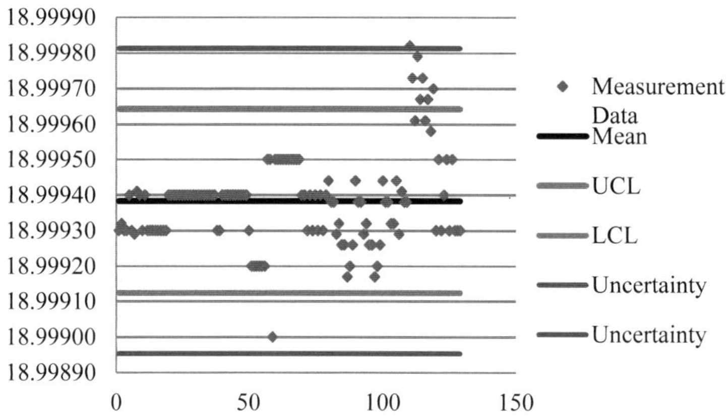

Figure 2. The Illustration of UCL, LCL, uncertainty and measurement data.

4. Conclusion

The quality assurance has been set as a recalibration of retained digital multi-meter that supporting Calibration and Measurement Capability (CMC) using the Multiproduct Calibrator (MPC). The quality assurance has been acquired more frequent by adapting KAN guide in term of electrical quantity. A typical calibration point of 19 kΩ using the MPC, both of LCL and LCL limits have been renewed for seven times. However, it is verified that the limits are still under the CMC estimation and do not change up to September 2017. The contributing factors include in measurement process such as a standard used, a digital-multimeter used, the implemented method and staff performance are observed to be consistent; however, the occurred variability is still in observation

References

[1] National Standardization Agency of Indonesia 2017 ".:: Laboratorium & Lembaga Inspeksi ::." [Online].Available: http://sisni.bsn.go.id/index.php/lembinsp/inspeksi/publik/1/X9/X9/2/X9/X9. [Accessed: 14-Nov-2017]

[2] International Standard Organization 2005 "ISO/IEC 17025 General requirements for the competence of testing and calibration laboratories *International Standard Organization* (Geneva: ISO Office) p 1–27

[3] ILAC 2008 Guidelines for the Determination of Calibration Intervals of Measuring Instruments *ILAC G24* (Silverwater: ILAC) 1–11

[4] Komite Akreditasi Nasional 2004 *Guide on Measurement Assurance* **33**

[5] López-tarjuelo J, Quirós-higueras J D, Bonaque-alandí J, García-mollá R, Juan-senabre X J and De Marco N 2015 What can statistical process control show us about ionization chamber stability? *Radiation Measurement*

[6] Sakamoto Y, Kaneko N H, Oe T, Kumagai M and Zama M 2011 Novel 100-Ω Metal Foil Resistor *IEEE Transaction Instrumentation Measurement* **60**(7) 2544–49

Acknowledgement
The authors express sincere gratitude to the Research Center for Metrology LIPI for providing devices. We also would like to thank to Dr. Sensus Wijonarko for the guidance. Lastly, we are very grateful for the support from the member of Electrical Metrology Subdivision.

IOP Conf. Series: Materials Science and Engineering **550** (2019) 012016 doi:10.1088/1757-899X/550/1/012016

Controlling pH and temperature aquaponics use proportional control with Arduino and Raspberry

O Supriadi, A Sunardi, H A Baskara and A Safei

Electrical Engineering Department of Pamulang University
Jln. Puspiptek Raya No.11 Buaran, Tangerang Selatan 15310 INDONESIA

E-mail: dosen01327@unpam.ac.id

Abstract. In planting plants using an aquaponic system. Plants are very dependent on nutrients found in water, pH and water temperature. pH and temperature become one of the parameters that determine the yield. pH must be maintained in conditions 6.5–8.0 and temperature must be maintained at 18°-30° Celcius. Therefore, an automation system designed in order to keep the pH and temperature always at a predetermined range. Automation system designed using ARDUINO and RASPBERRY Pi microcontrollers with proportional control method. From the results of experiments that have been carried out, it is found that the designed device can control the temperature condition in the water by activating Peltier elements and fans or heaters to maintain the temperature conditions in water with fastest response time is 25.5 seconds for pH and 473.5 seconds for temperatures reaching setting determined.

1. Introduction

1.1. Aquaponics

Amethod that combines of cultivating fish and farming in a limited environment is called aquaponics method, where this method is the development of hydroponic and aquaculture techniques that are constructed with a symbiotic principle of mutualism for plants and fish in the artificial ecosystem[1]. The benefit of aquaponics are: low in the consumption of water; growth of plants faster; vegetables are highly organic, bigger and healthier; reduction of use of artificial fertilizer; reduction in disposal of fish waste; reduction in land is required to grow the same crops; easier to set up; reduced damaged from pets and disease; and no weeding and bending down [2].

1.2. Water QualityParameters in Aquaponics cultivation system

Water quality is very important thing for aquaponics cultivation. There is five parameters that must be considered in aquaponics cultivation systems. The five most important water quality parameters are: Oxygen, pH (Hydrogen Power), Water temperature, Total nitrogen (ammonia, nitrite, nitrate) and Water hardness [3].From that five water quality parameters in aquaponics cultivation system, this paper only studied about pH and Water Temperature parameters. pH of the water in aquaponics cultivation systems acceptable is between 6.0 until 8.0 with the ideal level is 7.0 and water temperature is between 18° C until 30°C [2].

IOP Conf. Series: Materials Science and Engineering **550** (2019) 012016 doi:10.1088/1757-899X/550/1/012016

1.3. Problem in field
There are still many shortcomings of this aquaponics technique because the handling of plants must be more intensive than traditional planting using soil media [4]. One of the problems is pool water. Rarely replaced pool water will increase the pH and temperature in the water. It can make plants and fish eventually die because the turbid water is not well maintained [5]. The question is how to maintain pH andwater temperature at an acceptable condition in aquaponics cultivation system. So we need an effective, efficient way and bring a solution for agricultural and aquaculture actors.
Therefore, automation is carried out on aquaponics system cultivation by designing a device that can control pH and temperature in the water so that the harvest of fish and plants is as expected.This research aims to create a device that is able to maintain pH and temperature in ideal conditions so that fish and plants can develop well. This tool uses a mega Arduino microcontroller and raspberry pi, pH sensor, temperature sensor, with a proportional control system.

2. Methodology

2.1. Architecture of monitoring and control system
The design of monitoring systems and controls on aquaponics is conducted by controlling the pH and temperature in water as the main media for aquaponics cultivation. The monitoring and control system is divided into three parts: the aquaponics system, data acquisition, and control equipment.The aquaponics monitoring and control system are shown in figure 1.

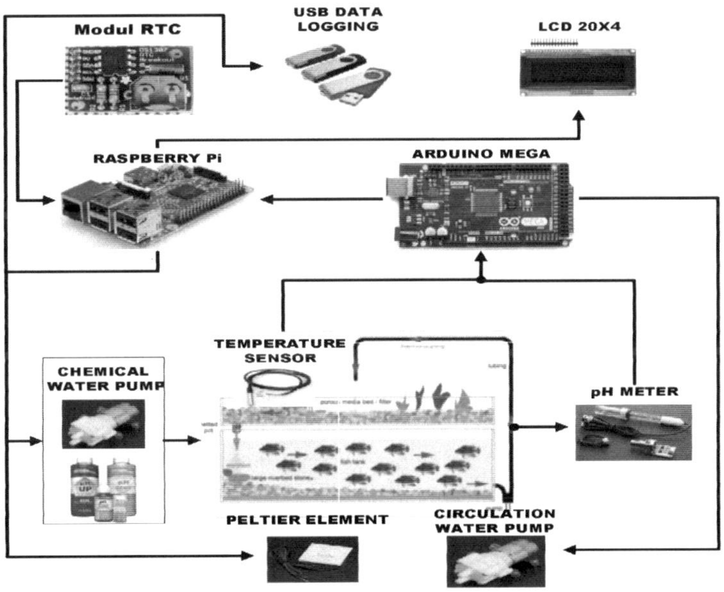

Figure 1. Block diagram of aquaponics monitoring and control system.

Figure 1 shows that the aquaponics system is a parameter in the aquaponics environment especially about the temperature and pH in the water. Data acquisition is about how to collect real-time environmental information such as pH and temperature. Data acquisition process uses two sensors: a pH sensor and temperature sensor.Both of these sensors are used to detect real-time pH and temperature aquaponics environment and the data obtained will be compared to a pH and temperature setpoint that is good for plants and fish.
Control equipment is the control and management center of the whole system, which manages all datasheet. The control equipment use is microcontroller Arduino Mega and Raspberry Pi with a proportional control system to control circulation water pump, chemical water pump and Peltier

IOP Conf. Series: Materials Science and Engineering **550** (2019) 012016 doi:10.1088/1757-899X/550/1/012016

element.To make it easier monitoring and data logging about pH and temperature in aquaponic environment use LCD display and USB.

2.2. Flowchart Aquaponics monitoring and control system

The working principle of the aquaponics monitoring and control system is shown in figure 2.

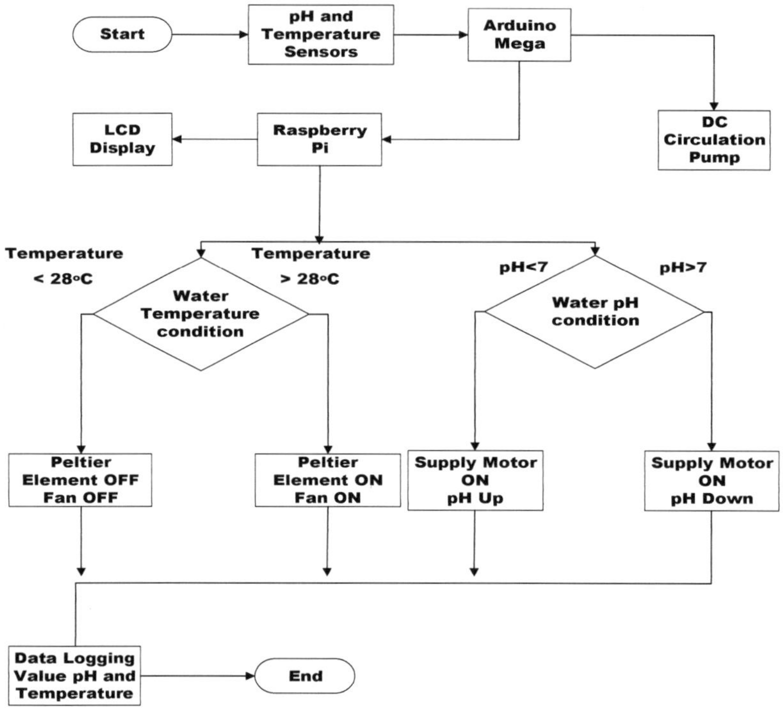

Figure 2. Flowchart of Aquaponics monitoring and control system.

The working principle of aquaponic control and observation systems begins with signals by pH and temperature sensors. Both of these sensors will send data regarding the conditions in real time to the aquaponics environment to the Arduino Mega and Raspberry Pi microcontroller. The output signal from Arduino Mega will be passed to the circulation pump to circulate the water in the aquaponics environment and to the Raspberry Pi. Raspberry Pi output will be delivered to the LCD display to inform the temperature and pH of the aquaponic environment in real-time so that it can be seen directly by the observer. In addition, the output of Raspberry Pi is also delivered to the chemical water pump and Peltier elements. The chemical water pump will work when the pH level in the aquaponic environment is not less than 7 or over than 7by pumping the pH solution (pH-Up or pH-Down) into an aquaponic environmentso that the pH corresponds to the setting value (SV) which is at pH = 7. The chemical water pump will supply pH solution (pH-Up) when pH level in the aquaponic environment is less than 7 and will provide pH solution (pH-Down) when pH level in the aquaponic environment is over than 7.While the Peltier element will work when the temperature is more than 28°C and will turn on the fan and Peltier element will turn off when the temperature is less than 28°C. Each real-time condition about pH and temperature will be saved in USB data logging.

3. Result

In designing the pH and temperature control at Aquaponics, several analyzes were carried out to find out whether the system was working properly or not. An analysis is carried out namely pH sensor analysis, temperature sensor analysis, proportional control system analysis on DC water pump, network communication analysis, overall system analysis.

3.1. pH sensor analysis

pH sensor analysis is conducted by comparing the measurement results of the pH sensor with the measurements result of the digital pH meter using a pH solution (pH-Buffer). In this analysis, six random samples were tested at pH 2, 4, 7, 9, 10 and 13. The purpose of this analysis was to determine the deviation of pH sensor readings against other comparable devices namely digital pH meters so that error correction can be done reading through the Arduino program. Sensor reading error data for the comparator is shown in table 1.

Table 1. The result of the error value pH sensor.

pH Liquid Buffer	pH Meter Digital	Litmus Paper	E201-C-9 pH Sensor	Error Value Sensor pH to pH Meter Digital	Sampling Time (Menit)
2.00	1.77	2.00	1.938	0.20	10
4.00	4.03	4.00	4.135	0.11	10
7.00	7.00	7.00	7.028	0.12	10
9.00	9.27	9.00	9.211	0.09	10
10.00	10.03	10.00	9.889	0.14	10
13.00	13.06	13.00	13.11	0.16	10

From table 1, it was informed that the measurement error of the pH sensor that had been compared with the measurement results by the digital pH meter obtained a minimum pH sensor measurement error 0.09, an average sensor pH measurement error 0.15 and a maximum sensor pH measurement error 0.20. Chart of the sensor pH measurement error is shown in figure 3.

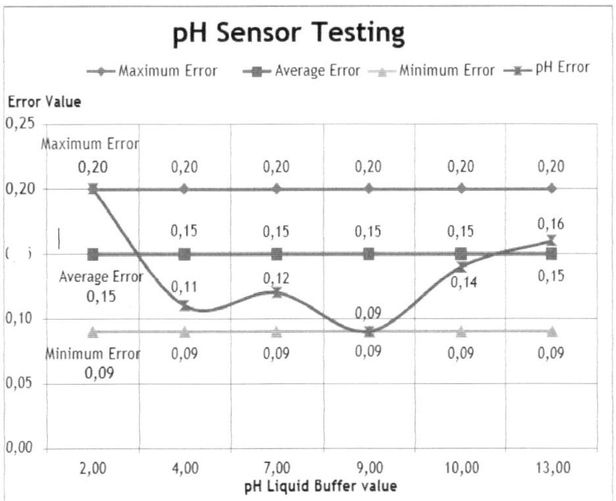

Figure 3. The result of the error value pH sensor.

3.2. Temperature sensor analysis

Temperature sensor analysis is done by comparing the measurement results of the temperature sensor with the measurements result of the digital thermometer using water. In this analysis, the test is carried out on several temperature values, six samples were tested at temperature 0°, 10°, 20°, 30°, 40° and 50°C. The purpose of this analysis was to determine the deviation of DS18B20 temperature sensor readings against other comparable devices namely digital thermometer so that error correction can be done at the Arduino program. Sensor reading error data for the comparator is shown in table 2.

Table 2. The result of the error value temperature sensor.

Water Temperature	Digital Thermometer	Mercury Thermometer	DS18B20 Temperature Sensor	Error Value Sensor pH to pH Meter Digital	Sampling Time (detik)
0.00	0.10	0.00	0.37	0.27	100
10.00	10.80	10.00	10.58	0.38	100
20.00	20.40	20.00	20.56	0.16	100
30.00	29.50	30.00	29.37	0.13	100
40.00	40.90	40.00	40.72	0.18	100
50.00	50.50	50.00	50.25	0.25	100

From table 1, it was informed that the measurement error of the DS18B20 temperature sensor that had been compared with the measurement results by the digital thermometer obtained a minimum DS18B20 temperature sensor measurement error 0.13, an average DS18B20 temperature sensor measurement error 0.23 and a maximum DS18B20 temperature sensor measurement error 0.20. Chart of the DS18B20 temperature measurement error is shown in figure 4.

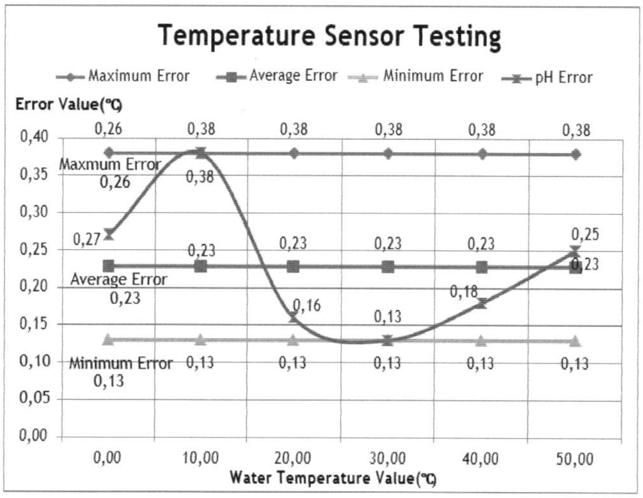

Figure 4. The result of temperature sensor error value.

3.3. Proportional control system analysis

Control system testing is done by setting the minimum setting value (SVmin) and setting the maximum setting value (SVmax) at pH equal to 7. When the pH is below SVmin, raspberry pi will instruct the chemical pump to supply pH solution (pH-Up) so that the pH value on aquaponics system increases to 7 according to the setting value, whereas if the pH value is above the SVmax value, the raspberry pi will instruct the chemical pump to supply pH solution (pH-Down) so that the pH value of the aquaponics system drops to 7 according to the setting value. Proportional control system response test to the pH is shown in table 3. Table 3 informs thatwhen pH value less than 7chemical pump pH-Up will turn onand when pH value over than 7 chemical pump pH-Down will turn on. Response time when switch from less than 7 until chemical pump pH-Up turn on is 1 second. Chart of chemical pump response to pH value is shown in figure 5.

IOP Conf. Series: Materials Science and Engineering **550** (2019) 012016 doi:10.1088/1757-899X/550/1/012016

Table 3. Proportional control system response test to the pH.

Water Temperature (°C)	pH Value	Motor Speed pH-Down (%)	Motor Speed pH-Up (%)	Time
28.69	6.62	0.00	5.48	13.50.42
28.69	6.62	0.00	5.48	13.50.43
28.69	6.62	0.00	5.48	13.50.44
28.69	6.48	0.00	5.48	13.50.45
28.69	6.40	0.00	7.44	13.50.46
28.69	6.21	0.00	8.61	13.50.47
28.69	6.12	0.00	11.35	13.50.48
28.69	6.26	0.00	12.52	13.50.50
28.69	6.56	0.00	10.57	13.50.51
28.69	6.86	0.00	6.26	13.50.52
28.69	7.00	0.00	1.96	13.50.53
28.69	7.37	0.00	0.00	13.50.54
28.69	7.82	5.28	0.00	13.50.55
28.69	7.89	11.74	0.00	13.50.56
28.69	8.10	12.72	0.00	13.50.57
28.69	7.64	15.66	0.00	13.50.58
28.69	6.95	9.20	0.00	13.50.59
28.69	2.11	0.00	0.78	13.51.00
28.69	0.01	0.00	69.86	13.51.01
28.69	2.21	0.00	99.80	13.51.02
28.69	8.10	0.00	68.49	13.51.03
28.69	12.30	15.66	0.00	13.51.04
28.69	13.97	75.73	0.00	13.51.05
28.69	13.97	99.61	0.00	13.51.06
28.69	13.60	99.61	0.00	13.51.07
28.69	9.37	94.32	0.00	13.51.08
28.69	7.21	33.86	0.00	13.51.09
28.69	6.99	2.94	0.00	13.51.10
28.69	7.40	0.00	0.20	13.51.11

3.4. Network communication system analysis

There are 2 parts of communication network testing, namely testing of I2C and serial communication networks. The I2C network is used to display data from the sensor to the LCD display. The I2C test results are shown in figure 6.

Figure 7 shows serial communication testing by sending sensor data from Arduino to Raspberry Pi. Sensor data transmitted from the pH sensor and the temperature sensor used, then the data is recorded to USB as a data logger.

IOP Conf. Series: Materials Science and Engineering **550** (2019) 012016 doi:10.1088/1757-899X/550/1/012016

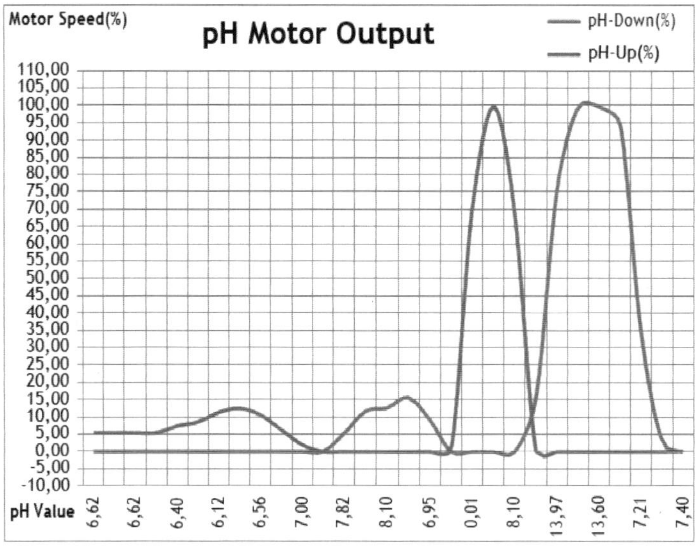

Figure 5. Chemical pump response to pH value.

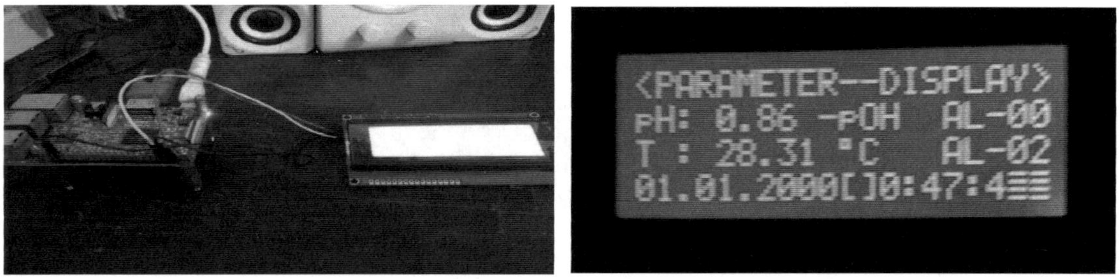

Figure 6. I2C communication testing to LCD display.

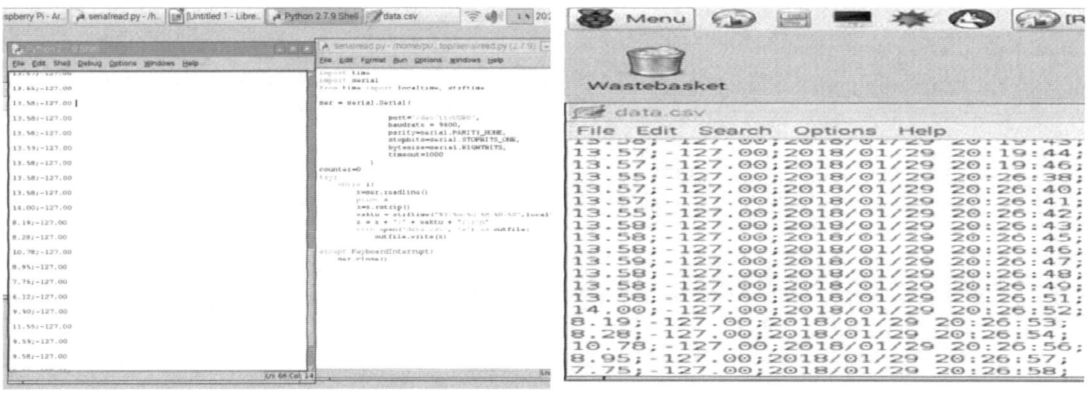

Figure 7. Serial communication and data logging testing.

3.5. Overall system analysis

Overall system analysis aims to see the response time of the control system that has been designed, by knowing the response that is given, it can be known the success of the system designed to answer the problem formulation, in table 4 shows the changes conditions in the device currently used for testing in prototype form (after) and the previous device in the form of the original scale (before).

Table 4. Device comparison testing.

Item	After	Before
Water Volume	±50 Liter	±10 Liter
Water pH Setting	7.5/6.0	7.5/6.0
Temperature Setting	29/27°C	29/27°C
Scale of real-time	1:20	1:4

Table 5. Final data of test results.

Data	Steady State Time	Maximum Overshoot Time	Peak Time	Error Max	Error Average	Error Min	accuracy	precision
pH-Up	29	14	10	0.2	0.15	0.09	Yes	No
pH-Down	22	9	7	0.2	0.15	0.09	Yes	No
Temperature Up	138	0	0	0.38	0.23	0.13	Yes	Yes
Temperature Down	759	0	0	0.38	0.23	0.13	Yes	Yes

From table 5 can be known the success of the prototype made on a scale of 1:20 from the scale of the original tool, taking into account all the errors and weaknesses that exist in the prototype makes the success rate of the tool cannot reach 100%. So the success rate that can be achieved is 80% for pH control and 80% for temperature control.

4. Conclusion

Based on the design, implementation and testing of the device it can be concluded thatactual information about pH and temperature of the water can be seen directly through LCD display, pH sensor and temperature sensor has worked properly. It is shown by the error percentage of calibration measurement sensor. Error pH sensor 1.07% and error temperature sensor 0.89%. The device can automatically control pH condition by operating pH liquid supply pump if pH conditions are outside setpoint and can control the water temperature conditions by activating Peltier elements and fans or heaters to maintain temperature conditions in the water with fastest response time is 25.5 seconds for pH and 473.5 seconds for temperatures reaching setting determined.

References

[1] Somerville C, Cohen M, Pantanella E, Stankus A, Lovatelli A 2014 *Tec. Paper. Small-scale aquaponic food production. Integrated fish and plant farming* (Roma : FAO) p. 288
[2] Akil K, Sivasakthi M S and Maha K P 2016 *Aquaponics – the Future Farming April Edition* (Florida: Raider Review Online Newspaper)
[3] Somerville C, Cohen M, Pantanella E, Stankus A and Lovatelli A 2014 *Water Quality for Plants. In: Small-scale aquaponic food production - Integrated fish and plant farming* (Roma - FAO) 83–102
[4] Ravan T and Shingate S 2016 *Aquaponics Based* **41** 8–13
[5] Filep R M, Diaconescu S, Marin M, Badulescu L and Nicolae C G 2016 *Curr. Trends Nat. Sci.* **5**(9) 6–9

IOP Conf. Series: Materials Science and Engineering **550** (2019) 012017 doi:10.1088/1757-899X/550/1/012017

Hybrid Billing Parking System (HBPs) for service quality improving

N Herlina[1], Herianto[2], N Hiron[3], Firmansyah M S Nursuwars[4] and H Mubarok[5]

[1, 2]Civil Departement, Universitas Siliwangi. Tasikmalaya, Indonesia 46115
[2, 3]Electrical Departement, Universitas Siliwangi. Tasikmalaya, Indonesia 46115
[4, 5]Informatic Departement, Universitas Siliwangi. Tasikmalaya, Indonesia 46115

Email: ninaherlina@unsil.ac.id[1], herianto@unsil.ac.id[2], hiron@unsil.ac.id[3], firmansyah@unsil.ac.id[4], husni.mubarok@unsil.ac.id[5]

Abstract This project provides a concept of computer-based Hybrid Billing Parking System Service that could work both online and/or offline. We propose Hybrid Billing Parking System (HBPs) in which this system could work and keep the data of billing parking transaction saved digitally. The main goal of this research is design a new concept of database and computer network which includes operational procedure, unique barcode for Hybrid Billing Parking System (HBPs). We have successfully created Hybrid Billing Parking System (HBPs) which has advantages on managing system, for evaluation and financial report and also continuing parking service system. The result of this reseach that the concept that has been designed has two benefit aspects. From energy perspective, HBPs consume less energy and from simplicity procedure perspective, HBPs can be designed into universal architecture of parking system, but the unique code of barcode of each customer will be different.

1. Introduction

Parking service is a crucial part of improving customers' satisfaction. Human-based parking system has weaknesses in accuracy and stability. This case wouldn't be happened on computer-based parking service system. However, it has weaknesses on external trouble, for instances, malware virus, inter-system coordination failure, missing data caused by electricity or mechanism disturbance, etc. making that parking service system doesn't work properly.

One of the virtues of parking management is billing parking system. In this study, Hybrid based Billing Parking System (HBPs) is conveyed. It takes major function since the system of billing and payment commonly uses full-automatic system, or man-less system. However, man-less system is not necessarily as a solution because it highly depends on the machine communication reliability involved in the parking system by considering to incoming vehicle identification sensor, parking gate, time of parking duration, data storage of parking payment authentication which finally become a bill for parking payment. All of these cycles could work properly as long as there is no internal or external trouble.

Even though man-less system is the most widely applied system in the modern parking service system nowadays, but in case of there is system failure (malware), then the invoice billing couldn't be accessed from the server where every billing parking data is processed and stored. In such condition the conventional system becomes the solution, where HBPs is operated by manpower system, or parking billing uses manual system. The transition of man-less to man-power system becomes crucial problem since the billing data couldn't be accessed from the server while parking customers need service immediately.

The HBPs we propose in this paper as a solution to address the problems faced by the parking billing system in general, as problem tha found in [1], which believed that the suitable parking cost decision will increase the efficiency of billing management system. Focusing on the case in [1], HBPs provide flexible parking billing system solutions where the HBPs system can automatically switch from online man-less based prosedure to offline man-less based prosedure. As a result, there is synchronization in the billing system to the system of billing parking.

HBPs has two integrated mode systems that work both online and offline system. Online man less system in which the communication of billing parking system between entrance gate and exit gate involves the server. On the other side, Offline man less system doesn't involve the server. Although in the form of offline system, the server doesn't miss the billing data. On the other side, offline man less system is where the system of billing parking between entrance gate and exit gate communicate without data from server.

2. Related work

The main problem in the automatic parking system is how to create a tight and accurate connection between customer and the sensor. Based on the study done by P. Christiansen, et al [1] showed that the technique of accurate billing payment system affects to the system efficiency of parking management, so that many researchers have offered solution to make the parking system more effective and efficiency. Intelligent parking techniques in which sensors communicating at close range such as Radio Frequency Identification (RFID) have been proposed on [2], while RFID technology in parking systems that produce very fast memory [3], Parking Guidance Information system in parking service known as (PGIs) is widely proposed by researchers, including in [4-6]. The parking control system using a programmable logic controller (PLC) is believed to provide a more reliable engine communication di parking system [7] reservation based [8], automatic parking system NFC based [9], [10], parking system WSN based [4] with ZigBee [11], identification of vehicle serial number technology by image processing [12-16].

Many researchers claim that the intelligent parking system is required to have the characteristics of a timeless service system, accuracy, reliability, security and able to provide security management, great reporting system, and vehicle detection at the parking area [5], survey system for smart parking system has been develope in [17] or intelligent system in wireless communication [18-19].

Those empirical studies have been discussed constitute to parking system related to the smart system to improve service quality of parking for the customers but remember that smart system highly depends on the ideal situation, which is the electricity and machine process run properly. On the other hand, manual service, semi manual service, or the real manual become another alternative way to be considered in order to keep parking service still be enjoyed by the customers in any condition and without any risk to the parking service vendor.

The combination system between automatic, semi-automatic, to the full manual service as called hybrid-based parking service system as this research proposes focuses on the manner of how identity data of vehicle can be identified both offline and online system.

3. Result and discussion

In this study, the hybrid-based parking system includes entrance gate station, exit gate station, the server, unique barcode in the parking ticket (figure 1), with the server as the core of data transaction. HBPs as proposed in this paper includes entrance station, exit station, and server unit. Exit gate station is equipped by independent database that is used as temporary data manager (batching). This database is used to batching process, which is temporary saving storage that works when the system is offline. The transaction of data in the exit gate station would be sent to the server if the system is back online.

In online condition, the transaction data is sent to the server via entrance gate station and also exit gate station. While in offline condition, the transaction of data is only storage temporarily in exit gate station. Offline system causes the entrance gate station not able to connect to the server, so that the

IOP Conf. Series: Materials Science and Engineering **550** (2019) 012017 doi:10.1088/1757-899X/550/1/012017

connection media is changed by the unique code in the parking ticket (figure 2). This unique code then links the information about billing parking that should be paid by the customers in the exit station. this unique code is fetched in the parking ticket in the form of barcode, so, when the system is verifying the vehicle in the exit gate station, the vehicle entered could be seen whether the system online or offline, as a result, the transaction data won't be lost. However, offline procedure highly depends on the unique code in the parking ticket.

Principally, HBPs works like a parking system in general, but billing parking system works even though the system is offline, and the transaction data is saved in the temporary exit gate station before it is saved in the system when it gets back online. By using data from the entrance gate station, the system then does validation to the exit gate to print billing parking ticket. In figure 1 shows the flowchart of HBPs in the entrance gate vehicle, starting from the customers push the button of entrance gate ticket, the system validates the status whether it is online or offline. If the system condition is offline, then the receipt received by the consumer is printed using offline procedure, meaning the printer prints the receipt not based on command from the server, then the system will open the entrance gate and let the vehicle enter the parking lot. While on condition of online system, the system would check the server status first, if there is failure between entrance gate computer station to the server, then the offline procedure would be applied. If the server is online, the data of transaction would be stored to the server, and then the system let the vehicle parks.

Figure 2 shows HBPs procedure of vehicles exiting from the parking area. Started from scanning parking ticket, the system would verify the ticket to see if the ticket has the unique code that indicates the ticket is printed at the time of the system offline or online. Again, the exit gate station will check the status of the system. If the system is online, then the transaction data would be stored directly to thein server after billing parking calculated, if the system is offline, the transaction data would be temporarily stored in the data base of exit gate station. And then if the system returns online, all of the data of transaction would be automatically sent to the server. This procedure keeps a lot of data and helps the manager of financial or executor for financial evaluation.

HBPs could be designed by the same architecture of general billing parking, the unique code on the barcode of the customer makes them different each other. This would be a serious concern if the customer loses the parking ticket, so that a strategy is required to avoid parking ticket lost

We offer a concept of barcode which is suitable to the HBPs. Every vehicle goes through the entrance gate station would get the barcode. This barcode functions as a tool of billing parking, by using unique code in the barcode. Exit gate station would be able to do payment transaction even though the system is offline. Figure 3 shows the concept of barcode using unique code whether the vehicles get in while the system is online or offline (figure 4). In the figure 3, it is illustrated that the online system is marked 01. While in the offline condition, it is marked 09 (figure 4). The barcode also contains code to indicate which entrance gate the vehicles come. This code is needed to know the information of entrance. This is needed as supporting information of improving service or evaluation of parking service.

The network of HBPs between between HBPs's server to entrance gate station server and exit gate station server. While the server communicates to the station occurred in a long distance. We recommend using the fiber optic based communicationd system, and using the Shielded Twisted Pair (STP) for in chase short distance.

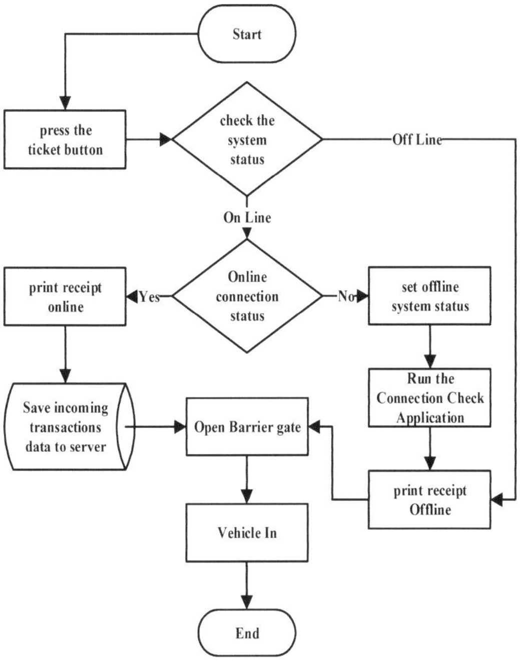

Figure 1. Flowchart of Hybrid Hybrid Billing Parking System (HBPs) entrance validation of vehicle.

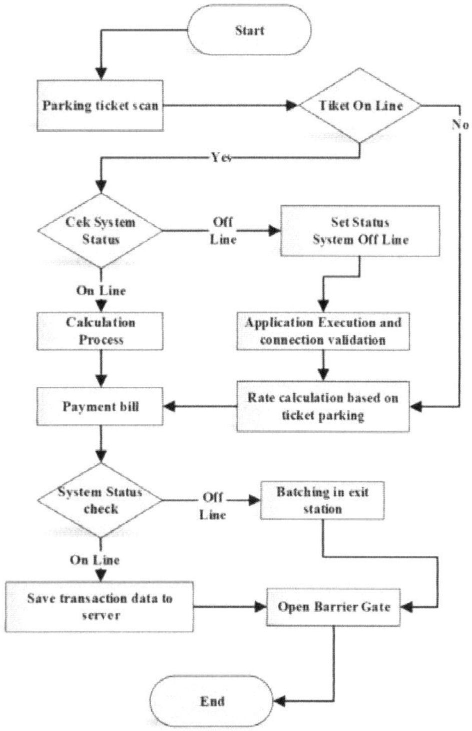

Figure 2. Flowchart of Hybrid Hybrid Billing Parking System (HBPs) exit validation of vehicle.

Figure 3. HBPs barcode design with unique code in online mode.

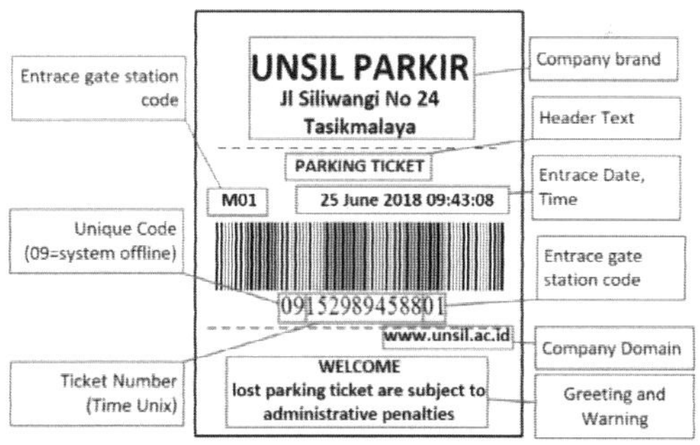

Figure 4. HBPs barcode design with unique code in offline mode.

IOP Conf. Series: Materials Science and Engineering **550** (2019) 012017 doi:10.1088/1757-899X/550/1/012017

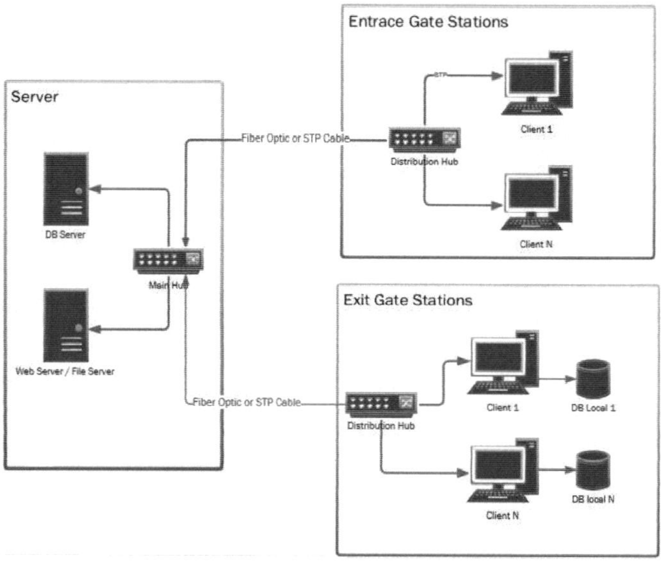

Figure 5. Architecture design for Hybrid Billing Parking System (HBPs) in online mode.

Figure 6 is a designed applied in HBPs. In general, architecture is designed into several local networks, namely entrance gate stations, exit gate stations, and server node. If the distance between the server and the entrance gate is remote, it is recommended to use fiber optic or Spanning Tree Protocol (STP).

In the online condition, entrance gate stations and exit gate stations directly get connected to the server. Billing and transaction data transfer are stored to the server (figure 5). Digital unique barcode is sent directly to server. Meanwhile, offline condition (figure 6) is assumed that communication of entrance gate stations and exit gate station are disconnected with the server. In offline condition, barcode with unique physical code are transferred to entrance gate station to the exit gate station through the consumer.

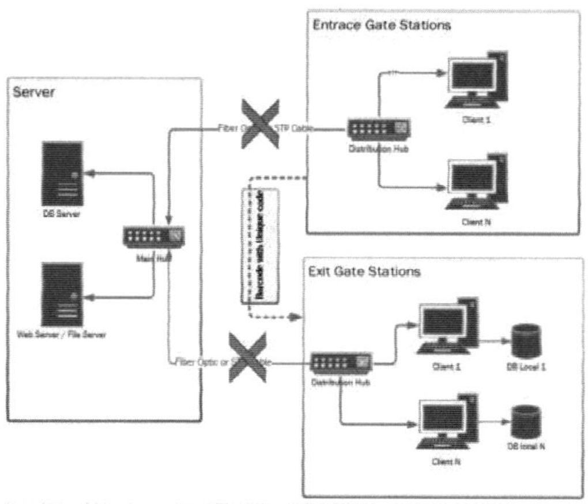

Figure 6. Architecture design for Hybrid Hybrid Billing Parking System (HBPs) in offline mode.

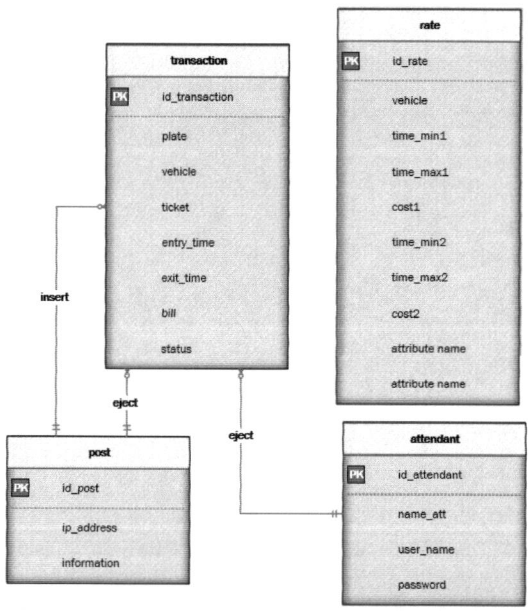

Figure 7. ERD Hybrid Billing Parking System (HBPs).

Figure 7 is Entity Relation Diagram from HBPs. ERD consists of transaction cable, attendant cable, rate table, post table. Transaction cable serves to store activities log system along the running system. The table is responsible for delivering incoming and outgoing transaction from parking or consumer vehicles. The attendants table is a table to keep the ideas of officers guarding the exit only. This table is responsible for providing the identity of the officer. Table rate is a table that provides the applicable rate which also shows prevailing price to be used as the basis for calculating the parking bill. Table post is a table that provide idea post at the entrance or exit.

This project provides an in-depth description in providing solution by keeping payment transaction data or parking billing is safely stored without being disturbed by machine communication status.

4. Conclusions

The proposed Hybrid Billing Parking System is a solution to improve parking service especially in the billing and parking payment. HBPs contributes to the concept of thinking that the operational system of an online based parking service does not guarantee that the service could work properly consistently, therefore, it backup system is required as a support system of the main system in offline condition.

The offered HBPs has the strength in the management process, for evaluation and financial report, also for continuously parking service system. From energy perspective, HBPs consumes less energy, because it could work without huge backup power system. As long as the database in the gate station could work, the report of transaction could be saved.

HBPs can be designed by using the same architecture in general system parking but the unique code that exist in the barcode of each customer makes it different. This case becomes crucial because the problem would occur if the customer loses the parking ticket, therefore, a strategy is required to save the parking ticket. Surely, this is the future work that can be done for further research.

References

[1] Fadeyev D 2016 *Transp. Res. Procedia* **20** 193–99
[2] El-hageen H M M, Ibrahim K, Ata M, Chesoh A and Jaafar H 2017 *Int. J. Sci. Eng. Res.* **8** 557–62
[3] Mukajkar R M, Phatale A P and Kulkarni P H 2011 *Int. Conf. on VLSI, Communication & Instrumentation (ICVCI) (Kerala)* (New York: IJCA) p 14
[4] Patil M and Bhonge V N 2013 *Int. J. Emerg. Technol. Adv. Eng.* **3** 188–92
[5] Ranjini R and Manivannan D 2013 *Int. J. Eng. Technol.* **5**1763–67
[6] Pala Z and Inan N 2007 *Proc. of the 1st RFID Eurasia Conference (Istanbul-Turkey)* (US: IEEE) p 121
[7] Sarayu S, Rajendra S S and Bongale V V 2013 *Int. J. Sci. Eng. Technol.* **2** 857–60
[8] Patil M and Sakore R 2014 *Int. J. Sci. Eng. Res.* **2** 21–6
[9] Al-ofeishat H A and Rababah M A A A L 2012 *IJCSNS* **12** 93–9
[10] Radhakrishnan A 2015 *J. Eng. Res. Appl.* **5** 56–9
[11] Asyrani H *et al* 2013 *Int. J. Eng. Technol.* **5** 3282–300
[12] Tian Q, Guo T, Qiao S, Wei Y and Fei W 2014 *J. Multimed.* **9** 774–80
[13] Rashid M M, Musa A, Rahman M A, Farahana N and Farhana A 2012 *Int. J. Mach. Learn. Comput.* **2** 93–8
[14] Karunamoorthy B, Sureshkumar R and Jayasudha N 2015 *Int J. Adv. Res. Comput. Eng. Technol.* **4** 85–90
[15] Al-Kharusi H and Al-Bahadly I 2014 *World J. Eng. Technol.* **2** 55–67
[16] Choudhary E S, Thakare S A, Raut S D, Jadhao K M and Tiwari M V 2016 *Int. J. Innov. Emerg. Res. Eng.* **3** 1–5
[17] Shaikh F, Nikhilkumar B S, Kulkarni O, Jadhav P and Bandarkar S 2015 *Int. J. Innov. Res. Sci. Eng. Technol.* **4** 9933–39
[18] Hiron N, Andang A and Setiawan H 2016 *Int. J. Futur. Comput. Commun.* **5** 163–6

[19] Hiron N and Andang A 2016 *2nd Int. Conf. on Science in Information Technology (Balikpapan-Indonesia)* (US: IEEE) p 250

Acknowledgments
This research is a part of an external research program and is supported by research funding from KEMENRSITEK-DIKTI at the scheme "Applied Product Research for the second year. Thanks to LPPM-PMP Siliwangi University for supporting this project.

IOP Conf. Series: Materials Science and Engineering **550** (2019) 012018 doi:10.1088/1757-899X/550/1/012018

Investigation of ultrasonic sensor type JSN-SRT04 performance as flood elevation detection

A Andang, N Hiron, A Chobir and N Busaeri

Siliwangi University, Jln. Siliwangi No. 24 Kota Tasikmalaya 46155, Indonesia

Email: andhangs@unsil.ac.id

Abstract. To measure water surface elevation, the measurement technique with ultrasonic technology has been recently being applied to various sectors, such as automotive, robotic and industrial requirement. This research aims to create water surface elevation measurement with JSN-SR04T ultrasonic sensor, which has 0.5 cm measurement resolution. Arduino-Uno microcontroller with a microprocessor-Atmega238 type was used as a measurement data processing system. The system test was conducted in a river where the system was placed under the bridge, and then the elevation water surface levels were captured by the sensor. After that, the results were displayed on the LCD screen. The test was executed in 37 testing scenarios of the condition of the river water surface. Then, the measurement results from the system were compared to the manual measurement. The results showed that the value of measurement error was 0.75%. This finding is in line with the datasheet which indicates that WSN-SR04T sensor can provide a maximum resolution of 0.5 cm.

1. Introduction

Unpredictable weather condition results in unforeseeable and uneven distribution of rainfall. Flooding becomes a threat that can cause considerable losses. Hence, several human efforts to control flooding have been made. For instance, through the improvement of rainfall prediction techniques such as by using multilayer neural network with backpropagation algorithm [1], hydrological analysis, hydraulics analysis and flood prevention analysis [2], early detection system by monitoring the river water level using microcontroller and SMS technology [3], and using ultrasonic sensor PING [4].

Detection of river water level by using pressure ratio in the water principle technique based on Atmega328 microcontroller and pressure sensor BMP085 has been successfully done [5].

River level detection technique using ultrasonic sensor is currently widely used because of its high accuracy which reduces analysis errors. While ultrasonic sensor PING has an accuracy of 1 cm and an effective measurement range of 3 cm-3 m [4], the range of the JSN-SR04T type ultrasonic sensor is from 25 cm-4.5 m so that this type of sensor can be placed in a higher place and also safer from the submerged condition.

The JSN-SR04T type ultrasonic sensor is equipped with the IO port TRIG at the minimum signal level of 5µs. The JSN-SR04T sensor will send signals with a frequency of 40 KHz, and then it will process the return signals. Signals with high status will come out of IO port ECHO. The difference between the output and returned signals is determined by using an equation.

The process of determining the distance based on the difference of the output and returned signals is performed by using equation (1), where D is the distance between the water surface and the sensor (meter). HLT stands for High-Level Time, while SS means Speed of Sound that is equal to 340 m/s.

$$D = \frac{(\text{HLT})\ x\ (SS)}{2} \qquad (1)$$

Arduino Uno is a circuit board with Atmega-328 type processor and 28 pins. On the circuit board, there are 14 digital inputs/outputs (six outputs for PWM), 6 analog inputs, 16 MHz ceramic crystal resonator, USB connector, adapter socket, and ICSP header pins. The difference between Arduino Uno and Arduino board one of which is the use of special chip as the USB-to-serial FTDI driver [6].

Figure 1. Arduino-Uno circuit board and JSN-SR04T sensor [6].

Figure 1 is Arduino-Uno circuit board and JSN-SR04T ultrasonic sensor, while IO pin configuration on Atmega-328 IC is as shown in figure 2. The Arduino-Uno block diagram is presented in figure 3.

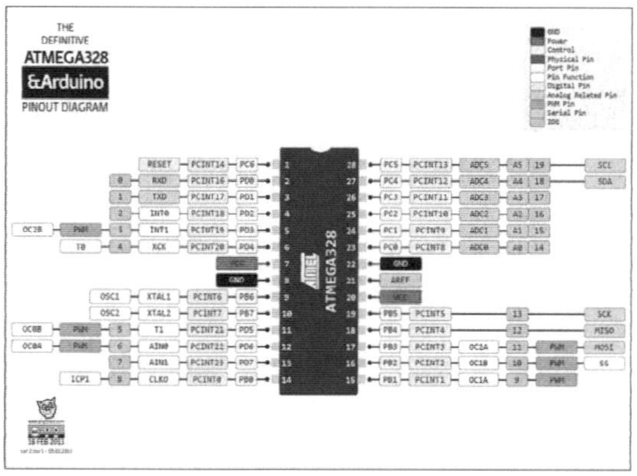

Figure 2 Atmega-328 IC pin configuration [7].

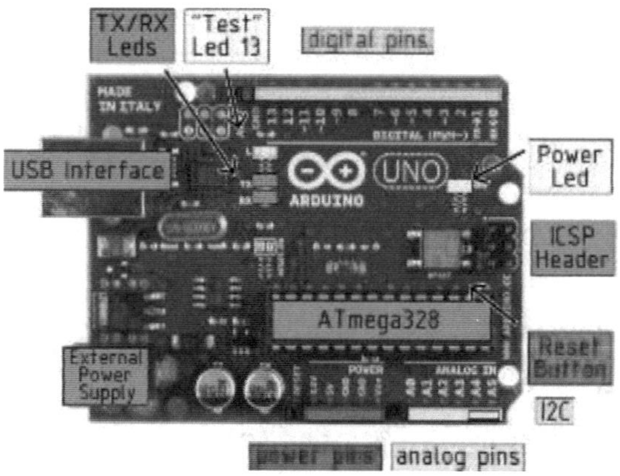

Figure 3. Diagram block of Arduino-Uno [6].

Ultrasonic is equal to sound waves with frequencies above 20 kHz [8]. In general, they have the same characteristics as sound waves. It has the properties of bouncing, propagating on solid medium and air with low energy. Thus, these waves are suitable for distance measurements both in air and water [9].

Figure 4 illustrates the working principle of the proximity sensor with the ultrasonic waves that is by transmitting a particular wave and then calculate the time when it is received back by the sensor [10].

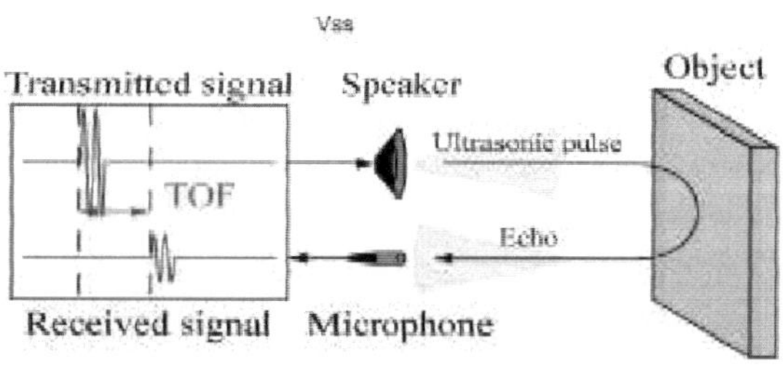

Figure 4. The principle of distance measurement with ultrasonic waves [7].

IOP Conf. Series: Materials Science and Engineering **550** (2019) 012018 doi:10.1088/1757-899X/550/1/012018

The sensor works by sending eight steps ultrasonic signal with a minimum frequency of 40 kHz when the positive pulse voltage is applied to the trigger pin for 10μs. Furthermore, the reflected signal that had been returned was received by the Echo pin (figure 5). The distance measurement of the object reflecting the signal can be performed using equation (1) that is by identifying the difference between the times of sending and receiving the reflected signal.

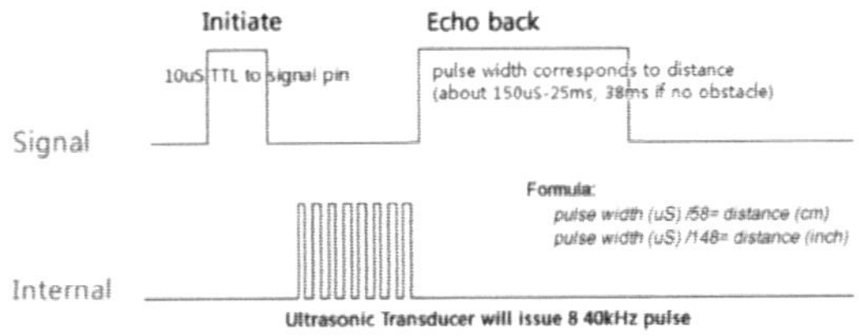

Figure 5. Timer system on the HC-SR04 sensor.

2. Methods

2.1. Hardware and system configuration

Figure 6. is a system architecture in which the sensor interacted by sending a comparison of the sent signal to a received signal in the processor. Then, the final result of the measurement was displayed on the LCD in meters unit.

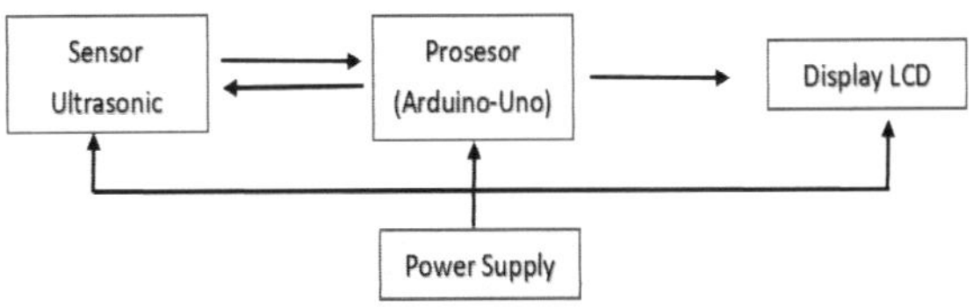

Figure 6. System architecture.

The sensor pole design process and the measurement were conducted as illustrated in figure 7. The sensor system was set at an altitude of about 4m from the surface of the river water.

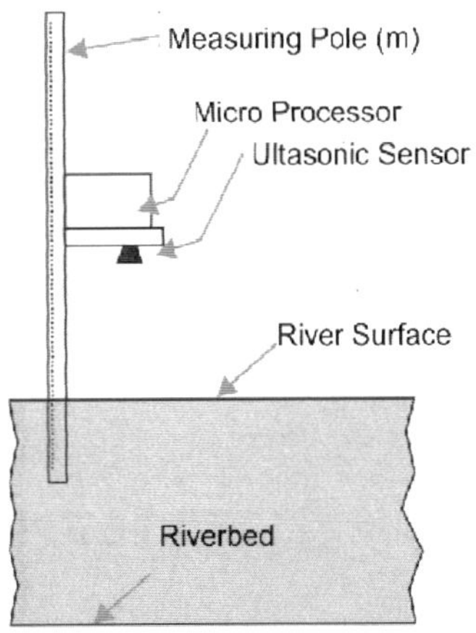

Figure 7. Model of the system and illustration of the system implementation.

Figure 8. is the design of communication configuration between JSN-SRT04 ultrasonic sensors and Atmega-238. The system consisted of poles equipped with the manual measurement instrument, processor units equipped with Arduino-Uno and LCD, and ultrasonic sensors that was directed to the water surface.

2.2. Testing Procedure

The system testing, that was performed as shown in figure 8(a), was a scenario in which the river water was in rising or tidal conditions. Meanwhile, in figure 8(b), the river water was receding. After the measurement results had been displayed on the LCD, subsequently, they were compared to the measurement in centimeters (cm) unit

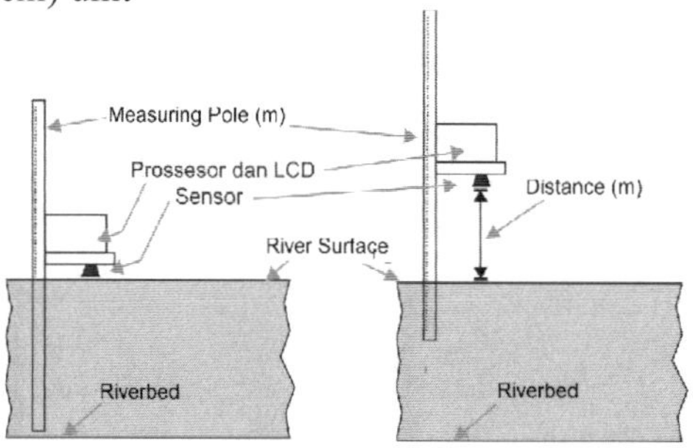

(a). High tide condition (b). Low tide condition

Figure 8. Test scenario at high tide and low tide level condition.

Then, to determine the level of error that occurred during the test,a validation of test results were performed by using the following equation:

$$Error\ (\%) = \frac{\text{System Measurement} - \text{Manual Measurement}}{\text{Manual Measurement}}\ x\ (100) \quad (2)$$

Schematic sensor communication with Arduino is demonstrated in figure 9. The TX (Echo) pin on the sensor, which served as the signal sender, was connected to pin 13, while the RX (Trig) pin, as the signal receiver, was connected to pin 12 on the Arduino board. The Arduino board supplied a working voltage sensor of 5V.

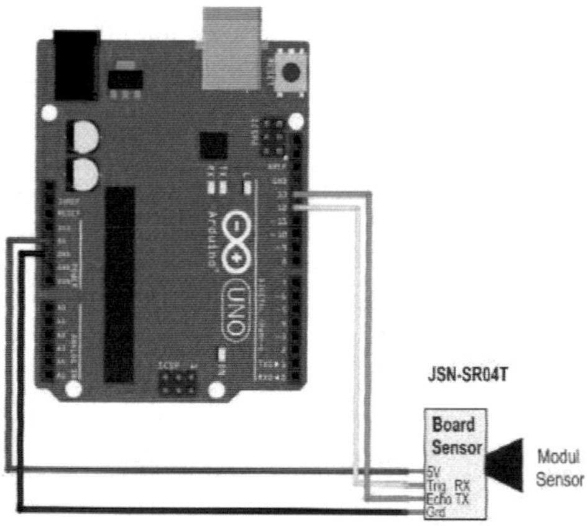

Figure 9. Wiring Diagram of communication between sensor and Arduino-Uno.

Figure 10. Sensor testing JSN-SRT04.

2.3. Software configuration

Figure 11. shows the software configuration where the initial stage was the variable and pin declaration. This stage was intended to state what variables would be used as well as the data type and the initial value. To connect with the ultrasonic sensor, the ports used on the Arduino Breakout pins were pin 1 and pin 13, while for ports connected to the LED pin was pin 8 (figure 11). Figure 12 is the result of sensor testing that had been connected with the Arduino-Uno board.

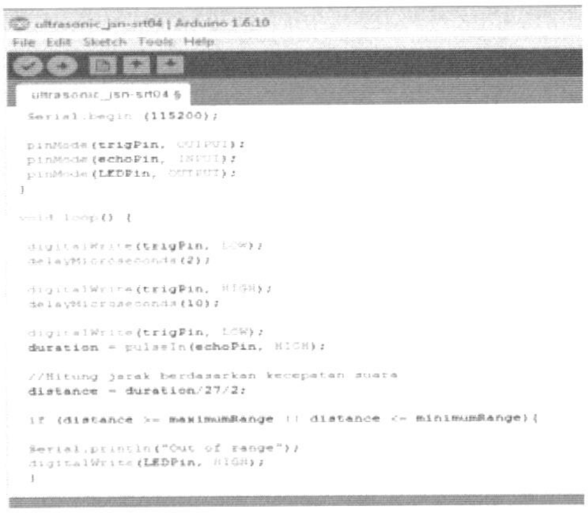

Figure 11. The syntax of communication program between sensor and Arduino-Uno.

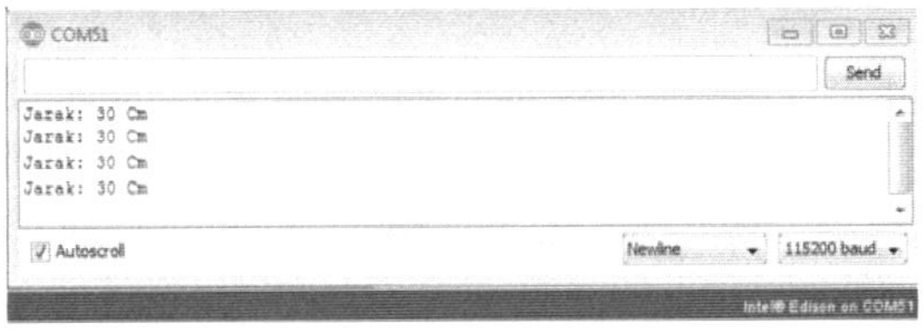

Figure 12. Distance testing validation in arduino programme.

3. Results and discussion

The system testing was undertaken in the river in Tasikmalaya city of Indonesia. The system which has been equipped with standalone power supply, in the middle of the river water through the bridge. The test scenario corresponds to figure 8, with the assumption of the distance of 0 meters to 2 meters. At 0 meters, it represents the condition of the water level rising close to the sensor, while at a distance of 2 meters, it exemplifies the surface condition of the low tide. After the distance measurement results had been displayed on the LCD, they were compared with the manual distance measurement results. Not all of the photos of the test are presented in this paper.

After the data had been collected, then the validation of the error test was performed to calculate the error value of the test results. This validation aimed to determine to what extent the accuracy of the proximity sensor toward changes in surface elevation of the actual water level after being compared with the manual measurement.

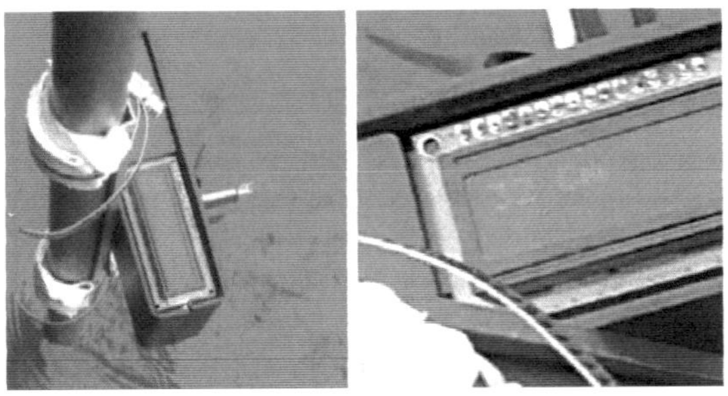

Figure 13. Water level testing at 32cm.

Figure 14. Water level testing at 50cm.

Figure 15. Water level testing at 70cm.

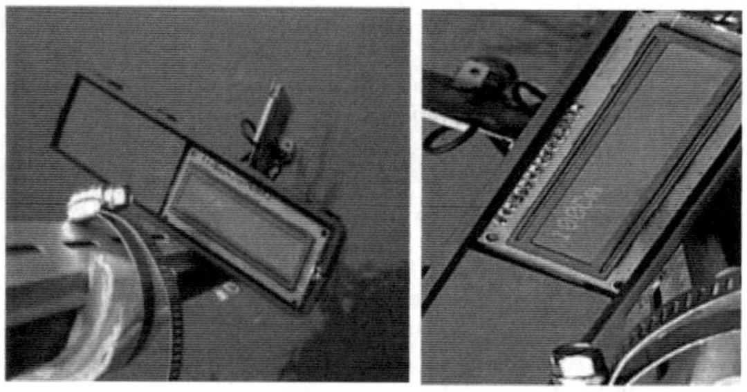

Figure 16. Water level testing at 100 cm.

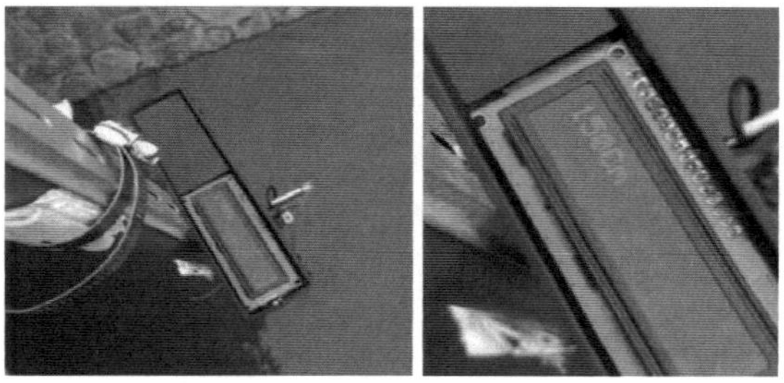

Figure 17. Water level testing at 150 cm.

Figure 18. Water level testing at 200 cm.

The result of system testing in river depicted in figures 13-18. Measurements are made sequentially from the height of the river water level at 32 cm to 200 cm. After a comparison between the measurements of the system and the measurement results manually, it was obtained that the accuracy of the measurement system is quite good, although the measurement value change due to the water surface is not fixed.

Figure 19. Validation of measurement result by measuring pole.

The test results suggested that the difference in value between the measurement by using the system and the manual measurement was less than 0.5 cm. This is in accordance with the resolution of the sensor JSN-SRT04 contained in the sensor datasheet that is 0.5 cm.

Table 1 shows the results of comparison between the measurement by using the system and the manual measurement of 37 experiments. The measurement error value was obtained by using equation (2) that is 0.75 present (less than 1). This result illustrates that the measurement results of the system are acceptable

Figure 20 is a graph of comparison between the measurement by using the system and by the manual technique. In the test, it was also found that at the time of measurement by the manual technique, the water surface was uneven or wavy as a result of water flow. In addition, the manual measuring medium (a metric tape measurer) also swayed. Therefore, it can be concluded that natural interference and human error are possible to contribute to the measurement by using the manual technique.

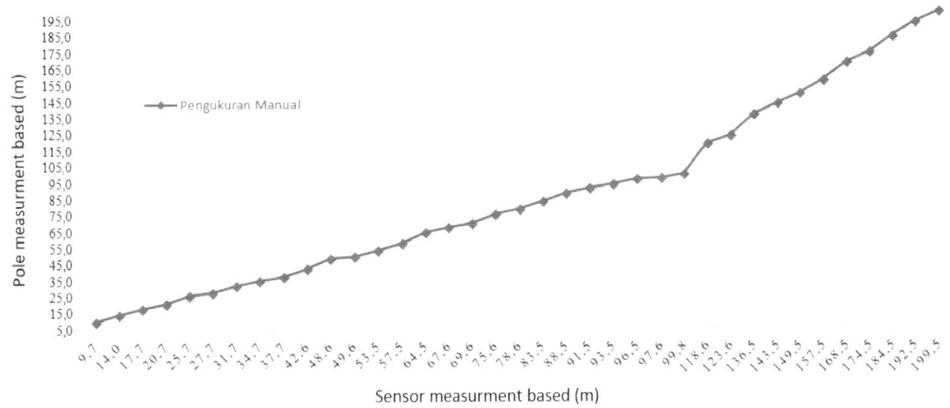

Figure 20. The comparison result of the measurement system to manual measurement.

4. Conclusions

Flood elevation detection system based on Arduino-Uno with sensor JSN-SRT04 has been successfully made and tested. For all water surface condition has been investigated and the error measurement was found 0.74%. compared with manual measurement. It was also found that at the time of measurement by the manual technique, the water surface was uneven or wavy as a result of water flow. In addition, the manual measuring medium (a metric tape measurement) also swayed. Therefore, it can be concluded that natural interference and human error were possible to contribute to the measurement by using the manual technique

The communications between Arduino-Uno and sensors with 37 experiments went successfully with no lag (lag) on the process of data processing by Atmega328. In the tests with the manual technique, natural interference and human error, which happened due to the wavy water surface and the water flow, caused the manual measuring media also swayed.

References

[1] Yuniar R J, Rahadi D and Setyawati O 2013 *Jurnal EECCIS* **7** 65–70
[2] Dewandaru G G T and Lasminto U 2014 *Jurnal Teknik Pomits* **3** C35–40
[3] Sulistyowati R, Sujono H A and Musthofa A K 2015 *Prosiding Seminar Nasional Sains dan Teknologi Terapan* **III** 49–57
[4] Saleh K, Fauziyah, Hadi and Freddy 2013 *Prosiding Semirata FMIPA Universitas Lampung* 511–15
[5] Indarto B, Sunarno H, Fahrudin M and Rahmat D B 2015 *Jurnal Fisika Dan Aplikasinya* **11** 120–26
[6] Sudhan R H, kumar M G, Prakash A U, Devi S R and Sathiya P 2015 *IJIREEICE.* **3** 27–9
[7] Atmel Coorporation 2015 *ATmega328P: 8-bit AVR Microcontroller with 32K Bytes In-System Programmable Flash [DATASHEET]* (CA 95110 USA: www.microchip.com)
[8] Kaur M and Pal J 2015 *IJSRD.* **3** 503–05
[9] David J and Cheeke N 2002 *Fundamentals of ultrasonic waves* (Florida, USA: CRC Press)
[10] Prawiroredjo K and Asteria N 2008 *JETri.* **7** 41–52

Acknowledgment

This research is a research funded by Siliwangi University through Internal research Program of Higher Education 2017. We are grateful to KEMENRISTEK DIKTI, LPPM-PMP of Siliwangi University for its support.

IoT forensic: Optimizing Raspberry Pi for investigation on the smart home network

N Widiyasono[1], I K G D Putra[2] , I A D Giriantari[3] and M Sudarma[4]

[1] Siliwangi University, Siliwangi Street, Tasikmalaya 46115, Indonesia
[2,3,4] Udayana University, PB Sudirman Street, Denpasar 80234, Indonesia

Email: nur.widiyasono@unsil.ac.id

Abstract. Cybercrimes can occur by utilizing existing technology, and any such crime leaves a trace on the system, i.e. data digital evidence. Therefore, it takes a scientific way to analyze the digital evidence on the forensic of such cybercrime. Raspberry Pi is one of the systems that can be developed and functioned as a device used for investigation on the network. The device can capture data on layer 5 (session layer) and investigation process using NIST (National Institute of Standard Technology) method. Raspberry utilization is not only in the process of capturing the data alone but can be used for Mobile Wireless Hacking. Like the two blades of the device has a benefit according to the needs of its users. This device can be further developed for the investigative process by mobile law enforcers, by making some modifications so that they can be taken wherever they go. The contribution of this research is developing this tool for investigation process on network service where after system able to do wireless hacking then data collection at layer 5 is done automatically during the specified timeframe, then the system will auto shutdown by itself, besides modification of battery as energy source so that this tool can be used in mobile.

1. Introduction

Cybercrime is a term that refers to criminal activity with a network computer as a tool, target, or place of crime, such as fraud (online auction, check, car credit, identity, child pornography) so that cybercrime can be defined as a crime that uses the computer illegally. Cybercrime activities have electronic evidence that is electronic and requires forensic management to be analyzed. Digital Forensic is this case isa management of cybercrime.

Cybercrime is done through the internet network media, criminals perform the action by using weaknesses of the system and Internet network infrastructure to get important information. Forensic management and handling attacks on Internet networks are called Network Forensics. Network Forensics is an activity carried out to collect digital data evidence and analyze cyber crimes committed by perpetrators and present the report in the trial.

Certain IoT application services are connected to cloud computing or IoT ecosystems connected through existing networks to enable security holes. IoT benefits for the industry are very large but have the potential for unique security challenges given the growing cybersecurity threats and will affect the implementation of IoT. Enhancing IoT security systems and developing effective digital forensics techniques for investigating sources of attack and obtaining digital evidence data is important so that they can be used in the trial process. The heterogeneity of IoT devices and the absence of standardization become IoT Forensics challenges, besides requiring special applications that have the ability to collect digital data evidence. Application capabilities are several stages of the investigation process such as data collection, examination, analysis and reporting [1]

Interoperability and IoT connectivity standards as well as IoT security, privacy, availability and resilience issues have been conducted, although IoT security issues are still open to review, and studies of digital forensics to investigate cybercrime on IoT are also more important [2].

IoT can connect humans with various services, sensors, and other objects. IoT is currently used in several applications such as smart transportation, smart healthy and smart sensor networks [3]. IoT development leads to cloud-based services and is a business opportunity that increases the number of smart devices [4]. The IoT environment has challenges to security issues including privacy, access control, data storage, and communication security [5]. Caution Investigations are also directed at the use of devices, sensors, and every byte connected to the IoT ecosystem. This problem causes the growth and spread of vulnerable and insecure IoT devices [6]. Conventional user-based security architecture is not used on IoT networks [7]. so in carrying out the ecosystem investigation phase IoT requires special tools, techniques, and procedures such as to secure IoT networks, collect data, maintain data and analyze digital data evidence.

IoT devices are designed to work passively and independently [8], thus requiring a broad search process, even though they can be found, but for digital forensics there is no way to do digital evidence data collection and documentation [9], in addition to methods for imaging IoT devices Forensics is so limited that it ignores ethical considerations to find evidence of devices running in a multi-tenancy environment [10].

Smart home infrastructure usually uses local area networks including shared access points with routers to connect to existing devices, in addition to using smart plugs. The smart plug has several functions, such as the wireless signal strength is much stronger, reduces cable problems, controls several smart devices at the same time with a touch of a button [11]

IoT Forensic is a digital forensic activity conducted to gather digital data evidence within the Internet of Things (IoT) device. The process of collecting digital data evidence requires tools such as raspberry pi but by default raspberry pi requires installation and configuration in advance that is adding the operating system *Kali-Linux RPI, Wireshark, Ethercap, Aircrack-ng,* and programming, besides those devices are not can be done in mobile so that require change to intake of power source in the form of battery. Therefore, this research aims to develop an optimized raspberry Pi for investigation on the smart home network.

2. Method

Process flow stages of research can be described in figure 1.

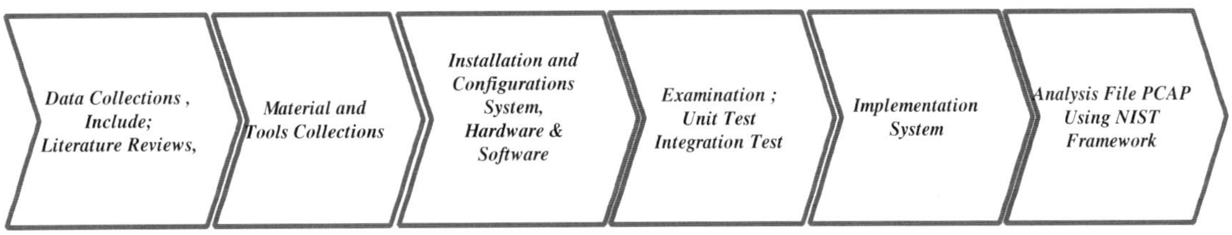

Figure 1. Research process flow.

The necessary data collection on theories relating to this research such as Raspberry Pi, networking, digital evidence, data acquisition and how the system is built, with literature study techniques that collect the necessary data from journals, books, internet, articles, as well as discussions with counselors during the consultation process.

The collection of tools and materials is any collection of necessary materials that support such as hardware and software. The following of hardware and software required are shown in Table 1.

Table 1. The hardware and software needed for the development of Raspberry Pi.

No	Hardware / Software	Function
1	*Raspberry Pi 3 B*	Device for system control
2	MicroSD 16GB	As Storage Operating System, coding and other data.
3	*USB Wireless Adapter (TP-Link TL-WN722N)*	Media for wireless access
4	*USB Drive / Flash disk*	External storage for (.pcap)
5	5v power supply adapter	Power source
6	*Box Casing*	Packing all hardware
7	*LED (red, yellow, green, orange)*	Set the system work process
8	4 Resistors 330 Ω	Electric current inhibitor to led
9	4 pieces Keys: 1 Large keypad, 3 Small keys	Device point indicator indicator
10	Cable Jumper	Raspberry Pi connector with LED and button
11	Image Kali Linux Rpi (.img)	Operating System for Raspberry Pi special Forensic / Cracking
12	Bash Script	Script or coding to control the system
13	Python	Script for controlling Raspberry Pi to network access point, LED, and Button
14	Aircrack-ng and crunch	Tools for cracking passwords
15	Ettercap (Text version)	Tool for sniffing
16	MD5 & SHA Checksum Utility	Tool for hash testing on pcap files
17	Wireshark	Tools to perform pcap file analysis

Installation stages and system configuration was prepared to illustrate the workings of the system before designing the design of hardware and software.

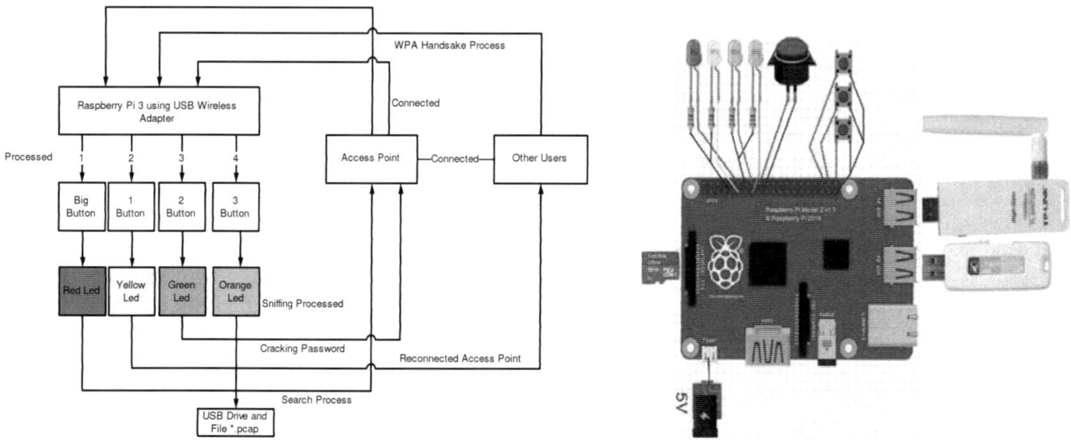

Figure 2. System block diagram.

Figure 2 describes the overall system block diagram. Having input and output to run the process, Input is run using the buttons, and the output of a lighted LED indicates the process of the system. Raspberries act as systems that receive input, process, and output. Raspberry Pi interacts with the access point and other users get pcap files and last stored on USB Drive.

The detail of system operation are as the following: start the process by pressing big button, with the red indicator indicating that it is searching the access point, pressing the button 1 with the yellow indicator light, stopping the access point search process, another user search to perform "WPA Handshake "button, pressing button 2 with green indicator to stop other user search process, and cracking password process, if cracking process too long it can be stopped by pressing button 3, and function button 3 to do sniffing.

3. Results

The test of the system was done by implementation of the system in the field. Selection of access point and user, and WPA Handshakes. The main process that is done in the process of selecting access point and user to be targeted or target to go to stage cracking. The trial of Raspberry tool developed is done on the access point available in the environment of Siliwangi University. The test results in the form of file "*. Pcap" which then analyzed the file by using the NIST method.

Table 2. The result of system testing.

Test	Access point name	The other users connected with the same Access Point	The process of WPA Handshake with other users	Cracking Access Point password	Sniffing the Network (Access point)	Get the digital Evidence (*pcap*)
1	Wakwekwok	Yes	Yes	Yes	Yes	Yes
2	NUABi	No	No	No	No	No
3	HANAWA	Yes	Yes	Yes	Yes	Yes
4	MyPublicWifi	Yes	No	No	No	No
5	MyPublicWifi	Yes	Yes	Yes	Yes	Yes

Table 3 shows that Raspberry Pi 3 has many advantages so that it can be optimized as a tool to get digital data evidence on IoT devices.

Table 3. Comparison of Raspberry Pi3 , Windows IoT, and android things.

No	Parameter	Rasspberry Pi	Microsoft IoT	Android Things
1	Operating System	Raspbian Jessie, ARM-Based Linux OS, Raspberry Pi Media Centers, RISC OS Pi, Plan 9, Retro Gaming Suites, FreeBSD, Chromium OS, Windows 10 IoT Core, Android 7.0 Nougat, AROS: AmigaOS Remake	Windows 10 IoT	Google Brillo
2	License type	Open Source	Closed Source	Open Source
3	Programming Tools	Raspbian 8.0., BlueJ, an IDE for Java., Geany, an IDE for almost any language, Greenfoot, another IDE for Java, Mathematica, serious math on an Rpi, Node-RED - Wiring for the IoT, Python, a modern, flexible language for the Rpi, Scratch, a stepping stone to programming.	Windows Code	LeakCanary. AIDE. Vysor,Stetho, B4A, One of "Speed-Injecting" Android App Development Tools.,Android Studio. GenyMotion, Visual Studio with Xamarin, Takt, Unreal Engine, InstaBug, Eclipse, Source Tree, Fabric, Codota

4. Conclusion

The optimization has been successfully conducted to raspberry device so that it can be used to conduct a mobile investigation for a smart home environment using Access Point to connect between IoT devices. This system capable to obtain digital evidence data of PCAP extension equipped with RDP help, to access point search stage, find other users connected to same access point, WPA Hand11shakes, password cracking, password found, connect to access point, sniff, obtains "* .pcap" files and then performs the analysis and the final report makes a withdrawal.

IOP Conf. Series: Materials Science and Engineering **550** (2019) 012019 doi:10.1088/1757-899X/550/1/012019

References

[1] Harrell C 2010 *Journey Into Incident Response* available at: http://journeyintoir.blogspot.co.id/2010/10/overall-df-investigation-process.html
[2] Giaretta A S B 2016 *IEEE Trans. Inf. Forensics Secur.* **11** 665–76
[3] Zarpelão B B, Miani R S, Kawakani C T and de Alvarenga S C 2017 *J. Netw. Comput. Appl.* **84** 25–37
[4] Chifor B.-C, Bica I, Patriciu V.-V and Pop F 2017 *Future Gener. Comput. Syst.* **86** 740–49
[5] D'Orazio C J, Choo K-K R and Yang L T 2017 *IEEE Internet Things J.* **4** 524 – 535
[6] Alaba F A, Othmana M, Hashem I A T and Alotaibi F 2017 *J. Netw. Comput. Appl.* **88** 10–28
[7] Kent K, Chevalier S, Grance T and Dang H 2006 *Guide to Integrating Forensic Techniques into Incident Response* (Gaithersburg: National Institute of Standards and Technology Special Publication) p 800
[8] Wang L, Peng D and Zhang T 2015 *IJSH* **9** 173-82
[9] Ambrosin M, Anzanpour A, Conti M, Dargahi T, Moosavi S R, Rahmani A M and Liljeberg P 2016 *IEEE Micro* **36** 25–35
[10] Harbawi M and Varol A 2017 *5th International Symp. on Digital Forensic and Security (ISDFS)* (Tirgu Mures, Romania: IEEE) p 1
[11] Venčkauskas A, Damaševičius R, Jusas V, Toldinas J, Rudzika D and Drėgvaitė G 2015 A *IJESRT* **4** 460–77

IOP Conf. Series: Materials Science and Engineering **550** (2019) 012020 doi:10.1088/1757-899X/550/1/012020

Measuring effectiveness of control of information security management system based on SNI ISO/IEC 27004: 2013 standard

A P Aldya[1], S Sutikno[2] and Y Rosmansyah[3]

[1] Informatics Engineering Siliwangi University, Indonesia
[2,3] School of Electrical Engineering and Informatics Institute Technology of Bandung, Indonesia

Email: aldy@unsil.ac.id

Abstract. One of the keys to the successful implementation of information security management in an organization is the selection and implementation of an information security management system control that is good and in accordance with the needs of the organization, the information security management system control can be adopted based on ISO/IEC 27001: 2013 standard document. To ensure the success of information security controls, it is necessary to measure the effectiveness of each control applied. SNI ISO/IEC 27004: 2013 is a standard that provides guidance on the development and use of measures and measurements to assess the effectiveness of controls and control groups in the information security management system as stated in the ISO/IEC 27001 standard, but to do the measurement process, required objects and measurement attributes and metrics, which are not explained in detail in the ISO ISO/IEC 27004: 2013 standard. This study aims to assist in measuring the effectiveness of information security management control by generating the flow of steps in determining the object and measurement parameters and the metrics used based on the provisions contained in the ISO ISO/IEC 27004: 2013 standard.

1. Introduction

Managing information security is as important as managing the core of the business itself. Nowadays, business organizations are becoming more concerned about the security of their information, as well as legal and audit institutions expect every organization to follow certain security rules compliance in line with the implementation of information security management infrastructure in its business. Therefore, build and implement appropriate information security infrastructure is one of the priority needs for most business organizations.

In this context, what is meant by 'information' that includes all forms of data, knowledge, documents, communications, conversations, messages, recordings, and images. Since the information, information systems, and computer are all interrelated, most of the information security means protecting all related assets including hardware, network, Internet connectivity, software, databases, the data (both data stored or data sent and received), file, directory, file reports, telephone, fax, documents, etc., such that only the authorities and legitimate users can access the information [1].

Practice in the field indicates that the problem of information security often caused by bad management. The comparison itself is 30% on the technical side and 70% on the management side. If the technique is the key to information security problems, then management is the core [2]. Therefore, we need to implement good information security management system. In the information security management system standard set forth by ISO/IEC 27000 standard series, the approach presented in the process of ISMS standard is based on the principle that is commonly known as the Plan - Do - Check - Act (PDCA):

a. Plan - set goals and make plans (analyze the situation of the organization, define the overall objectives and set targets, and develop plans to achieve them);
b. Do - implement the plan (do what was planned to be done);
c. Check - measure (measure and monitor the extent to which the implementation is done to meet the planned objectives); and
d. Act - to improve and enhance the activity of (learn shortage to boost activities in order to achieve better results).

As has been described above that one of the principles approach to the process of the ISMS (check) where the process is made by measuring the implementation of information security management systems, one of them is a measure of the effectiveness of the implementation of information security management system controls.

2. Literature Review

2.1. Measurement of information security management system

Security management, such as computer technology management, sales management, production management, and other functions of a company or organization well, should be measured, because something cannot be measured then it will not be evaluated. The evaluation is done to determine whether the measures taken for the organization are correct and also whether the resources provided are adequate [3].

Here is the definition of measurements according to some sources of reference:

a. According to the great dictionary of Indonesian word measurement meaning process, how, or deed measure.
b. According Atzeni and Lioy [4] measurement that is a process that aims to obtain quantitative values of attributes or qualitative in the real world. Attribute real world is any property of an existing entity, may be abstract or concrete.

While the definition of the measurement of information security management systems according to the document ISO/IEC 27 004: 2013 [5], which is the process of getting information about the effectiveness of the ISMS and control using measurement methods, measurement functions, analytic models and decision criteria.

Security management measurements help calculate and measure the effectiveness of the ISMS processes and controls. Measurement facilitates decision making and improve performance and accountability. Measurements can also help managers decide where to place controls and identify unproductive control [6].

A good measurement includes the following qualities: "SMART" (Specific, Measurable, Attainable, Repeatable, and Time-Dependent). Characteristics metrics, including accurate, precise, valid, and true, meaningful, reproducible, objective and unbiased, and is able to measure progress toward the goal, can be measured consistently, cheap to do, expressed as cardinal numbers or percentages and use at least one unit of measure, and contextually specific. [6-7].

2.2. Information security metrics

Metrics can provide an overview for the organization, to prioritize threats or vulnerabilities and risks generated. The metric is a tool designed to facilitate decision making, improve performance, and accountability, so that can be aligned with the goals and objectives of the organization [8].

Here is a list of definitions of metrics:

a. Measure: A variable whose value, its main purpose is to measure the data in order to facilitate insight.
b. Metric: measure or unit of measure that is designed to facilitate decision-making and improve performance and accountability through collection, analysis, and reporting of relevant data is proposed.
c. Metric: broad category of tools used by decision makers to evaluate the data. A metric is a system of related measures that facilitate quantification of some particular characteristics. In simple terms, the metric is a measurement that is compared with the scale or benchmark to yield useful results.
d. Security metrics: measurement of computer security standards.

Metrics can be used to improve the effectiveness of information security. The metric is a system parameter or quantitative assessment and periodic way of the process to be measured, along with procedures for carrying out the measurement and interpretation procedures for assessment from the standpoint of prior assessment or comparable. Metrics are usually devoted to the subject area, in this case the metric is only applicable in a particular domain and cannot directly refer to or interpreted outside of it [9-10].
Mostly of scientists agree that there are at least two approaches are important in the development of metrics. The first type is a mathematical or formal approach and the second is the definition of pragmatic [10-11].
Generally, security metrics can be considered as a part or an extension of the system or the organization's information security management program. Security metrics also have an important relationship with risk management. It can be said that many decisions are supported by security metrics in risk management decisions, because the ultimate goal of all activities of security management is a security risk management. Therefore, metrics can complement certain risk management activities, directly contribute input to the analysis and the overall ability of the organization to deal with the risks faced by facilitating continuous improvements to security [12-14].

2.3. Information security controls
Information Security controls are a set of information security policies that must be defined, approved by management, published and communicated to employees and relevant external parties [15].
Table 1 shows the control area, and the amount of control contained in the document of ISO/IEC 27001: 2013 Annex A.

2.4. Information security measurement model
Here's a picture of the information security measurement model described in ISO/IEC 27004: 2013

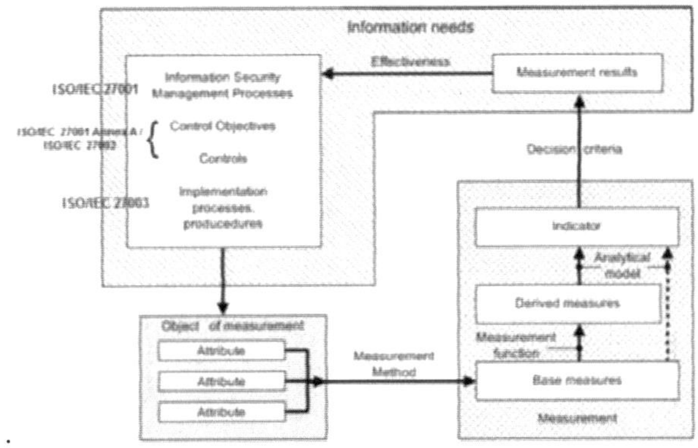

:

Figure 1. Model of information security measures described in ISO/IEC 27004: 2013 [5].

Table 1. Area of control and total control of the ISO/IEC 27001: 2013 Annex A.

Clause number	Area Control	Control Amount
5	Information security policies	2
6	Organization of information security	7
7	Human resource security	6
8	Asset management	10
9	Access control	14
10	Cryptography	2
11	Physical and environmental security	15
12	Operations security	14
13	Communications security	7
14	System acquisition, development and maintenance	13
15	Supplier relationships	5
16	Information security incident management	7
17	Information security aspects of business continuity management	4
18	Compliance	8
	Total Amount of Control	144

Information security measurement model is the structure that connects the needs of the information with the relevant measurement objects and attributes. The object of measurement can be planned or implemented process, procedures, projects and resources
The information security measurement model illustrates how the relevant attributes are measured and converted into indicators that provide a basis for decision making.

3. Method and results

In this research, the development of methods of measurement and metric measures, the development of methods of measurement carried out to facilitate the measurement, especially for ease in determining the measurement object and its attributes, and to assist in making the metric measures.
Development of measurement methods and metric measurement performed in this study was developed and adopted from the measurement model contained in the ISO standard of ISO/IEC 27004: 2013.

3.1. Determining the measurement objects and attributes

The main stage in measuring and making information security controls metric is to determine the measure of the measurement object and its attributes, the following is a flowchart that outlines the image stages in determining the measurement object and its attributes:

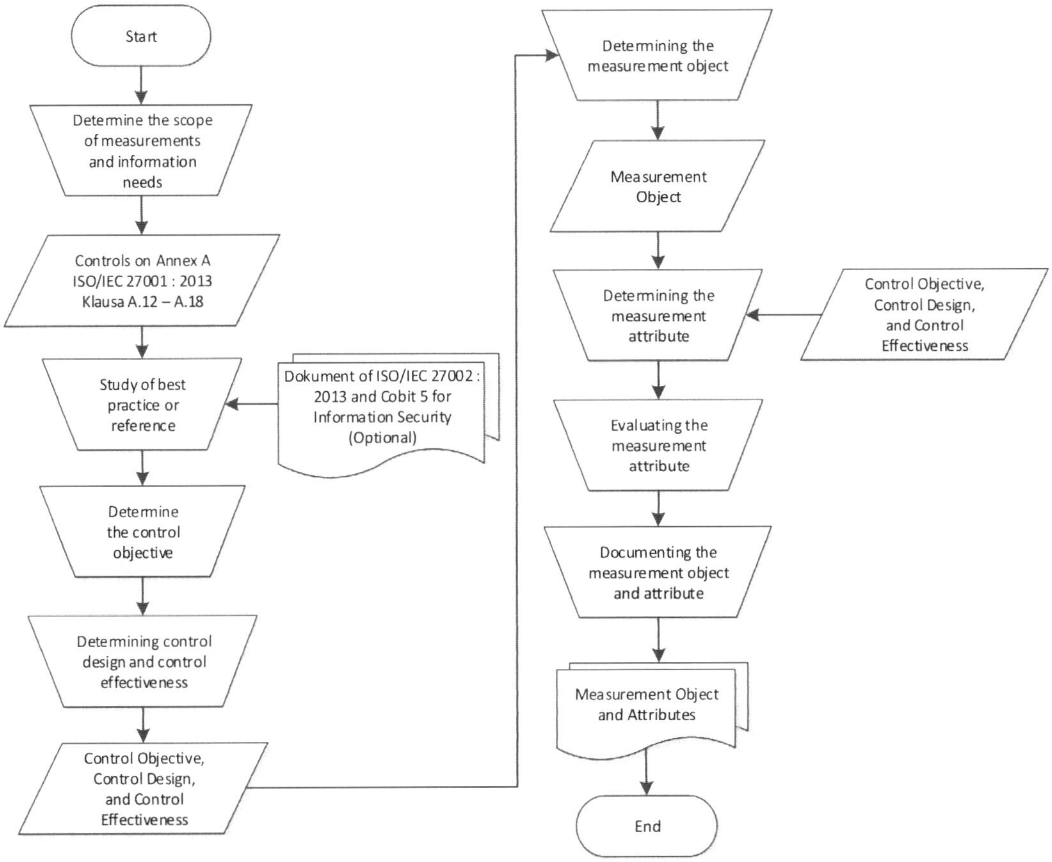

Figure 2. The steps to determine the measurement object and attribute.

3.1.1. Determine the scope of measurements and information needs. The first phase to measuring information security management system is determining the scope of measurement and information needs. The broad scope of the measurement depends on the capabilities and resources of the organization, the scope of the organization's initial measurement activities will be limited to elements such as certain controls, information assets are protected by a specific control, special activities for the information security with highest priority. Each measurement's construction must comply with at least one information's need.

The following activities can be carried out to identify the needs of the relevant information [5]:

a. Check-up the ISMS and processes, such as:
   1. ISMS policy and objectives, control objectives and controls;
   2. Laws, regulations, contracts and organization of information security requirements;
   3. The results output from the information security risk management process, as described in ISO/IEC 27001.

b. Prioritize information needs that identified based on criteria, such as:
   1. Risk management priorities;
   2. Capabilities and resources of an organization;
   3. Stakeholders' interests;
   4. Information security policy;
   5. The information necessary to meet legal requirements, regulations, and contracts;

6. The value of information in relation to the cost of measurement;

c. Select a subset of information that should be addressed in the measurement of the list of priority activities; and
d. Document and communicate the needs of selected information to all relevant parties.

All measurements that applied to the implementation of the ISMS control or control group should be implemented based on the needs of the selected information. Information needs to measure the effectiveness of controls is limited to controls contained in the document ISO/IEC 27001: 2013 Annex A, which controls contained in the document ISO/IEC 27001: 2013 Annex A is directly derived from and aligned with those listed in the ISO/IEC 27002: 2013.

3.1.2. Study of best practice or reference. Best practice is needed to assist in determining the measurement object and its attributes, best practice used in this research is the document of ISO/IEC 27002: 2013, but it also used another references that is COBIT 5 for Information Security document for other additional materials consideration.

3.1.3. Determine the control objective. Before determining the measurement object, control objectives need to be identified first, control objective is a statement of desired results or objectives to be achieved by implementing control procedures. The purpose of determining the control objective is to target the goal of every control on each clause, so that the measurement object can be achieved precisely in accordance with the target of control.

3.1.4. Determining control design and control effectiveness. In this research, the measurement itself is done to get the effectiveness of any control that implemented on the ISMS, therefore, need to be separated between the control design with control effectiveness measurement's selected so that the object can be in accordance with the requirements for the measurement itself. Control design is planning, or the description how a control is assembled or formed so that the controls can functionate in accordance with the expected goals or objectives, while the control effectiveness is an indicator of the success achieved by the control of the objective, or goals that have been set previously.

3.1.5. Determining the measurement object. Control objectives, control design, and control effectiveness that have been determined can be used to assist in determining the measurement object. The object that will be used by the measurement should be selected based on the priority information in accordance with the needs of the organization. Examples are included in the measurement object, i.e. [5]:

a. Products and services;
b. Process;
c. Applicable assets such as facilities, applications, and information systems as specified in ISO/IEC 27001: 2013 (Inventory of assets, A.8.1.1);
d. Business Unit;
e. Geographical location; and
f. Third-party services

3.1.6. Determining the measurement attribute. One object of measurement can have one, or more attributes that can be used to take measurements. To make it easier to determine the attributes, as well as the measurement object, it needs to be reconsidered, control objective, control design and control effectiveness of each of the control order attributes can be determined in accordance with the target control, and can be used for measuring the effectiveness of controls, measurement attributes also should be selected based on the priority information in accordance with the needs of the organization. Values that will serve as the basis of the relevant measure is obtained by applying the method of measurement in accordance with the selected attributes. This option should also ensure that:

IOP Conf. Series: Materials Science and Engineering **550** (2019) 012020 doi:10.1088/1757-899X/550/1/012020

a. A base measure is relevant and appropriate measurement methods can be identified; and
b. Valuable measurement results can be developed based on the values obtained and measures developed.

Characteristics of selected attributes determine the type of measurement method used to obtain the values that will serve as the base measure (e.g. qualitative or quantitative).

3.1.7. Evaluating the measurement attribute. Attributes that have been determined to be evaluated or reviewed to ensure that [5]:

a. Attributes for measurements have been appropriated; and
b. Data collection has been made to ensure the selected attribute can be used to perform effective measurement.

Only attributes that are relevant to the appropriate base measure should be selected. Although the selection of attributes should consider the level of difficulty in obtaining the attributes to be measured, so it is not just because the data is easily obtainable or attributes that are easy to measure.

3.1.8. Documenting the measurement object and attribute. Henceforth, the object of measurement and the attributes that have been evaluated shall be documented.

3.2. Developing metrics

After the measurement object and attribute obtained, the next step is to apply the measurement object and attribute it to the metric measure, stage of implementation of the measurement object and its attributes depicted in the figure flowchart as follows:

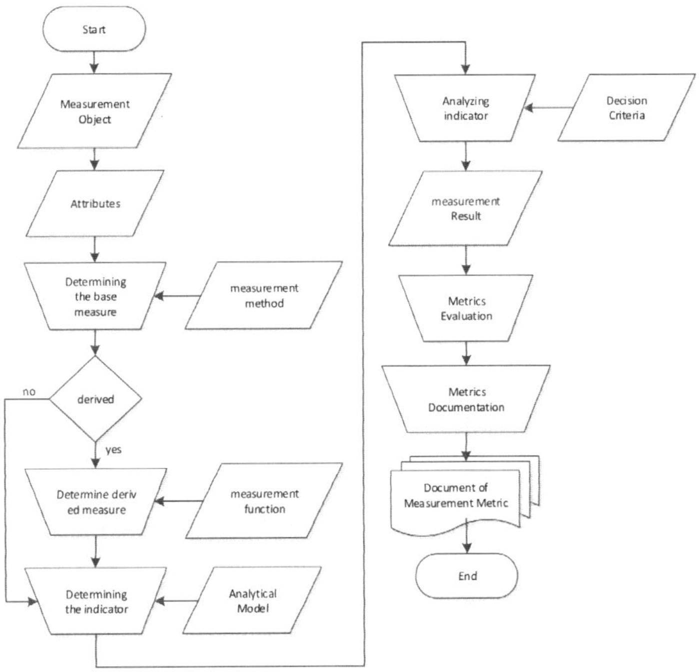

Figure 3. The steps to applying object attributes into the measurement and metrics measure.

3.2.1. Determining the base measure. Base measure is the simplest measure that can be obtained. Base measure is the result of the application of the method of measurement for the selected attribute of an object of measurement. A measurement object may have many attributes, but only a few that can provide useful value to be used as a base measure. A given attribute can be used for several different base measures. Measurement method is a logical sequence of operations used in connection with the scale to measure certain attributes. In the operation may involve activities such as counting occurrences or observing the passage of time. Given the measure of the base can be input for some measure of derivatives, but there is also a base measure that cannot be used as input for the derived measures, but can be directly used as an estimate or evaluation that results in the measurement's indicator.

3.2.2. Determine derived measure. Derived measure is a measure derived or developed from one or more base measure, based on measurement functions is used.Scale and the unit of derived measure depends on the scale and measure of the basic unit combined with the measurement function. Function measurement or calculation algorithm itself is used to combine two or more base measures to create the derived measure. Measurement functions can involve a variety of techniques, such as the average measure of the base, applying the base measure weights, or assign qualitative values for the measure of the base. Measurement functions can combine base measure that uses a different scale, as a percentage and qualitative assessment results.

3.2.3. Determining the indicator. Indicator is a measure that provides an estimate or evaluation of certain attributes, derived from the model analysis of information needs are defined. The indicator is obtained by applying the analytical model of the measure of the base or the derived measure. Scale and measurement methods influence the selection of the analytical techniques used to produce indicators.

3.2.4. Analyzing indicator. The results of measurement were developed by analyzing or interpreting the applicable indicators based on the criteria set out decisions, and should be considered in the context of the overall measurement goal of assessing the effectiveness of the ISMS controls. Decision criteria used to determine the need for further action or investigation, as well as to describe the level of confidence in the results of measurement. Decision criteria can be applied to a set of indicators, for example, to perform a trend analysis based on the indicators are received at different times.

3.2.5. Metrics Evaluation. Metrics that have been made subsequently evaluated, to ensure metric measures are made in accordance with the target effectiveness of the ISMS controls.

3.2.6. Metrics Documentation. Results of metrics that have been made and evaluated subsequently then documented, so it can be used as a reference in measuring the effectiveness of the ISMS controls.

3.3. Result of objective control, control design and control effectiveness, measurement objects and attributes

Here is an example of the control objective, control design and control effectiveness, as well as measuring objects and attributes that have been determined to control [A.12.1.1] Documented operating procedures:

Table 2. Examples of control objective, control design, and control effectiveness, as well as the measurement object and its attributes.

A.12 Operations security		
A.12.1 Operational procedures and responsibilities		
Objective: To ensure correct and secure operations of information processing facilities.		
A.12.1.1	Documented operating Procedures	Control objective: The whole operation is done in accordance with operating procedures Control design: Operating procedures are documented and available for users who need it. Control effectiveness: Operating procedures are executed by the relevant users Measurement's object: 1. Employees, 2. Operation procedure that shall be performed Attribute: 1. Employees who have access to the document operating procedures, 2. Executed operation procedure

3.4. Metrics result example

Here is an example of the measurement object and attributes applied into metrics, to control [A.12.1.1] Documented operating procedures:

Table 3. Examples of metrics.

Measurement construct identification	
Measurement construct name	Documented operating procedures
Numerical identifier	A.12.1.1
Purpose of measurement construct	To ensure the document is executed with an effective operating procedures.
Control/process objective	Clause [A.12.1.1] Documented operating procedures
Control/process	Clause [A.12.1.1] Documented operating procedures. Operating procedures shall be documented and made available to all users who need them
Object of measurement and attributes	
Object of measurement	1. Employees, 2. Operation procedure that shall be performed
attribute	1. Employees who have access to the document operating procedures, 2. Executed operation procedure,
Base measure specification	
Base measure	1. The number of employees who have access to the document operating procedures 2. The number of operational procedures performed, in accordance with the operating procedures documents
Measurement method	1. Count the number of employees who have access to the document operating procedures 2. Count the number of operational procedures performed in accordance with operating procedures documents
Type of measurement method	1. Objective; 2. Objective,

Scale	1. Numerical; 2. Numerical,
Type of scale	1. Nominal; 2. Nominal,
Units of measurement	1. Personnel; 2. Procedures,
	Derived measure specification
Derived measure	1. The percentage of the number of employees who have access to the document operating procedures 2. The percentage of the number of operational procedures performed in accordance with the operating procedures document.
Measurement functions	1. (Number of employees who have access to the operating procedures/total number of employees) * 100 2. (The number of operational procedures performed in accordance with operating procedures document/total operating procedures to be performed) * 100
	Indicator specification
Indicator	1. The bigger percentage of employees who have access to the document operating procedures is better, 2. The bigger percentage of the number of operational procedures performed in accordance with the operating procedures document is better.
Analytical model	1. Compare percentage of the number of employees who have access to the document operating procedures with the total number of employees who should have access to documents of operating procedures, 2. Compare percentage of the number of operational procedures performed in accordance with operating procedures document to the total number of operating procedures that must be met.
	Decision criteria specifications
Decision criteria	1. Percentage of employees who have access to the document of operating procedures should be above 85% 2. The percentage of the number of operational procedures performed in accordance with operating procedures documents should be above 85%
	Measurement results
Indicators interpretation	1. If the percentage below 85% it is necessary to be reviewed, 2. If the percentage below 85% it is necessary to control and better monitoring of the implementation of operating procedures.
Reporting format	1. Pie chart 2. Pie chart with furnished information about the reason for the procedure was not in accordance with the operating procedures document.
	Stakeholder
Clients for measurement	Information security managers and human resource development manager
Reviewers for measurement	Information Security Auditor or information security manager
Information owner	Operational managers and human resource development (HRD) managers
Information collector	Staff of operational and or information security staff
Information communicator	Information security manager
	Frequency/period
Frequency of data collection	Monthly

Frequency of data analysis	Monthly
Frequency of reporting measurement results	Monthly
Measurement revision	Reviewed every 1 year
Period of measurement	1 year

4. Conclusions

Based on the results of research conducted, the conclusion that can be drawn from this research are as follows:

1. Has been performed and resulted the development of methods of measurement, which can be used to help create measurement metrics of the effectiveness of information security management system controls.
2. Has applied the measurement object and attributes acquired in metric measures, which can be used to measure the effectiveness of information security management system controls.
3. Standard of ISO/IEC 27004: 2013, can be used to develop measure and measurement controls contained in document ISO/IEC 27001: 2013 standard (Annex A), or ISO/IEC 27002: 2013.

References

[1] M. Dey 2007 in *Proc. AFRICAN Conference* p 1-6
[2] Qingguo L and Wei Z *2009 Int. Conf. on Multimedia Information Networking and Security* (Los Alamitos: IEEE Computer Society) p 182
[3] Bilbao A and Bilbao E 2013 *47th Int. Carnahan Conf. on Security Technology (ICCST)* (Los Alamitos: IEEE Computer Society) p 1
[4] Atzeni A and Lioy A 2005 *Adv. Inf. Secur.* **23** 1-12
[5] ISO/IEC 2009 *International Standard ISO/IEC 27004 techniques-Information security management-Measurement* (Jenawa: International Organization For Standardization)
[6] Hong-li L and Ying-ju Z *2009 Int. Symp. on Computer Network and Multimedia Technology* (Los Alamitos: IEEE Computer Society) p 1
[7] R. Barabanov 2011 DSV Report series No 11-007
[8] Karabey B and Baykal N 2009 *43rd Int. Carnahan Conf. on Security Technology (ICCST)* (Los Alamitos: IEEE Computer Society) p 144-48
[9] Azuwa M P, Ahmad R, Sahib S and Shamsuddin S 2012 *Int. J. Cyber-Security Digit. Forensics* **1** 280-88
[10] Magott J and Woda M S 2008 *IJCCBS* **1** 161-77
[11] Brotby W K and Hinson G 2013 *PRAGMATIC Security Metrics: Applying Metametrics to Information Security* (New York: CRC Press)
[12] Barabanov R, Kowalski S and Yngström L 2011 *Information Security Metrics : Research Directions* (University of Stockholm)
[13] Bayuk J L 2000 *Information Security Metrics: An Audited-based Approach* (Washington, D.C : NIST and CSSPAB Workshop)
[14] Lundholm K, Hallberg J and Granlund H 2011 *Design and Use of Information Security Metrics Application of the ISO/IEC 27004 standard* (FOI-Swedish Defence Research Agency)
[15] ISO/IEC 2013 *International Standard ISO/IEC 27002 Information technology-Security techniqques-Code of practice for information security control*s (Jenawa: International Organization For Standardization)

IOP Conf. Series: Materials Science and Engineering **550** (2019) 012021 doi:10.1088/1757-899X/550/1/012021

Optimization design of irrigation water management to reduce field failure risk

A K Hidayat[1], I N Norken[2], M I Yekti [3], P Irawan[4] and A S Kosnayani[5]

[1, 4]Civil Engineering Department, Faculty of Engineering, Siliwangi University, Jl. Siliwangi No. 24, Tasikmalaya, West Java, Indonesia
[2, 3]Civil Engineering Department, Faculty of Engineering, Udayana University Jl.P.B. Sudirman, Denpasar, Kota Denpasar, Bali, Indonesia
[5]Nutrition Department, Faculty of Health Science, Siliwangi University Jl. Siliwangi No. 24, Tasikmalaya, West Java, Indonesia

Email: asepkurnia@unsil.ac.id

Abstract. The existing condition of the irrigation area after construction or the irrigation area in operation needs to be evaluated because in the planning of water availability and irrigation water needs many parameters-parameters which is the result of the analysis approach or indeed the data - the data that then changed. The changing conditions of the data will cause many impacts. The risk of field failure is one of impact if water availability is less than the initial estimate. This paper reports the analysis of the availability of water and irrigation water requirements is dynamic, in this study the availability of water is analyzed with 80% reliability. Irrigation water demand is based on regulation of planting schedule and reduction of field area with the intention of reducing the risk of greater field failure. The results obtained are alternative field area of optimum irrigation, regulation of planting schedule and operation of floodgates in the taking or building structure for the review. Field area and Regulation of planting schedule are analyzed by simulation. The mathematical model that will be produced is the optimum model of irrigation water management and mathematical model of field failure risk. This study can be an evaluation model especially for irrigated areas in dry field areas.

1. Introduction

Indonesia includes a country with more than one million hectares of irrigated field, measuring 6.8 million hectares [1]. Field planted in Indonesia of 18 million hectares, and irrigated areas covering 6.8 million hectares or 38%. In 2013 the width of rice fields in Indonesia covering an area of 8,112,103 hectares consists of irrigated rice fields of 4,819,525 hectares and irrigated rice fields of 3,292,578 hectares [2]. West Java Province has a rice planting area of 936,529 hectares and paddy rice harvested area for a year 1,979,799 hectares [3]. The decrease of paddy rice harvest area and paddy field production of the north coast of west java is 77,00 ha/year and 926,10 ton/year [4].

The irrigation problem is technically one of them is the water balance problem. Water equilibrium in irrigation is a condition where the availability of irrigation water is equal to the availability of water. Conditions in the field are 3 (three) possibilities: (1) Condition of water availability is greater than irrigation water demand, (2) condition of water availability equal to irrigation water demand and (3) condition of water availability less than irrigation water requirement. Field failure or crop failure in irrigation areas is more often caused by the condition (3) where there will be drought. Conditions (1) and (2) can sometimes lead to field failure or crop failures when drainage management is not working so the inundation causes the crops to break down and under conditions (2) field failures occur when water regulation is not working so water users scramble.

Research conducted on Cimulu Irrigation Area of 1546.20 hectares located in Tasikmalaya city. The extensive review of irrigation areas is relatively small for Integrated Water Management (IWM) to be more easily managed. Following the steps of the Integrated Water Management (IWM) approach in Indonesia, the EU, Brazil has shifted from a top-down approach to participatory water management involving various segments of society, that means participation and involvement of community segment will be easier if the scope is getting smaller [5]. Irrigation farm field simulations generally do not have a key interaction between the water requirements of plants and the water supply from irrigation systems [6]. Location Cimulu Weir located in the Village Tawangsari Tawang District Tasikmalaya City is located on 7° 19' 14.6" latitude and 108° 13' 17.4" longitude. Cimulu Weir lies on the Ciloseh river which is in the management of PSDA Center of Citanduy River Region. Cimulu Wei irrigates Cimulu Irrigation Area which is part of Citanduy Upstream Service Unit of PSDA Region Citanduy River Region. Irrigation Area Cimulu with irrigation area 1546.20 hectares consisting of 3 (Three) Working Areas Observers are (1) Working Area Observers Cimulu with 1008 hectares of rice fields, (2) Work Area Observers Dalemsuba with irrigated area 316.20 hectares (3) Cihanjang Observer Working Area with 222 hectares of the rice field area.

The real condition after the construction of the irrigation area needs to be evaluated, because in the planning of water availability and irrigation water demand there are many parameters which are the result of empirical analysis and approach. The condition of water availability in planning is also an empirical and statistical approach so that there is a possibility of real water availability condition less than irrigation water requirement. Failed field can occur either because of the availability of water is smaller and the availability of water is greater than the need for irrigation water. Therefore, it is necessary to re-analyze the post-structural irrigation area by arranging the operation of water gates in the building for and intake, the arrangement of planting schedule, and regulation of existing irrigation field. Research after completion of the structure becomes important because the condition of water availability is smaller than the needs of irrigation water will lead to disputes and seizure of water irrigation water user level so that the need for re-regulation. This model will produce an operating model of irrigation watergate and how much optimum field can be supplied. Mathematical model that will be produced is a mathematical model of optimization of irrigation water management and mathematical model of field failure risk.

Irrigation management research has been done using a technical approach [7]. These approaches are (1) research of irrigation water needs with observations in the field of evapotranspiration, percolation, evaporation, water discharge, soil physical properties. (2) research in the form of soil testing to find out soil texture and permeability coefficient by comparing the results in laboratory and observation of inflow-outflow discharge, percolation, evaporation and transpiration on the rice field. (3) research of water needs on the critical field through the bamboo system to know the physical condition and texture of soil in water absorption. (4) the provision of irrigation water supply system on the isolated field to determine the irrigation water supply system in the dry season by giving minimal administration and plant growth with the evaporation process system from the dew, and it turns out that some of these studies are still not optimal in finding the proper irrigation management. Research on the possibility of field failure from the existing water balance in the existing irrigation area, ie the post-structure or irrigation areas that are already operating or running which due to many factors, will certainly undergo many changes in the operation so that it is no longer in accordance with the time of planning, yet much researched.

This study offers a Dynamic system for the optimization of irrigation water management based on field failure risk, this system is an empirical system that is expected to predict the risk of field failure from the existing irrigation area, resulting in door operation from the intake door at the weir location and several buildings regulation. This system will produce the optimum area that may be irrigated with available water availability and if the irrigation system is forced to operate in accordance with available field or even increase the area of irrigation then it is predicted to risk field failure or crop failure. Thus this system can be used to reduce the risk of irrigation field failure. This system can be applied to irrigated areas with limited water availability or where water availability is less than irrigation water needs. The regulation is based on fixed parameters and the parameters change. The required parameters are planting pattern, type of rice variety, planting schedule, effective rainfall, water availability at door and irrigated area.

2. Methods

2.1. Research design

This review is a draft to review the post-construction irrigation area. The assessment can be applied to any existing irrigation areas that suffer from water shortages. Implementation in irrigation areas can be done at any point of water taking, eg in the intake. Research design can be seen in figure 1.

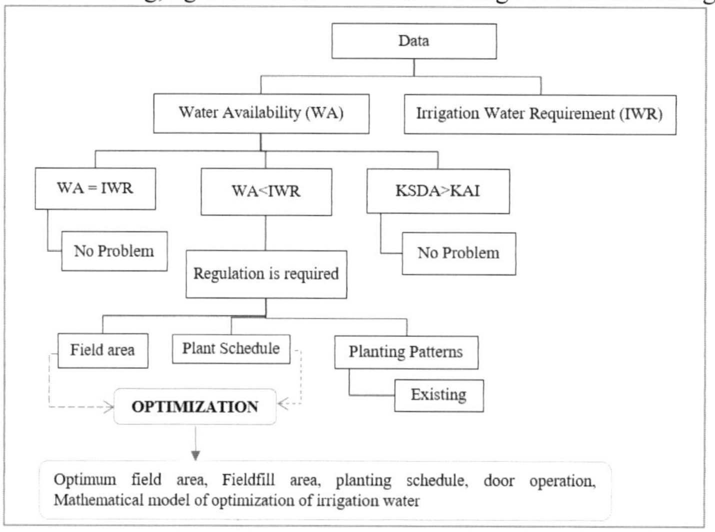

Figure 1. Research design.

2.2. Research stages

The stages conducted in research as follows:

- Preparation
  Preparation of data for analysis of irrigation water needs and water availability

- Analysis
  a. analysis of water availability
  b. analysis of irrigation water needs

- Factor analysis "k"
  There are 3 possibilities k = 1, k <1, k> 1, when k = 1 and k> 1 no problem

- When k < 1
  Regulation of planting schedule and/or reduction of irrigation area is applied

- Obtain mathematical model
  a. obtained optimum irrigation area and planting schedule
  b. operation of water gates and field area at risk of failure
  c. mathematical model of optimization of irrigation water management
  d. mathematical model of field failure risk

2.3. Data analysis

The data source is field data of Cimulu Irrigation Area located in Tawangsari Village Tawang District Tasikmalaya District. Climate data and climate data obtained from BPWS Citanduy. Existing data on cropping pattern and planting schedule are also obtained from the community, government apparatus, BPWS Citanduy, Citiguy Upstream SUP area of PSDA of Citanduy River Region in Tasikmalaya and Tasikmalaya. The data are collected from reports written in report forms consisting of 10 types of blanks. In the data analysis is limited some things as follows:

a. The existing irrigation water discharge is measured at the weir intake for the irrigated area under consideration and is considered to be repeated with the same pattern
b. The availability of river water discharge in the analysis for 80% dependent discharge
c. Analysis of irrigation water needs based on the semi-monthly planting period, so that in one year for three growing seasons there are 24 periods
d. Effective rainfall is assumed to be 70% and dependent rainfall 80%
e. Comparison of discharge for Irrigation Area in the first building is assumed to be proportional to the area of each irrigation area.
f. Variables in this study consist of (a) variables for the analysis of water availability are (1) the measurement of discharge in the intake, (2) the measurement of the discharge in the building for, (b) the variables for irrigation water requirements are (1), (2) effective rainfall, (3) irrigation area, (4) temperature.

The research materials are field data, climatology and climate data, secondary data of water availability and data of water demand analysis. Data on global cropping plan and tertiary plot data of Cimulu irrigation area. Data planting plan from PSDA will be analyzed and juxtaposed withexisting existing planting system in the community. The data were analyzed based on the diagram depicted in figure 2.

3. Results and Discussion

3.1. Shaping parameters design optimization

Irrigation is the use of water on the ground for the purpose of providing the necessary fluids for the growth of the plants. The water is derived from five sources, namely (1) precipitation; (2) atmospheric water other than precipitation; (3) flood waters; (4) groundwater; and (5) irrigation [1]. Irrigation area is the unity of territory or the expanse of field that gets water from one irrigation network. Water allocation is a concern in developing countries where water resources are limited and demand is greater with more parties [8]. Surface irrigation is an irrigation system that taps water directly in the river through a weir building or through a free intake building and then irrigation water is drained by gravity through the channel to the farm [9].

Optimization is a process to achieve an ideal or optimal result (an effective value that can be achieved). Optimization can be interpreted as a form of optimizing something that already exists, or design and make something optimally. Optimization of water use for irrigation field can be regarded as the efficiency of irrigation, also one of energy saving, water use efficiency and energy use efficiency are the two main focuses of the latest innovations on smart irrigation and smart grid [10]. The irrigation water distribution problem usually occurs in the furthest irrigation area of intake water. Lack of irrigation water is a frequent challenge in the furthest channel [5].

IOP Conf. Series: Materials Science and Engineering **550** (2019) 012021 doi:10.1088/1757-899X/550/1/012021

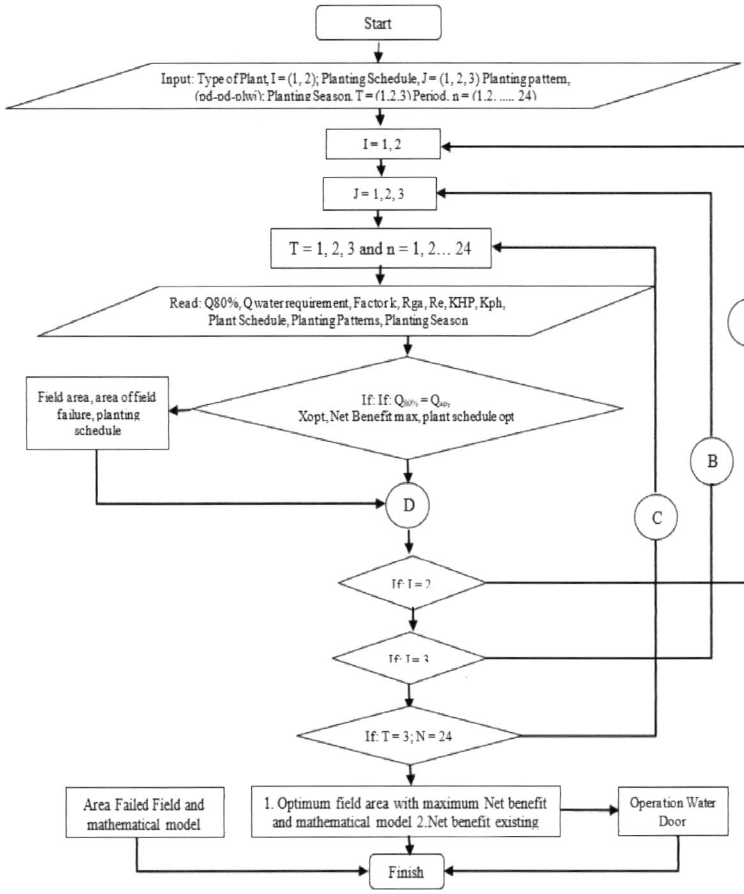

Figure 2. Flowchart of the data analysis.

Optimization of irrigation water is a very important component in irrigation water management [11], then it is said that optimization will be achieved if there is water balance between water availability and irrigation water requirement. The parameters of the optimization design-builder in this research are as follows:

a. Determine the alternative planting schedule before and after the existing planting schedule with a period of 15 (fifteen) daily so that can be made 24 planting schedule including the existing planting schedule.
b. Determine alternative cropping pattern and compare it with existing cropping pattern and cropping pattern which has been determined by PSDA Center of Citanduy River Region in Tasikmalaya and Tasikmalaya Regency.
c. Identification of parameters affecting the area of field failure, eg, planting schedule, available water debit, irrigation water needs, irrigation area, effective rainfall, and so on.
d. The operation of the sluice gate should be considered especially in the intake building, the 1st door building, and the door of the building for the review in order to minimize the risk of field failure,
e. Relation of planting schedule, available water debit, irrigation water requirement to field failure risk is built as the initial stage in making dynamic system design.

The objective function of this research is to maximize the profit of agricultural product (KHP) from irrigation area which can be compiled from the following equation:

IOP Conf. Series: Materials Science and Engineering **550** (2019) 012021 doi:10.1088/1757-899X/550/1/012021

$$(KHP)opt = \Sigma\, K_{ijT}\; A_{ijT} - RgL_{ijT} \tag{1}$$

Where:

KHP_{opt} = optimum agricultural yields (Rp)
K_{ijT} = profit per hectare obtained from each plant type i with planting schedule j in one planting season T (Rp./hectares)
A_{ijT} = decision variable (optimal area of each plant type i with planting schedule j on one planting season T (hectares)
RgL_{ijT} = cost of failure of planting field for each type of plant i with planting schedule j on one planting season T (Rp.)

Profit from Agricultural Product (KHP) is the result of multiplication of the profit/ha of each type of plant with planting area in one year of planting season (production price is considered fixed for each planting schedule). So that KHP is calculated with the following formula:

$$(KHP)_{iT} = K_{iT}\, A_{iT} \tag{2}$$

Where:

KHP_{iT} = benefit of plant i in planting season T (Rp.)
K_{iT} = plant profit per hectare during planting season T (Rp./hectares)
A_{iT} = plant area i in the planting season T (hectares)

The cost of field default risk is a function of the risk index of water failure, the risk of water failure, the reliability, the water discharge, the reliable water debit, the area of field and the profit per hectare.

$$RgL = \int * (IRga, Rga, Re, Q_{ap}, Q_{an}, A, K_{ph}) \tag{3}$$

Where:

RgL = The risk of field failure
IRga = Risk index failed water delivery
Rga = Risk of failure of water delivery
Re = Reliability
Q_{ap} = Discharge water debit
Q_{an} = Dependent Discharge
A = Field area
K_{ph} = The profit of plants per hectares

The equation approach for the cost of plant field failure risk for each planting season j in effective rainfall h% and planting area A is as follows:

$$RgL_{jhx} = IRga \,.\, Rga \,.\left(1 - (Re)^{\frac{Q_{ap}}{Q_{an}}}\right).A.(K_{ph})\, Rp \tag{4}$$

Where:

RgL_{jhx}= cost of failure of planting field for planting schedule T on effective rainfall h% and planting area A ha (Rp.)
IRga = risk index failed water delivery
Rga = risk of failure of water delivery
Re = Reliability
Q_{ap} = the need for irrigation water in the building (m^3/sec)
Q_{an} = Definitive of Mean discharge (m^3/sec)
A = Field area (hectares)
K_{ph} = profit per hectare (Rp/hectares)

3.2. Diagram design and relationship between parameter

Steps and phases of thinking are carried out as follows: (1) collecting data for analysis of the availability and need of irrigation water in the study sites; (2) analyzing the availability of water at the intake; (3) conducting effective rain analysis; (4) conducting irrigation water demand analysis for various planting schedule regulations; (5) conducting optimum analysis of irrigation water management by regulating the field area for various planting schedules; (6) analyzing optimization results to determine the best planting schedule and field area; (7) determining decision-making and inferring the extent of field failure that may occur. Optimization analysis of irrigation field irrigation, field failures and watergate operations can be performed on each building for and intake. For the test phase in this research, the simulation wbeas done on the building intake and building for the first, as in depicted in figure 3.

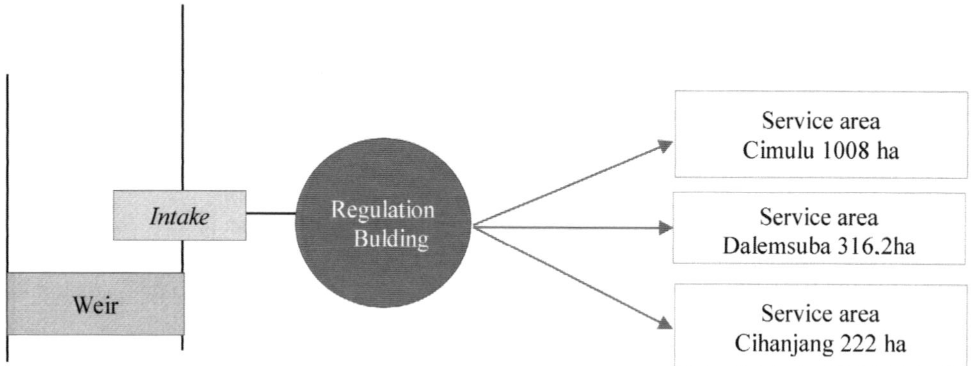

Figure 3. System simulation trial scheme.

Based on the results of the search for the dominant factors that affect the optimization of water management to reduce the risk of field failure, the initial design is obtained as can be seen inthe figure 4.

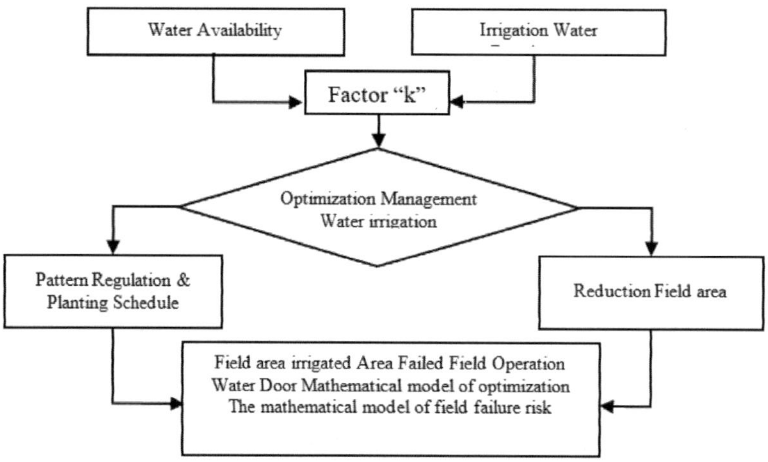

Figure 4. Initial design diagram.

Based on the results of the research steps, the initial design of optimization of irrigation water management to reduce the risk of field failure. The design shows the dominant parameters that affect the optimization of irrigation water management to reduce the risk of field failure. Further research needs to be done to present and perform simulation calculations as well as to determine the validity of the design and mathematical models for the various existing conditions created.

4. Conclusions

Based on the simulation result and some analysis from the mathematical model, it can be concluded that the optimum condition is reached at the time of water equilibrium condition (water balance). The condition of water equilibrium is defined as the total availability of water in the year is used and the same or close to the irrigation water requirement throughout the year. At optimum condition is reached, hence can get wide of irrigation area optimum that can be irrigated so can also know the area of irrigation field at risk of failure.

The design of this wake can analyze and compare the condition of global planting plan which is recommended by the government with the traditional farming condition of the farmer community and also offer an alternative planting system.

The value of agricultural production prices obtained is still assumed at the price of production per hectare so it is considered the same throughout the year, it needs to be analyzed further by including the time dimension for the price of agricultural production.

References

[1] Hansen V E *et al* 1986 *Dasar dasar dan Praktek Irigasi* (Jakarta: Erlangga)
[2] Billah M T 2015 *Statistik Lahan Pertanian Tahun 2009 – 2013* (Jakarta: Kementrian Pertanian)
[3] Purwana G 2015 *Jawa Barat dalam angka 2015* (Bandung: BPS Provinsi Jawa Barat)
[4] Ruminta 2016 Kerentanan dan risiko penurunan produksi tanaman padi akibat perubahan iklim di Kabupaten Indramayu Jawa Barat *Prosiding seminar nasional hasil hasil PPM IPB 2016* 62 –7
[5] El-Nashar W Y and Elyamany A H 2017 Value Engineering for Canal Tail Irrigation Water Problem *ASEJ.* xxx-xxx 1–9
[6] Winter J M *et al 2017* Intergrating Water Suply Constraints into Irrigated Agricultural Simulations Of California *Environ. Model. Softw.* **96** 335–46
[7] Supadi 2009 Model Pengelolaan Air Irigasi Memperhatikan Kearifan Lokal Thesis (Semarang: Universitas Diponegoro)
[8] Marselina M *et al* 2017 Model Prakiraan Debit Air dalam Rangka Optimalisasi Pengelolaan Waduk Saguling – Kaskade Citarum *J Teknik Sipil* **24**(1) 99–107
[9] Wiryawan A G P *et al* 2016 Efektifitas pengelolaan irigasi dengan sumur pompa guna meningkatkan pola tanam di Kecamatan Negara Kabupaten Jembrana *Jurnal Spektran* **4** 1 2016 88 – 96
[10] Nisiakou A *et al* 2016 Smart energy for smart irrigation Computer and Electronics *Agriculture* **129** 74–83
[11] Hidayat A K 2001 Model Optimasi Pengelolaan Air Lakbok Selatan Bendung Gerak Manganti (Yogyakarta: Universitas Gadjah Mada)

Optimization measurements on feeding machines with automated control system for aquaculture

N Busaeri[1], A Andang[1], Empung[2], N Hiron[1] and E Sumarsih[3]

[1]Electrical Departement, Universitas Siliwangi. Tasikmalaya, Indonesia 46115
[2]Civil Departement, Universitas Siliwangi. Tasikmalaya, Indonesia 46115
[3]Argiculture Departement, Universitas Siliwangi. Tasikmalaya, Indonesia 46115

Email: nundangb@unsil.ac.id

Abstract. This paper discusses about the measurement of optimization on feeder machine with automatic system control for aquaculture. This paper mainly aims to observe four main parts of feeding machine of aquaculture: major inlet, scale system, spiral tunnel, and throwing system. The observation done includes the energy of electricity usage of each unit of feeding weight which is processed, duration of process of each part, the exchange of rotor speeds, and the flow of feeding speed rate in gram per second. In this research, the design of feeding machine with automatic system based on Arduino Uno is conveyed. The feeds size used in this research are 2mm and 4mm. The measurement of optimization aims to consider the weight of the feeds, and the system of machine that works optimally. The result from the experiment releases that the machine designed is able to work optimally on 2kg feeds of each process by 70 seconds and 71 seconds, and 0.085Wh on energy of electricity usage.

1. Introduction

Time accuracy, quantity and quality of the fish feeds are the main key of successful aquaculture industry, especially in the large scale. The implementation of technology in the system of feeding is the appropriate solution since the establishment of public opinion that the involvement of technology improves the industrial productivity, especially in the large scale of aquaculture industry. Automatic system designed for feeding machine that can be programmed becomes an interesting case to be discussed, because the feeder machine produces more feeds from manual techniques [1].

Feeder machine developed by H. Wei (2017) provides a good technique on feeder system [2]. However, the weakness is that the machine should be placed above the fishpond. This affects to the container to get the steam from the fishpond evaporation directly and then create the feeds becomes humid. In this paper, we purpose the technique of making feeder machine that can be placed on the edge of the fishpond to avoid the humidity of the feeds. This is different from what have been done by Wei (2017), Atoum et al (2015) and Nirwan et al (2017) [2-4]. The proposed machine consists of several types of conveyer to flow the feeds from the container to the throwing system.

Besides the quality of feeds, a feeding method affects the fish growth in the pond [5], therefore, several studies discussing automatic technology-based fish feeding technique have been introduced by several researchers [2-4, 6-8]. However, conventional feeding-based techniques, namely integration of sheep stall also becomes a concern for fish farming [9], because it provides multiple benefits.

Arduino Uno is a control board and feasible to be implemented on an industrial scale, such as the main control board on the needs of wireless communication [10], Smart and Intelligent System [11], [12] as well as on pond fish feeder machine. The design of the feeding machine based on Arduino Uno had ever been purposed by M. Endebu (2016) where the technique offered is very simple for a small scale [4]. Feeding technique with an intelligent based control approach could decrease human labor costs [13]. The technique of throwing the feeds into fishpond is also purposed [2]. The need of power supplier for automatic feeder machine is also produced from solar panels [7].

In addition to automatic based feeding technique, visual approach using camera to detect the activities of fish in the pond which can then determine the amount of the feeds given [3] becomes a new technique in aquaculture. However, there are still barriers to perfect the proposed system [3]. The main objective of this paper provides the design of automatic feeder machine, and also the measurement of electricity usage on several parts of prototype designed. There are four main parts that will be observed: main inlet (Inlet A), the system scale (Inlet B), spiral tunnel (conveyer spiral) and the throwing system. The development included information of the electrical energy consume when the system was working, the duration of each process, the changes of speed rate of the motor or conveyer, and the feed flow rate flow in grams per second of each part observed.

2. Result and Discussion

2.1. Design

The automatic fish feeding system consists of six main parts: controller module, container, cylinder conveyer, the system of feeds weight validation, spiral conveyer, and throwing system. Figure 1 is the process of feeding machine purposed. It is assumed that the container is full, means that the feeds in the container is not empty, the process of system begins from the main inlet (inlet A) open or spin the conveyer, with conveyer diameter 4.5 inch and 6 parts. It is expected that the feeds would be assisted by gravity. Based on the figure 1, the conveyer of inlet A would always spin until the weight of the target attained, so that the conveyer of inlet A would stop and open the conveyer of B and also spiral conveyer. After the feeds reaches the end of spiral, then it will pass the throwing system, the feeds would be flown into the fishpond by the system of throwing.

Container is the place for feeds supply (figure 2). It is placed on the top and it should be in dry condition. The technique of keeping the container humidity is not discussed in this paper. In the main inlet (inlet A), there is a cylinder which is placed under feeds container (figure 2). The function of this main inlet is to move the feeds from the container into the scale system. The main inlet consists of cylinder conveyer (start fruit) with six parts (figure 3) and an external DC motor drive. The scale is the validator of expected feeds weight based on the instruction of controller board (figure 4). This system consists of three strong sensors and two parts of cylinder conveyer. The activator of conveyer uses DC motor with 12V. Strong sensors (figure 2) are as trigger to stop the conveyer from the main inlet and activate spiral conveyer.

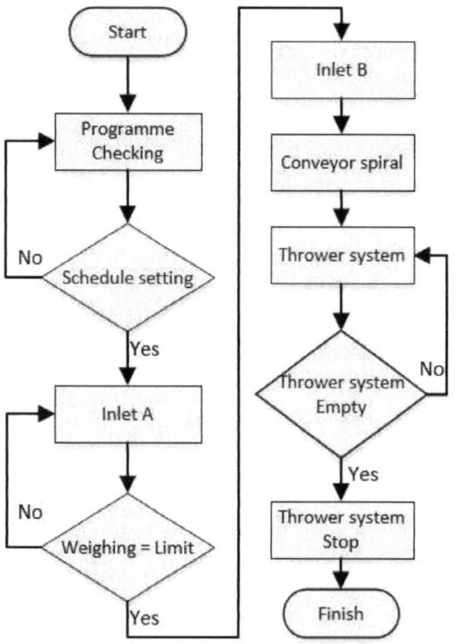

Figure 1. Workflow process of automatic feeder machine.

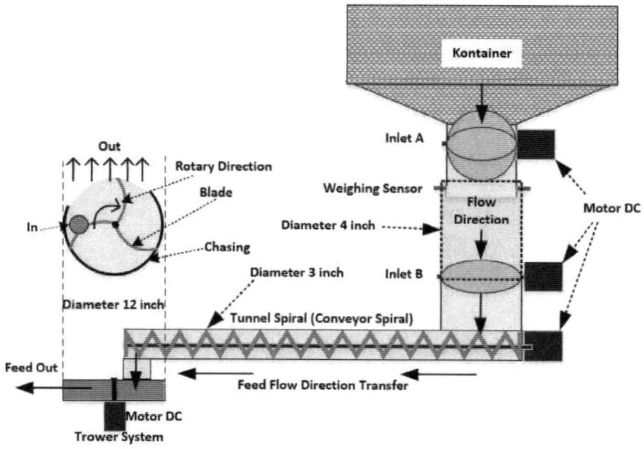

Figure 2. Design of automated feed control with rotary drive system.

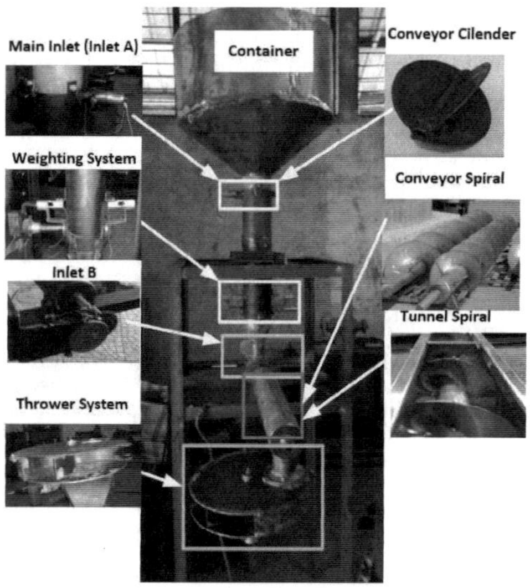

Figure 3. Automated feed control with rotary drive for large scale fish farms.

The spiral conveyer equipped by spiral conveyer serves to move the weighing system to throwing system. The length of the spiral tunnel is 70 cm. This spiral conveyer is activated by DC motor with 12V. This system functions to throw the feeds into the fishpond. This throwing system let the feeder machine to be placed on the edge of the fishpond. This throwing system is designed to spread out the feeds into fishpond. This system is activated by DC motor.

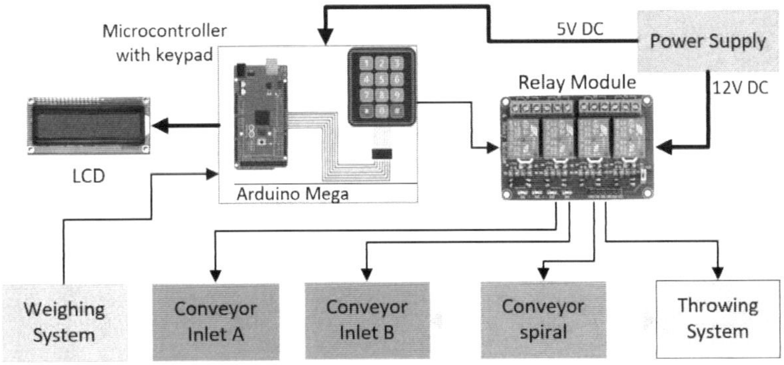

Figure 4. System architecture control.

Figure 4 shows the proposed automatic based feeder machine control architecture. Using Arduion Uno, as the main controller to handle four main parts of machine as output, which are inlet A, inlet B, spiral conveyer, and thrower system, LCD and keypad are used as input tool of feeding schedule program. The activator of those main parts uses DC motor induction with specification as shown in figure 5.

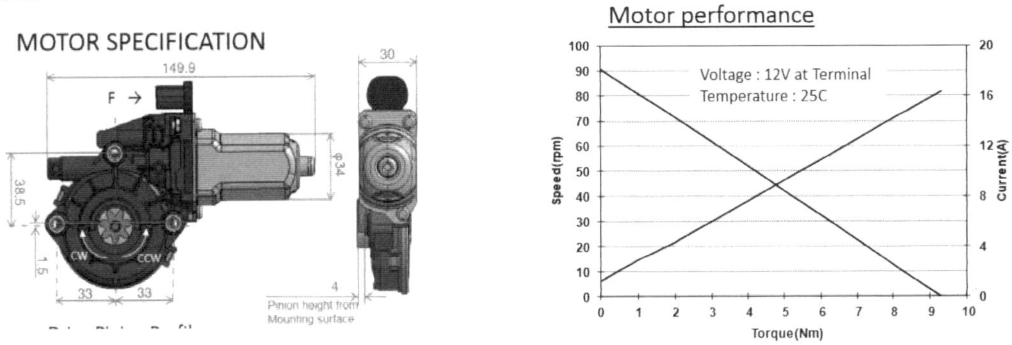

(a) (b)

Figure 5. The characteristic of DC motor rotary driver (a) motor specification; (b) motor performance.

2.2. *Testing*

The experiment is conducted with the feeds in 2mm and 4mm. Separately, the feeds with different sizes are put into the container and then left them through the cylinder conveyer at inlet A, and measure duration of the process, the energy usage, the changes of conveyer rotation speed, and the changes of feeds flowing in the conveyer in which 0.5kg feeds until 5 kg feeds with multiple 0.5 kg in each experiment. Each addition of 0.5 kg is tested three times. The measurement of duration process uses stopwatch. Meanwhile, the measurement of electricity energy usage uses Kyoritsu Kew 6315, and the changes of conveyer rotation speed uses tachometer.
There is no test on inlet B, because it only validates the weight of feeds that would be verified by the program. As a result, the energy usage of conveyer on inlet B could be ignored. On the other side, throwing system is activated by DC motor with a long process duration, so that it requires measurement of energy usage and the duration of the process that occurs on each unit.

2.3. *Measurement result*

From the result of experiment, it is obtained the measurement of duration and the energy usage in every multiple 0.5kg feeds. Duration of the process has character of energy usage which tends improved linearly based on the weight of feeds processed, but it is shown that in 2mm feeds and 4mm, the energy usage and the duration process increases drastically on the feeds weight greater than 2kg (figure 6).

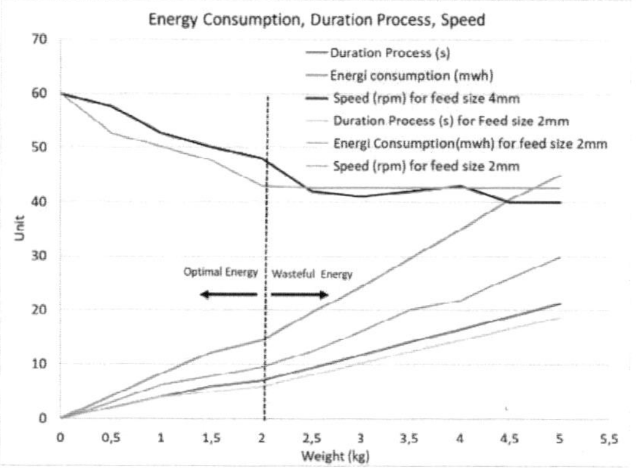

Figure 6. The result of duration process measurement and the change of energy in 2mm feeds.

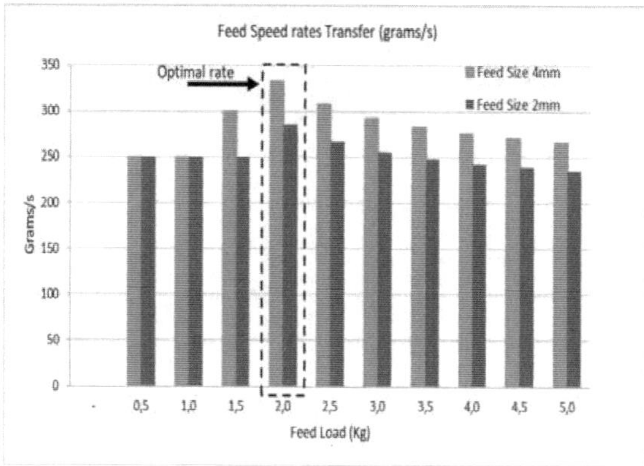

Figure 7. Feed Speed Transfer Rate in main inlet (inlet A) with feed size of 2mm and 4mm.

The size of the feeds also influences the energy of electricity usage. Figure 6 showed that 2mm feeds with weight 2kg consumes energy of electricity 0.0096Wh and 6 seconds in the duration of process. On the other side, 4mm feeds with weight 2kg consumes energy of electricity 0.014Wh and 7 seconds in the duration of process. The amount of the energy usage of 2kg or more feeds, then there will be an increase of electrical energy usage significantly, both for 2mm and 4mm. As a result, it can be concluded that the amount of inlet A is maximum of 2kg.

Figure 7 shows the result of feeds rate experiment in the conveyer of main inlet (inlet A). In 2mm feeds with 0.5kg to 5kg, it is obtained the exchange of feeds speed rate. The highest feed transfer rate is obtained at a load of 2kg which applies to feed sizes of 2mm and 4mm. Thus, it can be concluded that the optimal feed rate is at 2kg feed weight.

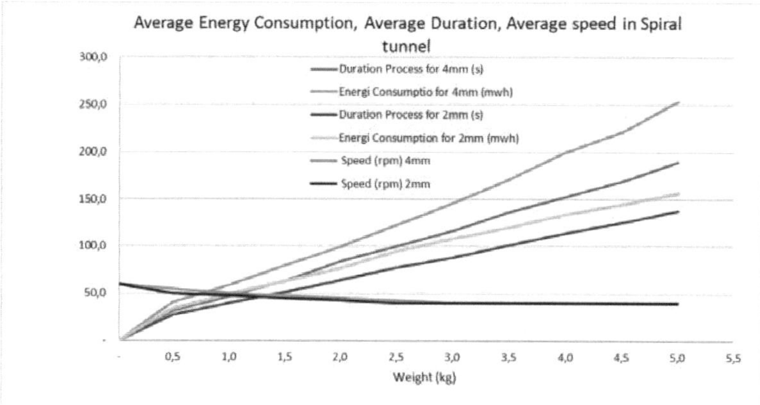

Figure **8.** The result of process duration measurement and changes of energy of 2mm feeds and 4mm on spiral tunnel

Figure 8 shows the result of measurement of electrical energy usage and duration of the processing time for each 0.5kg increase in spiral tunnel. Based on the result of measurement, it is known that the more size of feeds, the more the energy usage needed will be. For 2mm feeds with 2kg needs 64 seconds duration of process with the energy usage 0.0762Wh. On the other side, for 4mm with 2kg

needs 84 seconds duration of process with the energy 0.0994Wh. The speed of rotation is also reduced as the weight of feeds in the spiral tunnel.
Based on the exchanges of speed rate on the conveyer of figure 8, 2kg feeds with 45rpm speed rate of rotation is assumed to be accepted from 60rpm rotation with no burden. It means that by the regulation of 25% is still considered workable. The amount of lower rpm (less than 40rpm) would cause excessive heat on the motor and impact on reducing the life time of motor.

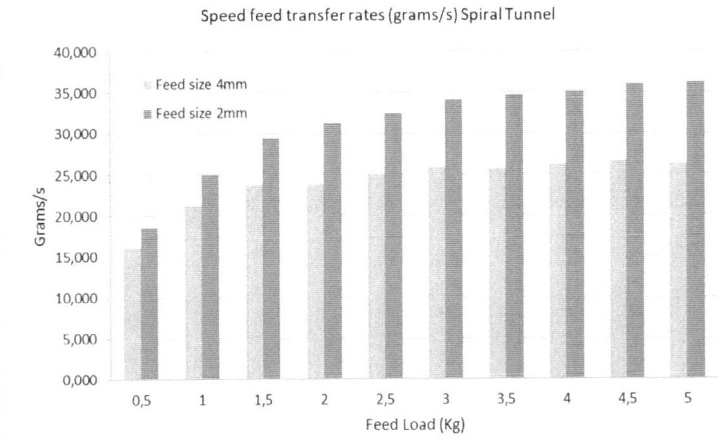

Figure 9. Feed Speed Transfer Rate in Spiral Tunnel with feed size of 2mm and 4mm.

Figure 9 shows the result of the experiment on feeding speed rate of spiral conveyer (spiral tunnel). At a feed size of 2mm feeds with a feed load ranging from 0.5kg to 5kg, it is obtained the change of the amount of feed rate against the time. The more feeds, the more speed rate of feeding in grams per second will be. However, there is stability on the feeds more than 3.5kg, there is 25 grams/s for 4mm feeds. On the other side in 4.5kg or more, there is 36 grams/s for 2mm feeds size.

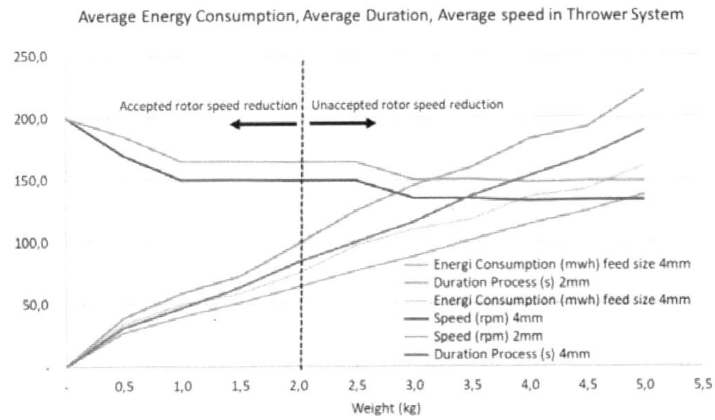

Figure **10.** The result of process duration and energy usage on the system of throwing in 2mm and 4mm.

Figure 10 shows the exchanges on the energy usage, duration of the process, and speed rate rotation of feed load given. At a maximum of 2kg feeds with 2mm, the energy usage is 74.8mWh and duration of the process is 64 seconds. While at a feed size of 4mm, the energy usage is 98.2mWh with 84 seconds of duration of process.

Throwing system has rotor speed rate on 2000rpm with no burden, while a 2kg load with 2mm feed size, the speed rate of rotation of rotor decreases to 1645rpm, meaning it decreases 11% at 4mm feeds, the decrease of speed rotor 25% or 1297rpm. This condition in 2mm and 4mm with maximum 2kg feeds is still accepted. Feeding loads above 2kg would affect to the decrease of rotor of throwing system significantly. This condition is unacceptable.

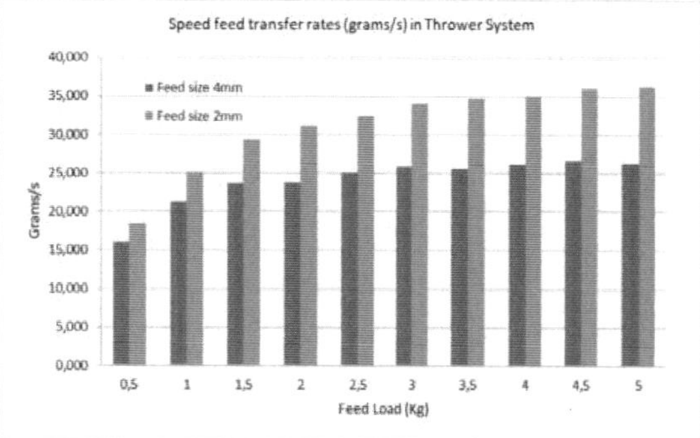

Figure 11. Feed speed transfer rate in thrower system with feed size of 2mm and 4mm.

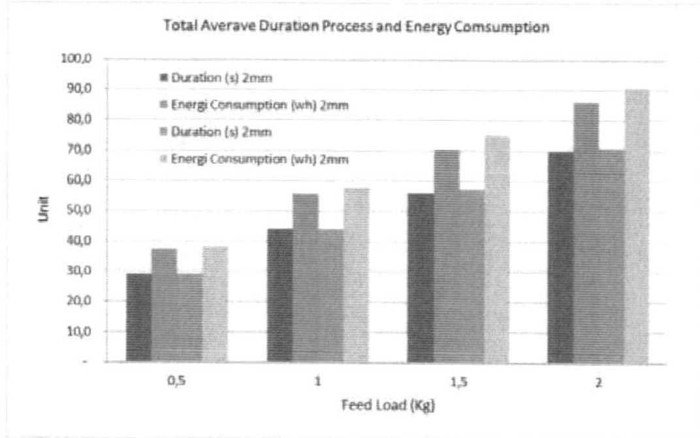

Figure 12. The electricity energy usage in total, duration of the process in 2mm and 4mm feeds

Figure 11 shows the result of experiment on feeds speed rate in throwing system. At a feed size of 2mm with a feed load ranging from 0.5kg to 5kg produces the exchanges on the amount of speed rate against time. The more feeds size, the more speed rate of the feeds in grams/s will be. Nevertheless, there is stability on the speed rate above 3kg. There is speed rate of the feeds in 34grams/s for 4mm. On the other side, there is feeds speed rate in 25grams/s for feed size of 2mm.
Figure 12 shows the total amount of energy usage, duration of the process for 2mm and 4mm feeds over all the process. In maximum 2kg with 2mm, total electricity energy usage is 85.8mWh or 0.085Wh. On the other side, in 4mm, total electricity energy usage is 90.7mWh or 0.0907Wh. Total duration of process in 2mm and 4mm in 2kg feeds is 70 seconds and 71 seconds, or 1.2minutes.
From the final result of several experiments, we concluded that the proposed design of automatic feeder machine has different performances at each feed load given. From several experiments, it is concluded that 2kg feed load for each process is the optimal amount. It means that no matter how much the feeds needed should be divided into 2kg per each process. It is assumed that the machine should flow the feeds 10kg into the fishpond, then the machine has to work 5 times of process with each process maximum 2kg.

3. Conclusions

The optimization of the machine designed is obtained. The machine could work optimally with 2kg feeds in each process, both for 2mm and 4mm feeds size. The optimization is the same in the energy of electricity usage. Another conclusion of this research is that the design of feeder machine could work normally and has less energy usage of each process. The machine only works for 2kg in maximum feeds. The motor of the main inlet should work immediately after the system of throwing is almost done, so that it is needed the speed rate of rotation on the spiral conveyer. On the other words, the speed rate of throwing system is very based on the duration of the process on the spiral tunnel. It means that the quicker the spiral tunnel works, the quicker the machine period works. Duration of the process of feeding machine for 2mm and 4mm is 1.2minutes with the total of energy usage 0.085Wh to 0.0907Wh or 0.9Wh for 10kg feeds. This design of the machine could be implemented in fish farm of catfish, shrimp, or fish farm in the offshore.

References

[1] Ogunlela A O and Adebayo A A 2016 Development and Performance Evaluation of an

Automatic Fish Feeder *J. Aquac. Res. Dev.* **7**(2) 7–10
[2] Wei H C et al Improvement of automatic fish feeder machine design *J. Phys. Conf. Ser.* **914** 012041
[3] Atoum Y, Srivastava S and Liu X 2015 Automatic Feeding Control for Dense Aquaculture Fish Tanks *IEEE Signal Process. Lett.* 1–5
[4] Nirwan S, Swarnakar R, Jayarajan A and Shah P 2017 The Developement of Automatic Fish Feeder System Using Arduino Uno *Int. J. Mod. Trends Eng. Res.* **4**(7) 64–8
[5] Rajeshkumar A and Balusamy M 2017 The effect of different fish feeding methods on growth performance and fish yield in composite fish culture system *J. Entomol. Zool. Stud.* **5**(6) 1514–18
[6] Hazwan M 2013 *Modeling and Control of The Fish Feeder System* Thesis (Malaysia: Universiti Tun Hussein Onn Malaysia)
[7] Ani D T, Cueto M G F, Diokno N J G and Perez K R R 2015 Solar Powered Automatic Shrimp Feeding System *Asian Pacific J. Multidiscip. Res.* **3**(5) 152–9
[8] Uddin N et al 2013 Development of an automatic fish feeder *Glob. J. Res. Eng.* **10**(1) 27–32
[9] Endebu M, Tugie D and Negisho T 2016 Fish growth performance in ponds integrated with poultry farm and fertilized with goat manure : a case in Ethiopian Rift Valley *Int. J. Fish. Sci. Aquac.* **3**(2) 40–45
[10] Hiron N and Andang A 2017 Wireless communication with batching method based on Xbee-PRO S2B module for sensing of wind speed in *Proceeding - 2016 2nd International Conference on Science in Information Technology, ICSITech 2016: Information Science for Green Society and Environment* (Balikpapan: Universitas Mulawarman)
[11] Hiron N, Andang A and Setiawan H 2016 Batch Processing Method in Machine to Machine Wireless Communication as Smart and Intelligent System *Int. J. Futur. Comput. Commun.* **5**(3) 163–6
[12] Hiron N and Andang A 2016 Wireless communication with batching method based on Xbee-PRO S2B module for sensing of wind speed in *2016 2nd International Conference on Science in Information Technology (ICSITech)* 250–3
[13] Zhou C, Xu D, Lin K, Sun C and Yang X 2017 Intelligent feeding control methods in aquaculture with an emphasis on fish: a review *Rev. Aquac.* 1–19

Acknowledgments
This research is a part of an external research program and is supported by research funding from KEMENRSITEK-DIKTI 2017-2019 at the scheme "Applied Product Research for the second year. Thank you to LPPM-PMP Siliwangi University for supporting this project.

Performance comparison of signed algorithms on *JSON Web Token*

A Rahmatulloh, R Gunawan and F M S Nursuwars

Department of Informatics, Siliwangi University
Jln. Siliwangi No. 24, Tasikmalaya 46155, Indonesia

Email: alam@unsil.ac.id, rohmatgunawan@unsil.ac.id, firmansyah@unsil.ac.id

Abstract. The system survive in this digital era is a system that can work on multiple platforms, the use of web services is one of the solution. Exchange data using JSON format and for the security of authentication using JSON Web Token (JWT). The importance of token-based authentication using JWT on web services can solve interoperability problems. JWT is stateless and can include information in the token authorization. JWT has several options for using algorithms namely HMAC, RSA, and ECDSA. Unfortunately, it is unknown which algorithm has better performance. Here, we directly tested the signing algorithm in the three algorithms seen from several parameters. The experimental results showed the use of HMAC algorithm produces an average value of token-generating time is 21.3 s, token size 109 bytes and data transfer token speed 91.2 s. It was considered that the HMAC had an excellent performance.

1. Introduction

Current technology trends are stateless systems and can be used multi-platform so that interoperability problems can be resolved. Interoperability problems can be solved such as using web services. Web service (WS) uses a service-oriented architecture (SOA) concept that provides services on networks such as the web, and these services contain well-defined business functions which can then be consumed by users in various applications[1, 2]. SOA which has better performance is the Representational State Transfer (REST) model [3-4]. REST is a standard web-based service architecture for data communication using HTTP protocol standards [5, 6]. However, REST is still weak on the security side [7], securing REST including securing data and data communication lines to protect confidentiality and integrity data [8].

Various techniques for secure the REST architecture have been carried out by several researchers, Hussain (2014) developing an application programming interface (API) on the web using ASP.NET based on the view controller (MVC) model, using basic authentication, claim based authentication, and token based [9]. Likewise with research conducted by Huang (2015) and Sahoo (2017) using token-based authentication [10,11]. Authentication based on ID claims has been done by Xinhua (2013) [12] while the Security Assertion Markup Language (SAML) technology was conducted by Anand (2017) [13]. Authentication is a standard aspect of information, authenticating one of the most critical parts in each application. Users can be authenticated to get resources using passwords, biometrics, token-based or through certificates [14]. For decades, cookies and server-based authentication are the most comfortable solutions for authentication in web-based applications. However, it will be different when handling authentication for multi-platform, the use of server-based authentication will be very complicated. Session-based authentication always makes the server overload, because the server must check the session all the time it reduces server performance. Unlike token-based authentication, there are no sessions created every time the user enters so this shows better performance than session-based techniques [15]. REST security using JSON Web Token (JWT) authentication has been conducted by Balaj (2017) [16].

Token-based authentication enables the credentials of the user's name and password to get tokens that allow them to retrieve certain resources. After a token is available, users can reuse the same token to access resources without using a username and password again [17]. The given token has been made by the server with a certain algorithm so that we can know the ownership of tokens and access rights owned by the user.

JWT is essentially part of the data signed in JSON format. JWT can be signed using an HMAC symmetric key algorithm or public and private key pair using RSA or ECDSA [18]. Research only compares the performance of the HMA SHA-256 algorithm with the SHA-512, there has been no performance testing of the algorithm options that JWT can be used [16]. Therefore, the focus of this research was to to test the algorithm performance on JWT and compare the performance of the three algorithms in the parameters of the speed of generating tokens, the size of tokens and data transfer tokens.

2. Method

2.1. Representational state transfer (REST)

REST is not a programming language, but rather a hybrid architectural style derived from several network-based architectural styles that are often applied in web-based services [5]. REST architecture is generally run via HTTP (Hypertext Transfer Protocol), involving the reading of certain web pages that contain an XML or JSON file. REST is not stateless and resource-oriented, everything in the REST architecture is a resource. Each request is independent, and the server does not store any request conditions. The Application Programming Interface (API) that follows the REST style is called the RESTful API. The RESTful API uses Uniform Resource Identifiers (URIs) to represent resources. Each data source is identified using a URI link. Some of the methods used in the REST such as the GET method is used to obtain resources, and the POST method is used to create new resources, the PUT method is used to update resources based on resources, and the DELETE method is used to delete resources or collections of resources. Examples can be seen in table 1.

Table 1. Example of a RESTful API.

Resource	Method			
	Get	Post	Put	Delete
/api/student	Get a list of all student	Create a new list of student.	Update a list of student	Delete all student
/api/student/123	Get a student by student's ID	Treat as a collection. Create a new student in it.	If a student exists, update the student. If a student does not exist. Create a new student.	Delete the student.

2.2. Token authentication

Token authentication or token-based authentication is a core element of scalable identity and authorization management. Authentication tokens are stateless, secure, scalable, and designed to ease users without overloading the server. Authentication is the process of identifying the user's identity. The traditional mechanism of an application will save the session ID on the server containing the user's identity from the session cookie. Unlike token authentication, it applies a modern approach designed to solve session ID problems, use tokens as a substitute for session IDs that can reduce server load, manage permissions, and as a new, better mechanism to support distributed or multi-platform and cloud-based architectures.

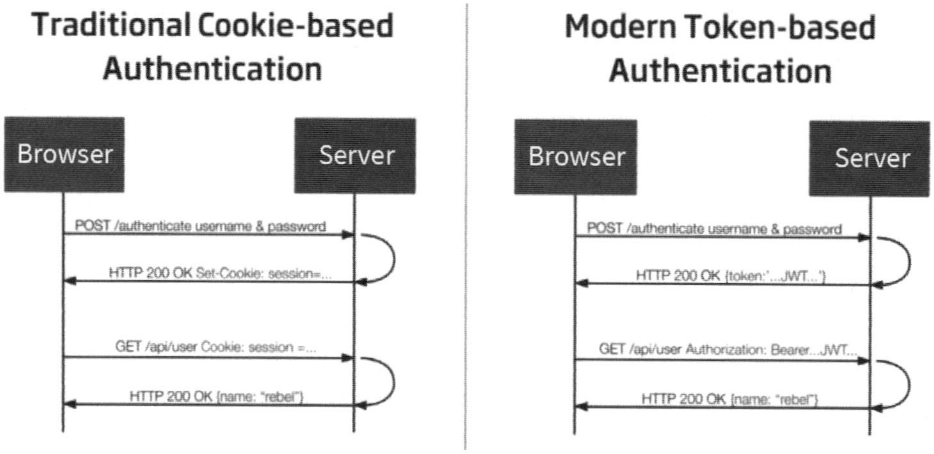

Figure 1. Difference between traditional and modern authentication tokens.

Figure 1 shows the authentication process between client and server, in modern authentication based on the user's credential token exchanged with a token that then the token can be used every subsequent request. Tokens are attached to users via cookies or stored in local storage or browser cookies.

2.3. JSON Web Token (JWT)

JWT or read 'jot' is a token in the form of a string consisting of three parts, namely header, payload and signatures used for authentication and information exchange processes [18]. Each part is separated by a dot symbol (.) as depicted in figure 2.

Figure 2. Three main elements.

JWT encodes claims to be sent as JSON objects that are used as load structures from JSON Web Signature (JWS) or as a plaintext structure of JSON Web Encryption (JWE), allowing claims to be digitally signed or protected with Message Authentication Code (MAC). According to how tokens are presented there are two types of tokens, Carrier Tokens, and Key Holder Tokens and according to the token purpose, there are two token schemes, Identity Tokens, and Access Tokens [19]. JWT works the same way as a password when the user successfully logs in, and the server will give a token stored in local storage or browser cookies. The token is used to access certain pages, the user will send the token back as proof that the user has successfully logged in. The JWT structure can be seen in figure 3.
In the Header section consists of two parts: the algorithm used and the type of token, then JSON is encoded Base64. The second part is the payload that contains the claim, and the claim is a statement about the entity which is usually additional user data and data. The payload is the same as the encoded header with Base64. The third part is the signature containing the encoded header, the encoded payload, the secret, the algorithm that has been defined in the header and the sign. Cryptographic algorithms for digital signatures and MACs at JWT can be seen in table 2.

Header: Algorithms and Token Types
{
"alg": "HS256",
"typ": "JWT"
}
Payload: Data
{
"iss": "#appbackend",
"user": "alam",
"pass": "rahasia"
}
Signature: results from Hash
{
"Base64-encoded(Header.Payload)" + "key" + "Algorithm"
}

Figure 3. JWT structure.

In the Header section consists of two parts: the algorithm used and the type of token, then JSON is encoded Base64. The second part is the payload that contains the claim, and the claim is a statement about the entity which is usually additional user data and data. The payload is the same as the encoded header with Base64. The third part is the signature containing the encoded header, the encoded payload, the secret, the algorithm that has been defined in the header and the sign. Cryptographic algorithms for digital signatures and MACs at JWT can be seen in table 2.

Table 2. Signing algorithm [20].

"alg" parameter value	Digital Signature or MAC Algorithm	Implementation Requirements
HS256	HMAC using SHA-256	Required
HS384	HMAC using SHA-384	Optional
HS512	HMAC using SHA-512	Optional
RS256	RSASSA-PKCS1-v1_5 using SHA-256	Recommended
RS384	RSASSA-PKCS1-v1_5 using SHA-384	Optional
RS512	RSASSA-PKCS1-v1_5 using SHA-512	Optional
ES256	ECDSA using P-256 and SHA-256	Recommended+
ES384	ECDSA using P-384 and SHA-384	Optional
ES512	ECDSA using P-521 and SHA-512	Optional
PS256	RSASSA-PSS using SHA-256 and MGF1 with SHA-256	Optional
PS384	RSASSA-PSS using SHA-384 and MGF1 with SHA-384	Optional
PS512	RSASSA-PSS using SHA-512 and MGF1 with SHA-512	Optional
none	No digital signature or MAC performed	Optional

The last stage of the three sections in figure 3 merged to produce JWT, with pseudocode (1).

$$\boldsymbol{Token} = f(Base64Encode) \sum_{n=\alpha,\beta}^{\infty} (header.payload.signature) \tag{1}$$

With tokens obtained from the server, users can access web service resources by entering tokens in the HTTP header in the authorization method. The JWT tokens example can be seen in figure 4.

*eyJ0eXAiOiJKV1QiLCJhbGciOiJIUzI1NiJ9.
eyJrZF9wZWdhd2FpIjoiUC0wMDYiLCJ1c2VybmFtZSI6InphbCIsImlhd
CI6MTUyNzM5NzIyMywiZXhwIjoxNTMwMDI2OTY5fQ.FqHaqcJUOg
XlSIg1Q7MuBhpqkWEqdGIAovzPOxqV2jg*

Figure 4. The example of JWT tokens.

2.4. Related work

The use of token-based authentication using JWT has already been investigated by Ethelbert (2017) which results in that token-based authentication using JWT is superior to the others seen from the Authentication mechanism, Access control mechanism, Compact & stateless token structure, Dual authentication / SSO support, and access control scalability [21]. Then in the study of Mestre (2017) that compares JWT performance with single and multiple tokens on one or several servers. The results show that the proposed system has a worse performance than those using a single token [22]. Research produces a comparison of the performance of SHA-512 better than SHA-256 in the HMAC hash algorithm alone [16]. While the general algorithm choice that is often used on JWT is HMAC + SHA256, RSASSA-PKCS1-v1_5 + SHA256 or ECDSA + P-256 + SHA256 but no one has yet tested its performance [18].

3. Results and Discussion

A simple simulation application has been created to find out the algorithm performance on JWT. Then the test scenario with experiment 50 times the process, starting from testing the time to generate the token, the size of the token generated and finally the data transfer speed of the token from the client request to the server until the token response is received by the client.

3.1. Generate token

The first test is to generate a token process 50 times and in table 3 only ten experiments are presented, but the results of the test are all poured into a graph in figure 5.

Table 3. Generate token processing time.

Testing	Signing Algorithm	Testing	Signing Algorithm
1	277762	114581	21171
2	271418	81652	21414
3	278841	86963	15092
4	252384	90251	14371
5	262182	126767	26075
6	263163	100349	17970
7	251739	86033	18839
8	243438	90439	51525
9	271581	101983	34792
10	262023	96329	22439
Avg	268290	95845	21322

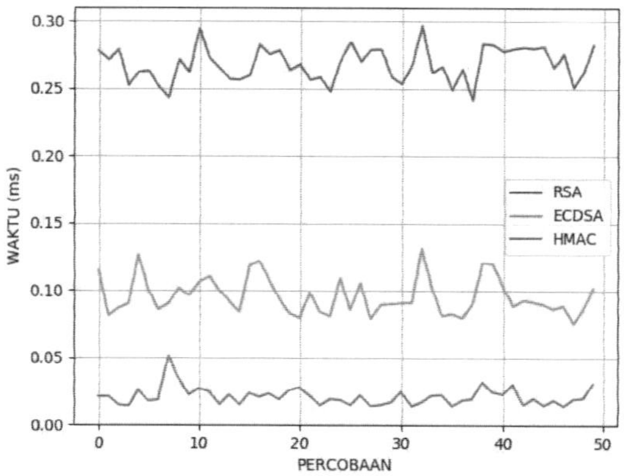

Figure 5. Comparison performance of time process generating for tokens using RSA, ECDSA, and HMAC algorithms.

3.2. Token size

The second test is testing the JWT token size, the results of the experiment 50 times can be seen in table 4 and graphical form in figure. 6.

Table 4. Token size.

Testing	Signing Algorithm RSA (bytes)	ECDSA (bytes)	HMAC (bytes)
1	164	164	109
2	164	164	109
3	164	164	109
4	164	164	109
5	164	164	109
6	164	164	109
7	164	164	109
8	164	164	109
9	164	164	109
10	164	164	109

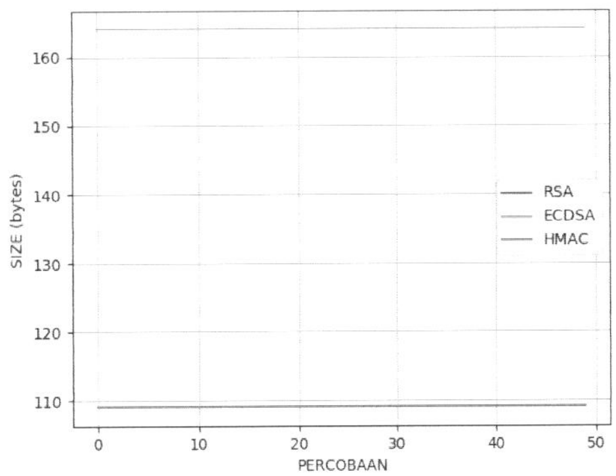

Figure 6. Comparison of tokens size using RSA, ECDSA, and HMAC algorithms.

3.3. Data transfer
Likewise, testing all three trials of transferring token data 50 times from client to server can be seen in table 5 and figure. 7.

Table 5. Data transfer.

Testing	Signing Algorithm RSA (ms)	ECDSA (ms)	HMAC (ms)
1	368412	190990	83660
2	339241	144385	91155
3	359306	151147	74678
4	311298	152766	77904
5	322892	218130	100363
6	319571	163614	78579
7	315724	147288	79581
8	301652	163899	138004
9	379141	202094	110295
10	320035	162519	93199
Avg	339676	173532	91180

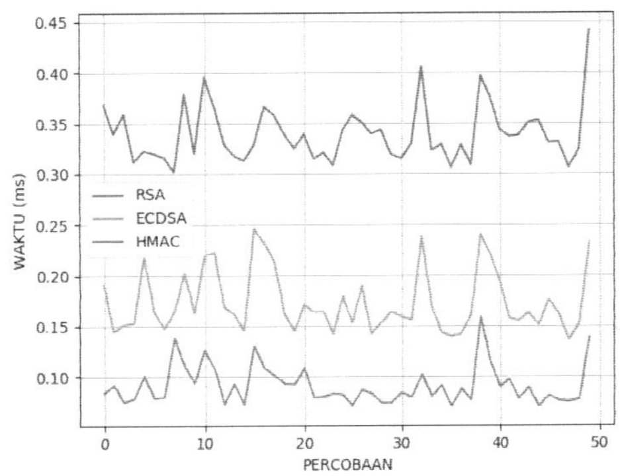

Figure 7. Comparison performance time of token data transfer using RSA, ECDSA, and HMAC algorithms

If we compare the performance of the token generating time (table 3 and figure. 5) we can see that the HMAC algorithm is superior to an average score of 21.3s, there is an increase of 349.5% when compared to ECDSA while with RSA an increase of 1158.3%. Viewed from token size (table 4 and figure. 6) HMAC produces 109 bytes token size when compared to ECDSA and RSA increased by 50.5%. The final test results (table 5 and figure. 7) in HMAC token transfers received an average score of 91.2s, while an increase of 90.3% in ECDSA and 272.5% in RSA.

4. Conclusions

The results of the research in this paper have been explained about the comparison of token-based authentication performance using JWT with several algorithms. The overall results show that the use of HMAC is superior to the parameters of time generate token, token size and token transfer speed.

However, there has been no discussion and testing when viewed from other parameters such as testing the security of tokens.

JWT is stateless so the server does not need to store data because data such as scope or authorization and a user can all be stored in JWT. This concept is very suitable if it is applied to a centralized authentication system that uses single sign-on. For future work, it needs to be tested regarding the mechanism for storing tokens on local storage (cookies), whether it is safe and how to save the correct token.

References

[1] Fielding R T and Taylor R N 2002 Principled design of the modern Web architecture *ACM Transactions on Internet Technology (TOIT)* **2**(2) 115–150

[2] Shofa R N, Aradea and Kurnia B B 2013 Penerapan Service Oriented Architecture (SOA) Dalam Pembangunan Web Based Learning, *Jurnal Penelitian SITROTIKA* 9(2) 221–19

[3] Rathod D M 2017 Performance evaluation of restful web services and soap / wsdl web services, *International Journal of Advanced Research in Computer Science* **8**(7) 415–420

[4] Memeti A, Imeri F and Cico B 2017 REST vs. SOAP: Choosing the best web service while developing in-house web applications, *Journal of Natural Sciences and Mathematics of UT* **2**(3) 63–8

[5] Fielding R T 2000 *Architectural Styles and the Design of Network-based Software Architectures* Dissertation (Irvine: University of California) 1–180

[6] Fielding R and Reschke J 2014 Hypertext Transfer Protocol (HTTP/1.1): Semantics and Content, *Internet Engineering Task Force (IETF)* (https://tools.ietf.org/html/rfc7231)

[7] Kumari V 2015 Web Services Protocol: SOAP vs REST *IJARCET* **4**(5)

[8] Kanmani K and Smitha P S 2013 Survey on Restful Web Services Using Open Authorization, *IOSR-JCE* **15**(4) 53–6

[9] Hussain M I and Dilber N 2014 Restful web services security by using ASP.NET web API MVC based *JISRC* **12**(1) 4–10

[10] Sahoo P, Janghel N K and Samanta D 2017 Securing WEB API Based on Token Authentication, *IJAECE* **4**(2) 1–4

[11] Huang X-W, Hsieh C-Y, Wu C H and Cheng Y C 2015 *NBIS '15 Proc. 2015 18th Int. Conf. Network-Based Information Systems* (USA: IEEE Computer Society) 601–6

[12] Li X 2013 The Design of Digital Campus Unified Identity Authentication System Based on Web Services *Applied Mechanics and Materials* **427–429** 2301–4

[13] Indu I, Rubesh-Anand M R and Bhaskar V 2017 Encrypted Token based Authentication with Adapted SAML Technology for Cloud Web Services *Journal of Network and Computer Applications* **99**(C) 131–45

[14] Pavlovski C, Warwar C, Paskin B and Chan G 2015 Unified Framework for Multifactor Authentication, *22nd International Conference on Telecommunications (ICT 2015)* 209–13

[15] Balaj Y 2017 Token-Based vs Session-Based Authentication: A survey 1–5

[16] Rahmatulloh H, Sulastri and Nugroho R 2018 Keamanan RESTful Web Service Menggunakan JSON Web Token (JWT) HMAC SHA-512, *Jurnal Nasional Teknik Elektro dan Teknologi Informasi (JNTETI)* **7**(2) 131–7

[17] Bhawiyuga A, Data M and Warda A 2017 Architectural Design of Token based Authentication of MQTT Protocol in Constrained IoT Device, *2017 11th International Conference on Telecommunication Systems Services and Applications (TSSA)*, 1–4

[18] Jones M, Sakimura N and Bradley J 2015 *RFC 7519: JSON Web Token (JWT)* (https://www.heise.de/netze/rfc/rfcs/rfc7519.shtml)

[19] Zheng K and Jiang W 2014 A Token Authentication Solution for Hadoop Based on Kerberos Pre-Authentication, *2014 International Conference on Data Science and Advanced Analytics (DSAA)* 354–60

[20] Jones M 2015 JSON Web Algorithms (JWA), Internet Engineering Task Force (IETF)

[21] Ethelbert O, Moghaddam F F , Wieder P and Yahyapour R A 2017 JSON Token-Based Authentication and Access Management Schema for Cloud SaaS Applications, in *2017 IEEE 5th International Conference on Future Internet of Things and Cloud* (Cornel University Library) p 6
[22] Mestre P, Madureira R, Melo-Pinto P and Serodio C 2017 Securing RESTful Web Services using Multiple JSON Web Tokens, in *Proc. World Congress on Engineering 2017* 418–23

Power analyzer based arduino-uno validation using Kyoritsu KEW 6315 and Hioki 328-20

Sutisna, I Usrah, N Hiron and A Andang

Electrical Departement, Universitas Siliwangi. Tasikmalaya, Indonesia 46115

Email: sutisna@unsil.ac.id; ifkar_usrah@yahoo.com; hiron@unsil.ac.id; andhangs@unsil.ac.id

Abstract. Online quality analysis technique of energy consumption is getting easier. It is because the sensor module and microprocessor are getting more simple, complex and possess high accuracy. This study offers an alternative to online electrical energy consumption monitoring technique by using Arduino. It is interesting to know the accuracy of the work of the Arduino-based energy quality monitoring system with high accuration measurments tool Kyoritsu power analyzer KEW 6315 and Hioki 328-20. The method of this research is the design of 1 phase electrical energy monitoring system by using current sensor type SCT-013 and voltage sensor in the form of transformator. EmonLib Library was used for processing voltage and current data in Arduino. The validation of measurement results was done by comparing the results of Arduino with Kyoritsu KEW 6315 and Hioki 328-20. The measurement reporting technique was displayed online based on the website. The results provide a clear picture of the measurement performance of the developed system, that the system has a quality of performance close to Kyoritsu type kew 6315 that against Hioki 328-20 with an average error below 5% is acceptable. The results of the measurement of electrical energy consumption had been successfully displayed in the form of a website-based energy report.

1. Introduction

The high demand for electrical equipment has caused the increasing use of non-linear loads. This increase of non-linear loads become one of the problems in electrical power quality [1]. Non-linear loads such as televisions, computer units, laptops, can generate high harmonics [2]. Poor power quality leads to high losses, sometimes leading to a voltage decrease in terms of electrical power users. Electrical equipment may be damaged due to voltage drop in the supply line [3].

Internet-based electrical power quality monitoring system based on internet technology becomes important because this system is not only accurate and fast but also can provide early warning of the decrease of electric power quality. The integration of internet technologies and the online monitoring of electric power quality makes it easy for human beings to make energy management decisions so that energy users can perform the routine behavioural evaluation of energy usage quickly and easily [4]. Therefore, online energy monitoring has the effect of changing people's behaviour in using electrical energy, to be more prudent in making decisions [5].

2. Methods

The purpose of this research is testing or validating the accuracy of energy measuring devices made with Arduino involving the voltage sensors transformer-based and current sensors SCT013 then comparing the quality of the measurement with a measuring device manufacturing Kyoritsu KEW 6315 and Hioki 328-20.

2.1. Current sensor YHDC SCT-013

YHDC SCT-013 is a non-invasive sensor used as a measure of energy consumption [6]. The working principle of YHDC SCT-013 is by sensing (induced) induction of electric current and then dividing the current value to the number of coils on the sensor [7]. YHDC SCT-013 is capable of reading current from 0 to 13A consisting of 18000-turn coils with Non-linearity ± 3% [8].

The YHDC SCT-013 current transformer is suitable to be used with the Open Energy emonTx [7]. Power quality variables, such as power factor and VAR power, can be determined from the phase change of the electric current. In this case, YHDC SCT-013 has an important role in dealing with such a task.

2.2. Electric energy consumption

Electric energy consumption consists of three primary variables as follows:

$$P = V_t \text{ x I x } \cos\theta \text{ (Watt)} \tag{1}$$

$$S = V_t \text{ x I (VA)} \tag{2}$$

$$Q = V_t \text{ x I x } \sin\theta \text{ (VAR)} \tag{3}$$

P is the actual power required by the load in (Watt). S is the real electrical power supplied by the primary generator to the consumer in Volt-Ampere (VA). Q is the power produced by the load which is inductive in Volt-Ampere Reactive (VAR). The active energy consumption of the load can be obtained from equation (4).

$$Wa = P \text{ x t (Watt} - \text{Hour)} \tag{4}$$

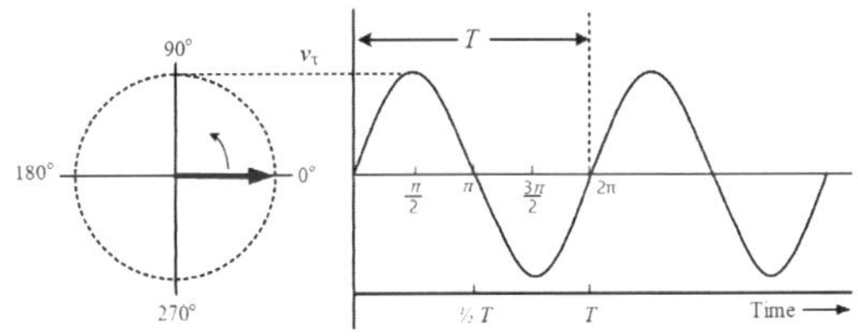

Figure 1. Sinusiodal of power characteristics.

Figure 1 illustrates the transformation of the sinusoidal waveform voltage (Vm) to the function of time into the form of angular frequency (ω). V (t) = Vm sin ωt. The function is repeated every 2π radian. Assumed that Vm sin ωt is a function of t and the period is T, the period can be expressed in degrees' form in (8), and θ is (ωt + θ).

$$f = \frac{1}{T} \tag{5}$$

$$\omega = 2\pi f \tag{6}$$

$$V_{(t)} = V_m \sin(\omega t + \theta) \tag{7}$$

There are some laboratory-scale power quality measuring instruments researches such as Kyoritsu KEW 6315 and Hioki 328-20. Kyoritsu KEW 6315 is a power quality measurement analyzer with measurement accuracy ± 0.2% rdg ± 0.2% f.s. (sine wave, 40 - 70Hz) [9], while Hioki 328-20 has Basic accuracy: ± 1.0% rdg. ± 5 dgt [10]. Both of these tools are reliable measuring instruments manufactured in Japan.

3. Related work

In this paper, we offer a conventional voltage sensing technique by using linear transformers as conducted by [11-14]. Meanwhile, the sensors utilized was YHDC SCT-013 as already used in [14] and

IOP Conf. Series: Materials Science and Engineering **550** (2019) 012024 doi:10.1088/1757-899X/550/1/012024

[15]. Arduino Uno microcontroller and internet shield were used as data processing and communication interfaces between Arduino-Uno [16-17]. The communication network from the system to the server performed by using the 3G protocol.

EmonLib Library, which was downloaded from openenergymonitor.org, was used as a method of power calculation algorithms in Arduino as it is believed to be more straightforward and easier [13, 18-21]. Some studies have also introduced voltage sensing techniques by using low-voltage sensors that can integrate directly with microcontrollers such as ZMPT101B [22-26]. There is also research introducing voltage and current sensing techniques simultaneously in one tool for example with PZEM 004 module which is more practical than ZMPT101B.

This research aims to conduct a system testing on load in 2 conditional modes. The first condition was an offline mode in which the testing was done on the power panel for 1 hour. Then, the measurement results were observed and compared with Kyoritsu KEW 6315 and Hioki 328-20. The second was an online mode in which the system was connected to the internet for 30 days. The test was conducted on residential buildings with a total power load of 2200VA.

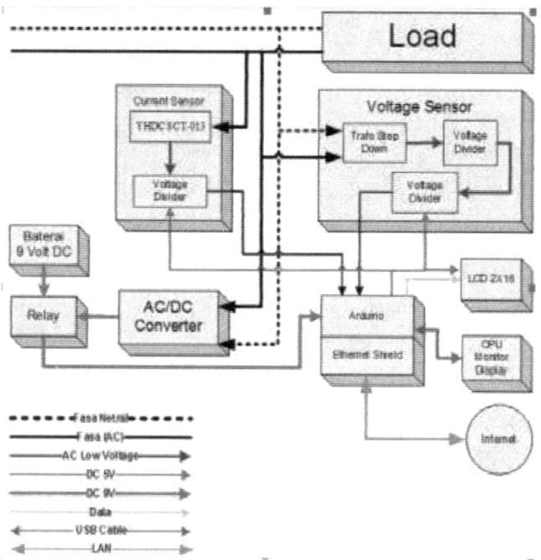

Figure 2. System architecture proposed.

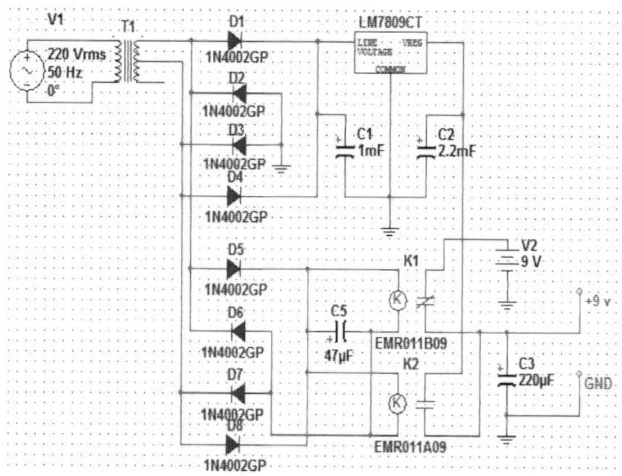

Figure 3. Circuit Scheme of backup power proposed.

Figure 2 is a system architecture, a complete architecture for energy monitoring with mobile web technology that has been discussed by Galera (2018) [27]. In this study, we have designed the architecture as shown in figure 2 where the dotted line is a neutral line, the black line is a single-phase AC 220V line voltage and the purple line is the low-voltage AC line. The red line is the DC 5V line voltage supplied from the main source and 9V supplied from backup power supply system. The yellow line is the digital data line, the blue line is the USB data path, the green line is the internet communication path using Local Area Network (LAN).

Figure 3 is a backup power supply scheme in which the input is obtained from a 220V main power and a 9VDC battery. If the main power supply is off, the power supply is switched from the battery without pausing.

IOP Conf. Series: Materials Science and Engineering **550** (2019) 012024 doi:10.1088/1757-899X/550/1/012024

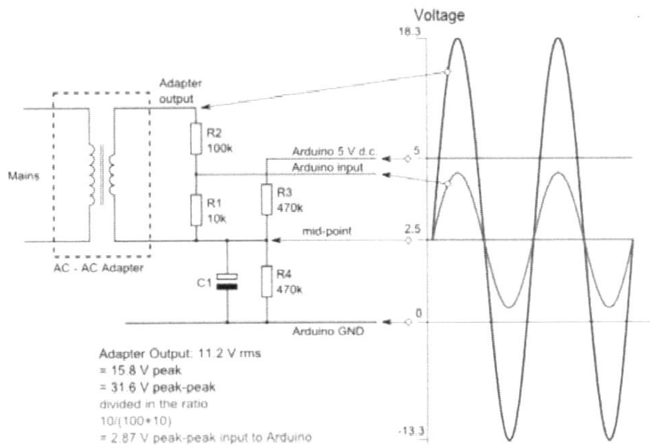

Figure 4. Line diagram of the voltage sensor using a step-down CT transformer [14, 15, 19].

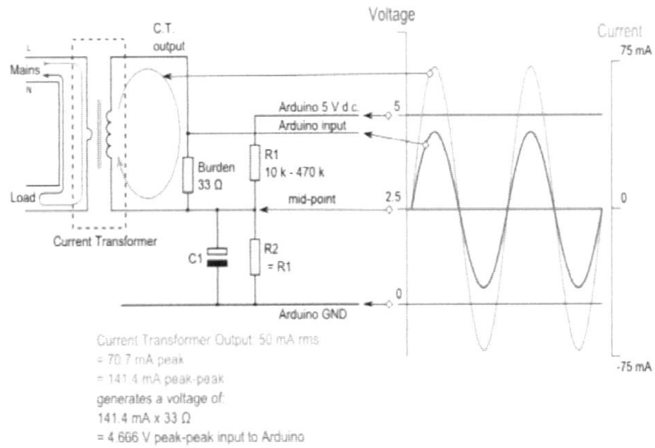

Figure 5. Line diagram of the current sensor using YHDC SCT-013 [14-15].

Figure 4 is the line diagram of the voltage sensor using a step-down CT transformer obtained from openeenrgymonitoring.org. The 220VAC input voltage was reduced to a voltage of 18.3VAC. The voltage value of AC as input to Arduino was determined by an equation (1) with the range between 0V and 5V. Then, a 100 μF capacitor serves to decrease the ripple voltage. The midpoint is the median value of the voltage dynamics of the target which will be processed by the Arduino in Pin 2. The voltage divider equation to obtain the Arduino input and midpoint values was done by using equation (1) with the current value connected to Pin 1 Arduino. The same is true for electric current sensors, where the YHDC SCT-013 sensor was the current transformer (figure 5) using the EmonLib Library function that is calcIrms to find the RMS current value [18].

$$U_{out} = \frac{R_4}{R_3 + 4} U_p \tag{8}$$

4. Results and discussion

The result of the tests that have been done is to prove that the voltage sensor with transformer and current sensor that has been carried by previous researcher [11-14].

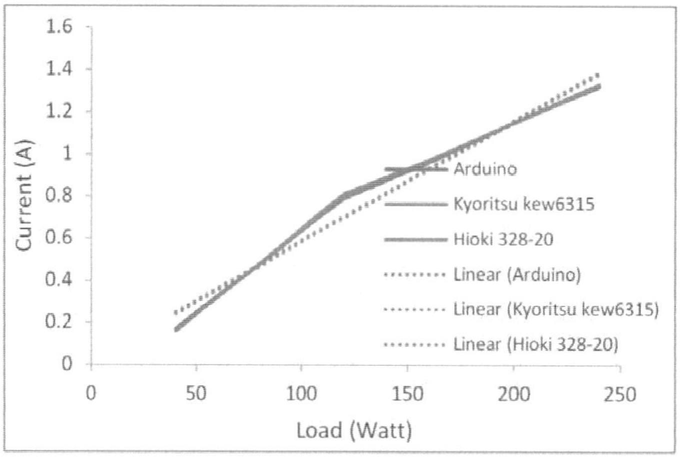

Figure 6. Testing of electrical current measurement system on linear loads.

The result of measurement test with linear loads in of heating elements was started from 40W to 240W with the power factor was assumed as 1 and then was compared with Kyoritsu KEW 6315 and Hioki 328-20 (figure 6). The Arduino-based measurement system showed results that resemble Kyoritsu KEW 6315 and Hioki 328-20. A nonlinearity occurred likely because there was a change in load resistance as a result of the heat that was held for a long enough period of time.

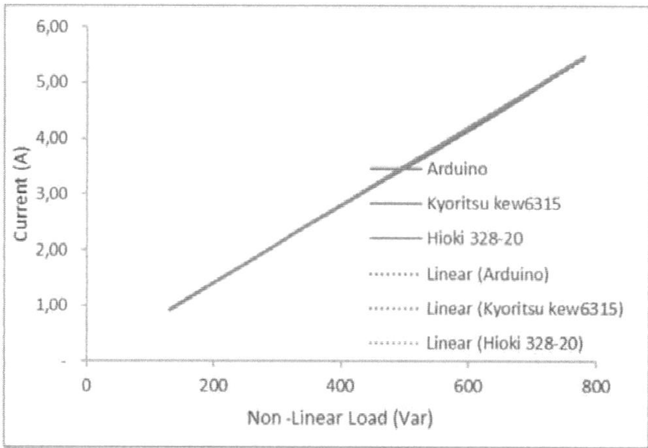

Figure 7. Testing of electrical current measurement system on Non-Linear Loads.

In the non-linear load test, some laptops from 130 Var to 780 Var with an average power factor of 0.65 were tested. The Arduino-based measurement system showed the same results as the measurement of electrical energy with Kyoritsu KEW 6315 and Hioki 328-20 (figure 7).

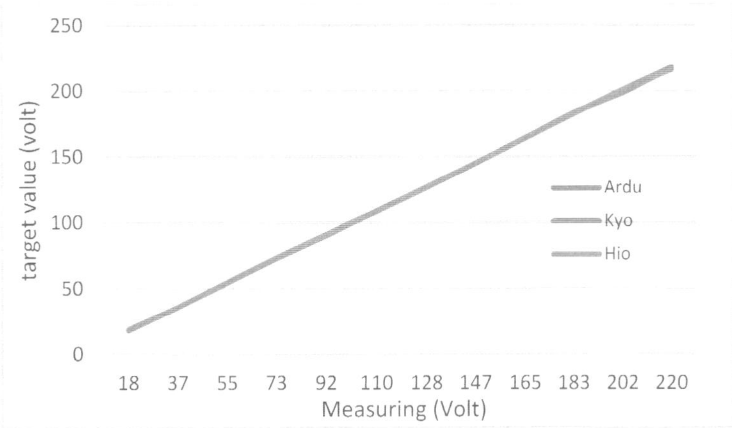

Figure 8. Testing of the accuracy of voltage measurement

The accuration result of measurement error of the system against Kyoritsu KEW 6315 was 0.5%, while against Hioki 328-20 was 0.26%, see figure 8.

4.1. Offline system test

In offline testing on electrical power panel (figure 9) where the power measurement system with Arduino worked properly. Communication between current sensors and voltage sensors against Arduino could run normally. Table. 1 demonstrates the measurement error value of the system after it was compared with the other measuring instruments. In terms of voltage and current variables, the error value of the comparator was below 1%, while the error value of power factor and active power variables was below 2%.

Table. 1. Comparison of Arduino-based error measurement values against the other measuring instruments.

Parameter	Ardu vs Kyoritsu (%)	Ardu vs Hioki (%)
Voltage (V)	0.24	0.17
Current (A)	0.37	-3.44
Active Power (W)	1.89	-2.19
Power factor	1.33	-2.69

Figure 9(a). Testing wiring procedure of the measurements on the power panel.

Figure 9(b). The comparison of measurement results.

IOP Conf. Series: Materials Science and Engineering **550** (2019) 012024 doi:10.1088/1757-899X/550/1/012024

Based on figure 11, the measurement results obtained between the system have been developed against the Kyoritsu kew 6315 and Hioki 328-20 as validator. Its found that they has the same characteristics in measurement. The average error of comparison between the Arduino and Kyoritsu for voltage measurements was 0.24%, while to the Hioki was 0.17% (figure 12). The average error of measurement of system based arduino against the Kyoritsu was 0.37%, while the Hioki was -3.44% (figure 12). The average error of comparison between the Arduino and Kyoritsu for power factor measurements was 1.33%, while against Hioki is -2.69% (figure 13). The average error of comparison between the Arduino and Kyoritsu for active power measurements was 1.89%, while the Hioki is -2.19% (figure 14). These results provide a clear picture of the measurement performance of the developed system, that the system has a quality of performance close to Kyoritsu type kew 6315 that against Hioki 328-20.

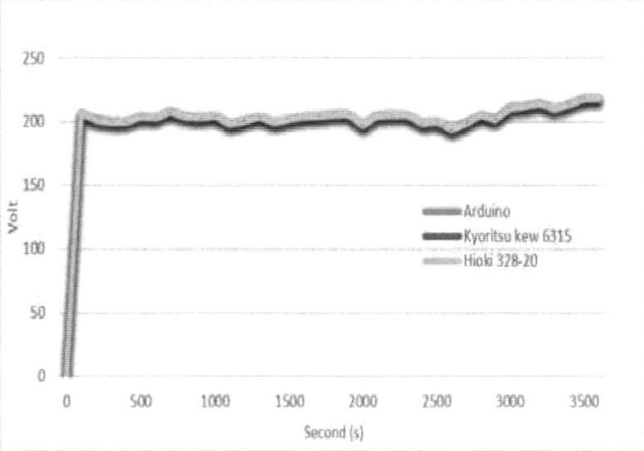

Figure 11. Voltage measurment value in comparison.

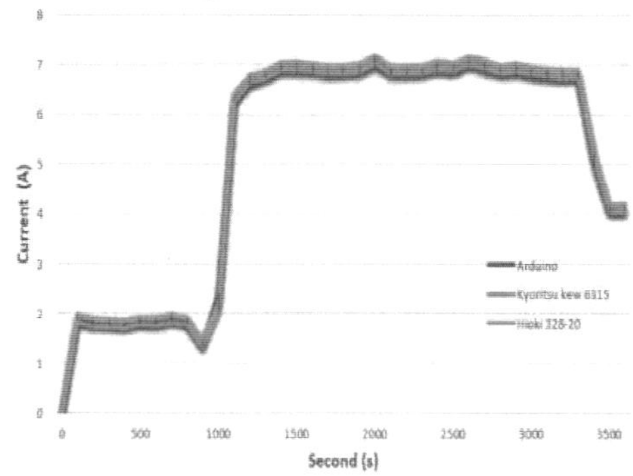

Figure 12. Current measurment capture in comparison.

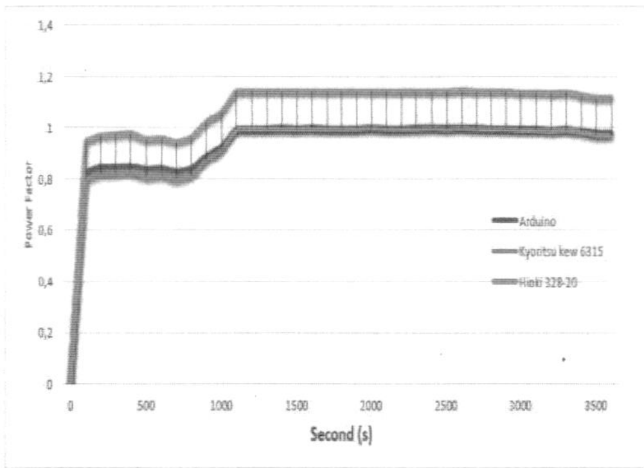

Figure 13. The power factor value measurements in comparison.

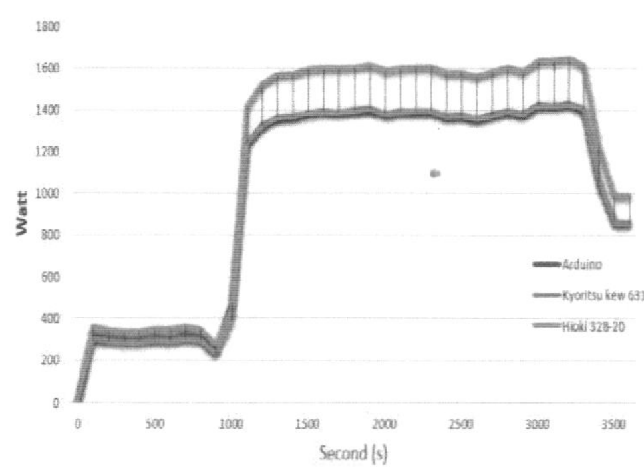

Figure 14. Active power measurment capture in comparison.

4.2. Online system test

The online system testing strongly depended on the reliability of the Internet network. Using a 3G communication network as a communication medium between the LAN Shield in Arduino and the server service had caused a delay in the data transfer from the sensors to the user display. The greater the delay in the data communications from the sensor to the server, the greater the measurement error occurred. This condition resulted in an inaccurate energy report although based on the offline test the measurement system showed accurate results. These results provide a clear picture of the measurement performance of the developed system, that the system has a quality of performance close to Kyoritsu type kew 6315 that against Hioki 328-20. The measurement system of electrical energy consumption that depends on internet service should be supported by reliable network quality. In other words, if the

IOP Conf. Series: Materials Science and Engineering **550** (2019) 012024 doi:10.1088/1757-899X/550/1/012024

electrical energy measurement system is implemented in the rural area, where the quality of the internet network is not guaranteed, a system failure might occur as if the whole system fails.

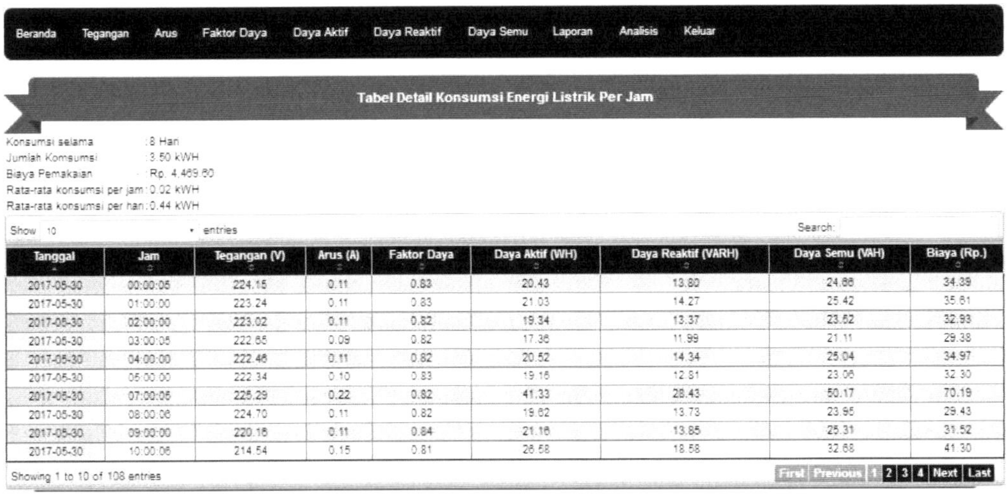

Tanggal	Jam	Tegangan (V)	Arus (A)	Faktor Daya	Daya Aktif (WH)	Daya Reaktif (VARH)	Daya Semu (VAH)	Biaya (Rp.)
2017-05-30	00:00:05	224.15	0.11	0.83	20.43	13.80	24.66	34.39
2017-05-30	01:00:00	223.24	0.11	0.83	21.03	14.27	25.42	35.61
2017-05-30	02:00:00	223.02	0.11	0.82	19.34	13.37	23.52	32.93
2017-05-30	03:00:05	222.65	0.09	0.82	17.36	11.99	21.11	29.38
2017-05-30	04:00:00	222.46	0.11	0.82	20.52	14.34	25.04	34.97
2017-05-30	05:00:00	222.34	0.10	0.83	19.16	12.81	23.06	32.30
2017-05-30	07:00:06	225.29	0.22	0.82	41.33	28.43	50.17	70.19
2017-05-30	08:00:06	224.70	0.11	0.82	19.62	13.73	23.95	29.43
2017-05-30	09:00:00	220.16	0.11	0.84	21.16	13.85	25.31	31.52
2017-05-30	10:00:06	214.54	0.15	0.81	26.58	18.58	32.68	41.30

Figure 15. Power consumption in daily report dashboard.

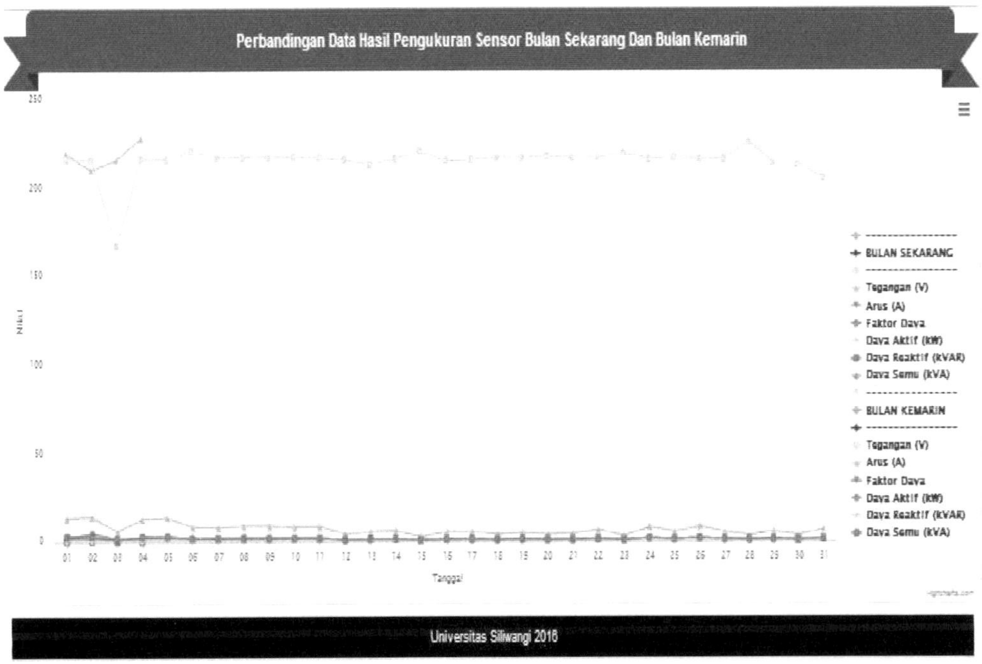

Figure 16. Power consumption in daily report chart.

5. Conclusion

Online energy monitoring system can be built by using Arduino which is cheaper and has good accuracy as well. The use of linear transformer-based voltage sensors depends on the quality of transformers. This study shows that the use of YHDC SCT-013 current sensors and voltage sensors with linear transformers resulted in good performance. It also reveals that communication through internet from sensors to server can be performed by LAN Shield module for Arduino. With an average error below 5% it can be said that an arduino-based energy measuring device that has been built can assume it is suitable for use as a measure of electricity consumption. The test results both on offline and online conditions demonstrated satisfactory results. The communication between the Arduino and the server can run well as long as the internet network is available. Although the test results showed satisfactory results, it is essential to know if the energy monitoring sensors use multi-sensors on a 3 phase power network with a single server.

References

[1] Yeotikar Ms A A and Mohod S B 2016 *IRJET* **3**(11) 581–4
[2] Hiron N, Andang A and Mubarok H 2015 Energy Information System (EIS) As Energy Mix Projections with Trend Analysis Approach for The Scenario Achievement Of The National Energy Policy In 2025, in *The 14 International Conference on QIR*
[3] Johnson D O and Hassan K A 2016 Issues of Power Quality in Electrical Systems *Int J Energy Power Eng* **5** (4) 148–154
[4] Rajurkar A, Shinde O, Shinde V and Waghmode B 2016 *Int J Adv Res Comput Commun Eng* **5**(5) 398–400
[5] Hiron N and Andang A Wireless communication with batching method based on Xbee-PRO S2B module for sensing of wind speed in *Proc - 2016 2nd Intl Conf on Sci Information Tech ICSITech 2016: Information Science for Green Society and Environment* 2017 p 250–3
[6] Pocero L, Amaxilatis D, Mylonas G and Chatzigiannakis I 2017 *HardwareX* **1** 54–67
[7] Wall R 2014 A Report on the properties of the Yhdc Current Transformer and its suitability for use with the OpenEnergy monitor **6** 1–10
[8] YHDC SCT-013-020 DATASHEET - Transformer 2017 [Online] Available: http://www datasheet4u com/pdf/SCT-013-020-pdf/1089415 [Accessed: 15-Jul-2017]
[9] Kyoritsu KEW 6315 | Power Quality Analyzer | PRODUCTS | KYORITSU [Online] Available: http://www kew-ltd co jp/en/products/detail/00100/ [Accessed: 15-Jul-2017]
[10] Hioki AC Current Clamp Meter | Digital Clamp On HiTester 3282 - Hioki 2017 [Online] Available: https://www hioki com/en/products/detail/?product_key=6359 [Accessed: 15-Jul-2017]
[11] Malagond G, Bagewadi G, Pooja L, Ronad B F and Jangamshetti S H 2015 Design and Development of Data Logger for Standalone SPV Systems *Int J Emerg Technol Adv Eng* **5**(5)
[12] Kokatanur A S and Rao S K 2015 *IJRET* **4**(4) 2319–21
[13] Murthy K V, Sriram P, Vignesh V, Suryaprakash C and Ranjankumar N 2017 Wireless Motor Monitoring and Controlling Through Visible Light Communication (Vlc) *SSRG Int J Electr Electron Eng* no March 76–81
[14] Gwil Glyn Trystan Kate Claire and Amy Learn | OpenEnergyMonitor *openenergymonitoring* [Online] Available: https://learn openenergymonitor org/electricity-monitoring/ct-sensors/interface-with-arduino?redirected=true [Accessed: 24-Apr-2018]
[15] M K Vishnu Murthy Sriram P Vignesh V Suryaprakash C and Ranjankumar N WIRELESS MOTOR MONITORING AND CONTROLLING THROUGH VISIBLE LIGHT COMMUNICATION (VLC) in *SSRG International Journal of Electrical and Electronics Engineering (ICRTECITA-2017)* 2017 no March pp 76–81
[16] Hiron N, Andang A and H Setiawan 2016 Batch Processing Method in Machine to Machine Wireless Communication as Smart and Intelligent System *Int J Futur Comput Commun* vol **5**(3) 163–6
[17] Hiron N and Andang A 2016 Wireless communication with batching method based on Xbee-PRO S2B module for sensing of wind speed in *2016 2nd International Conference on Science in Information Technology (ICSITech)* 250–3
[18] Putra L, Michael, Yudishtira and Kanigoro B 2015 Design and Implementation of Web Based Home Electrical Appliance Monitoring Diagnosing and Controlling System *Procedia Comput Sci* **59** Iccsci 34–44
[19] Rishav U and Tiwari M 2015 Automatic Power Factor Correction Using Capacitor Banks *IJIREEICE* **3**(1) 86–9

[20] Chishti M Z S 2018 Non-Intrusive Load Monitoring using Artificial Neural Network *Glob J Eng Technol Rev* **3**(2) 22–9
[21] Nataliani R, Nugroho F X and Mursito E 2014 Pembuatan Perangkat Keras dan Analisis Sub-Metering Konsumsi Energi Listrik *J OTOMASI KONTROL INSTRUMENTASI* **6** (1) 25–35
[22] Abubakar I, Khalid S N, Mustafa M W, Shareef H and Mustapha M 2017 Calibration of ZMPT101B voltage sensor module using polynomial regression for accurate load monitoring *ARPN J Eng Appl Sci* **12** (4)
[23] Saputra W R, Muid A, Rismawan T and J S Komputer 2016 Rancang Bangun Sistem Monitoring Gangguan Pada Gardu Listrik Menggunakan Arduino Dan Website *J Coding* **4**(2) 141–50
[24] David N, Anozie F N, Ebuka F O and Nzenweaku S A 2016 Design of an Arduino Based Wireless **7**(9) 466–9
[25] V S B V M Design Of Wireless Power Meter For Overbilling Prevention Using Arduino
[26] Mnati M, Van den Bossche A and Chisab R 2017 A Smart Voltage and Current Monitoring System for Three Phase Inverters Using an Android Smartphone Application *Sensors* **17**(4) 872
[27] Galera K M E and Llantos O E 2018 Mobile Web Energy Monitoring System Using DFRduino Uno *Procedia Comput Sci* **124** 706–13

Acknowledgments
This research is a part of an external research program and is supported by research funding from KEMENRSITEK-DIKTI at the scheme Applied Product Research for the third year Thanks to LPPM-PMP Siliwangi University for supporting this project.

RFID for nurse activity monitoring in the hospital's nurse call system with Internet of Thing (IoT) concept

F M S Nursuwars and A Rahmatulloh

Informatics Engineering, Universitas Siliwangi, Tasikmalaya, Indonesia

Email: firmansyah@unsil.ac.id; alam@unsil.ac.id

Abstract. The hospital nurse call system is a system found around a hospital bed that allows patients in health care to call nurses or other health care staff remotely about their need for assistance. When the button is pressed, the signal tells the staff at the nurse station via a voice signal, and usually the nurse or nurse assistant responds to the call. Today it has created many nurse call systems that are considered efficient both with the concept of wireless and cable. artificial equipment that has standard facilities and complete facilities. But this tool only emphasizes how patients call nurses, without thinking about how the quality of nursing services responds to each call. The purpose of this study is to develop an RFID-based nurse call system with the internet of things (IoT) concept. The system allowed nurses can respond to every call from the patient without the patient or family the patient comes to the nurse station, also the hospital executive can get a report on the quality of nursing services, in the form of reports on the number of calls, the time is taken by the nurse to respond to each call and report who the nurse responds to each patient's call. Thus, the information can be evaluated how the quality of service in the hospital.

1. Introduction

One of the essential things that patient need is Nurse Call system. It is intended for routine communication between the patient and the nursing staff, nurses immediately attend the patient as soon as they hear the call [1]. It is a must-have tool in the hospital for the patient, especially who are unable to move from their bed and not able to call someone loudly [2]. Nurse call system is needed because of the lack of ratio between the nurse and the number of patients, as well as other duties imposed on the nurse outside of her primary duties. According to the Institute for Healthcare Improvement, Nurses spend only between 20-30% of their time in caring for patients directly, the rest to do another task such as documentation and administration (www.amcomsoftware.com). These conditions resulted in a high workload of nurses that can lead to lower levels of patient satisfaction because nurse care to patients to be less optimal and nurse satisfaction that provides care [3].

The development of nurse call system already a lot in hospitals as an effort to improve nurse services, from wiring devices to wireless devices using xbee and Wi-Fi. Problems arising from existing tools, for nurse callers with cables, each button that is installed must use a cable drawn into the nurse room, so you can imagine the more number of rooms/bed of patients, the more cable that must be in the pairs [4]. Other problems of an existing nurse caller device, both wired and wireless devices, the installed tool only emphasizes how the patient or family of the patient calls the nurse and how the nurse knows which patient is calling, regardless of how the nurse's service quality responds to every call and did not consider how to give more time for nurses to take immediate action against patients because there is no need to document and administer the patient care process. Nurse's quality respond has been discussed in [5], but only discussed the duration of time spent by nurses in handling each patient's call without knowing who the nurse is handling it. Internet of Thing (IoT) generally refers to a network of interconnections between daily devices which are often equipped with the intelligence to support various kinds of control and monitoring. the ubiquity of the Internet will increase by IoT through integrating every device for interaction on network via an embedded system [5]. Nurse call system with Internet Of Thing (IoT) Concept has been developed by Aswin (2011), Sayekti (2013), Majumder (2016), Mahmud (2017) and Mardjoko (2017) [2, 3, 6-8]. The system Sayekti (2013) and Aswin (2011) not used TCP/IP for IoT Concept, it just used Xbee/ZigBee and Bluetooth. The limitation of the system Majumder (2016) the number of clients is limited to 36 clients per server node while the limitation of the system developed by Aswin (2011), Sayekti (2013), Majumder (2016) and Mardjoko (2017) no archive monitoring for system activity's logging. Another connection used for nurse caller communication uses rs-485 [4, 9]. While other studies emphasize how patients call nurses, such as make use of the patient's eye blink [10] and designing the human interface for patients [1].

Existing research except for Mahmud (2017) only emphasizes how patients call nurses and how to connect a client node to a server node, regardless of how the nurse's service quality looks like how long and who responds to every call[6]. But the limitation of Mahmud the system only monitors the duration of how long the response of the nurse without knowing who the nurse is responding to. By utilizing RFID in this paper the system can find out who the nurse is responding to.

With the Internet of Thing (IoT) technology and RFID is expected to solve installation problems on the nurse caller with the concept of cable. In addition, by using RFID to reset the patient's call, can provide information on how much time is spent from the patient start calling the nurse until the nurse takes direct action on the patient and who the nurse is doing the action. The purpose of this study is to develop an RFID-based nurse call system with the internet of things (IoT) concept.

2. Method

2.1. Block diagram of the system

Block diagram of the nurse call system is shown in figure 1. There are several types of nodes in the system. The server node is a Computer web server, the Web server is a software that provides data-based services using HTTP or HTTPS protocol from the client using a web browser application for data requests and the server will send data in the form of web pages and generally in the form of HTML documents [11]. It’s stored in the nurse station, any data that the web server receives will be uploaded to the database server. Data received web server can be seen on information systems that can be accessed from the PC Admin. In the implementation web server, db server and PC admin can be one Computer.

The client node is bed node and toilet node in each ward, client nodes are end node. This node is used by each patient. Each client node has a 2.4 GHz wifi module connected to the node server directly using the wifi access point. The limited range and limitations of wifi access point access capability can be helped by increasing the number of access points. Every 1 server node can handle simultaneously 253 client if using IP address class C, 65,533 client if using IP addresses class B and 16,777,213 client if using IP address class A. Network diagram of the nurse call system is show in figure 2. Data from the client node from both the bed node and the toilet node are the same, but the server will distinguish calls from the bed or toilet, whenever a patient presses a button from the toilet node, it indicates an emergency alert. And when the alert is generated from bed node it indicates a normal call for assistance or medication etc. the data that the client node sends to the server node is distinguished from the IP address of each client node. The client node has no time data, to determine when the patient calls and when the respond nurse is determined from the data receiving time by the server node. Each call made by the patient must be reset by tapping the nurse's RFID card to the client node, this process to determine how long and which nurse is responding to. Log data system activity can be seen in the PC admin information system.

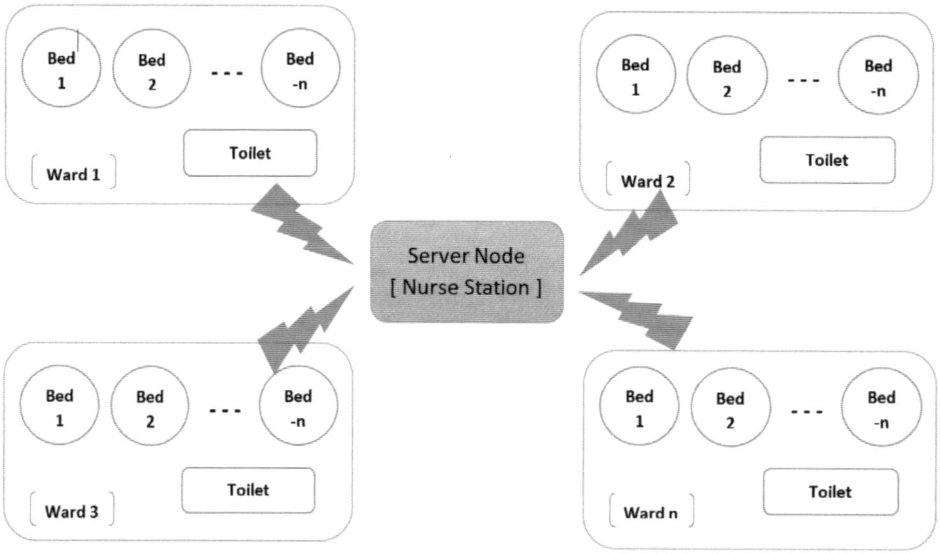

Figure 1. Block diagram of the complete system.

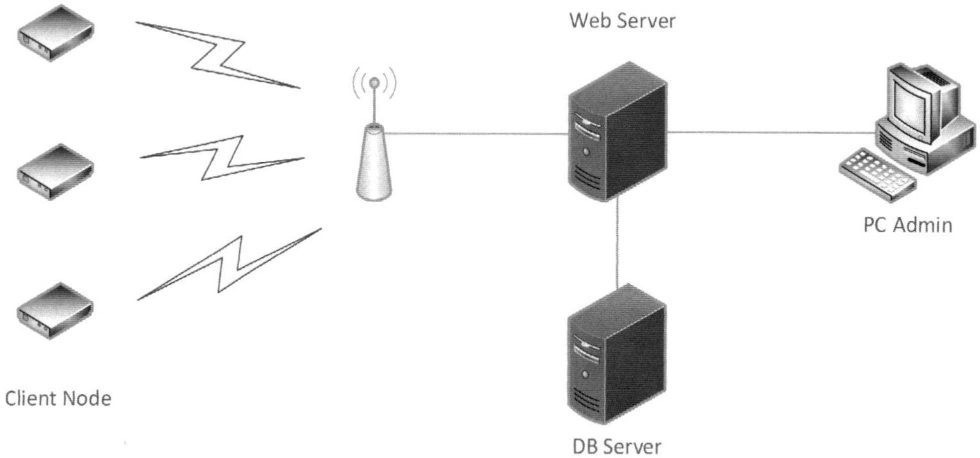

Figure 2. Network implementation.

2.2. Flowchart of the system

Flow diagram of the system is shown in figure 3. The system starts from patient push the button in client node. Client node will send call data status to the server node (web server). The data received by the node server will be identified from the client node where it came from. The identification is specified from the IP address of the sending node. The identification results determine where the position and type of client node. A call from the bed node will be considered a normal call and from the toilet, node will be considered an emergency call. Once the data is identified, the system will alert the nurse, there is a call from the patient, by displaying the space number, the bed/toilet number and the type of call on the monitor screen and the alarm sound on the pc speakers.

After a warning alert of the patient's call, the nurse approaches the patient's room, upon arrival, the nurse taps his RFID card to stop the call alert. Patient activity pressed call button and nurse stuck RFID card in stored on server. Log data can be viewed and analyzed by the executive of the hospital.

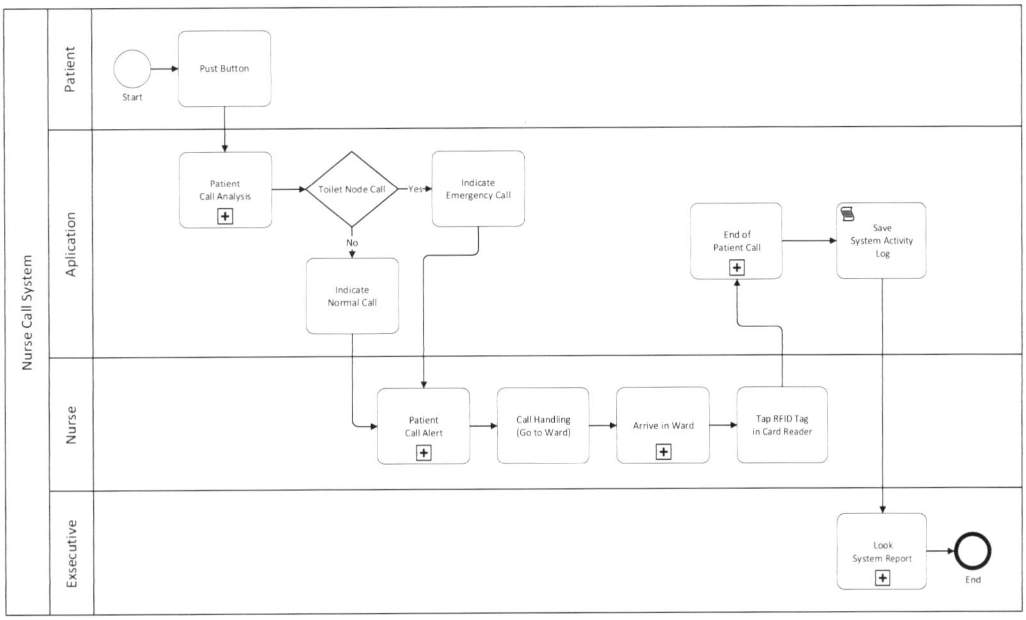

Figure 3. Nurse call system BPMN.

3. Results and discussion

3.1. Circuit analysis

Client node consists of several components. Among them, wemos D1 mini board is the heart of the system. Wemos D1 mini board consists of a ESP-8266EX microcontroller, ESP-8266EX is a microcontroller runs at 80MHz/160MHz clock speed. Wemos D1 mini board has 1 analog input, 1 micro USB Connector and 11 digital input/output pins that have interrupt, PWM, i2c and one wire supported in all pins except D0. It also compatible with Arduino, NodeMCU and MicroPython.

Figure 4 show each child node consists of Wemos D1 mini board, one RFID reader RS-522.1 Green LED for Network Status indicator, 1 RED LED for patient call status indicator, 1 momentary switch for patient button and relay module to communicate with room light signal node. Except for toilet node it also has buzzer module to give an audio signal for the patient family if something happens with the patient in toilet, it is shown in figure 5.

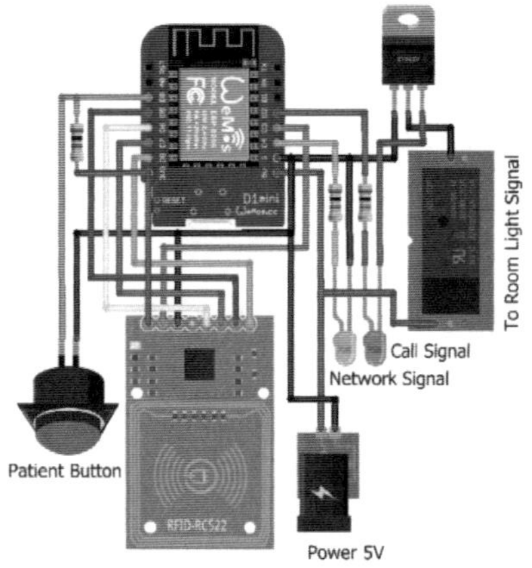

Figure 4. Circuit diagram of client node, bed node type.

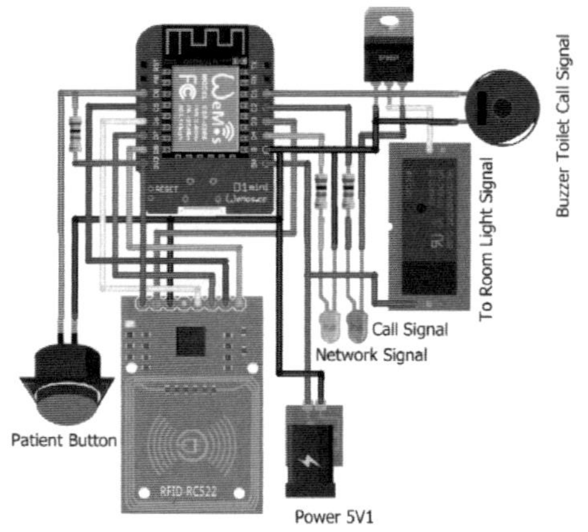

Figure 5. Circuit diagram of client node, toilet node type, it has a piezzo buzzer for sound signal.

3.2. Hardware implementation

The hardware created is a nurse caller tool on the client node. As in figure 6, the hardware is mounted a momentary button that can be in the patient's press if it requires nursing care. To reset the call, the nurse must tap the RFID tag using a 13.56Mhz RFID card with an RS-522 reader connected to the SPI method with the MCU. The MCU used is a Mini D1 wemos. The designed hardware could have added a Shield Battery module so that when the power fails, the system can still function or use a centralized UPS in the nurse's room as a power backup

Figure 6. Implementation of client node.

3.3. Output analysis

The data stored in the db server is what time the patient calls the nurse and what time and who the nurse responds to the call, figure 7 describes the log data displayed in table form with the content from which the patient is calling (Client Node), the time data the call is made by the patient (Call Time), the time data the nurse's response was done (Response Time) and who the nurse who handles the call (Nurse).

Data stored in the database can be used for executive information systems (EIS) that can display information about how often the patient calls, how long and how much time is needed by the nurse to respond to each patient's call, and who the nurse is responds to the call. So it is known how the performance of each nurse in doing treatment to the patient.

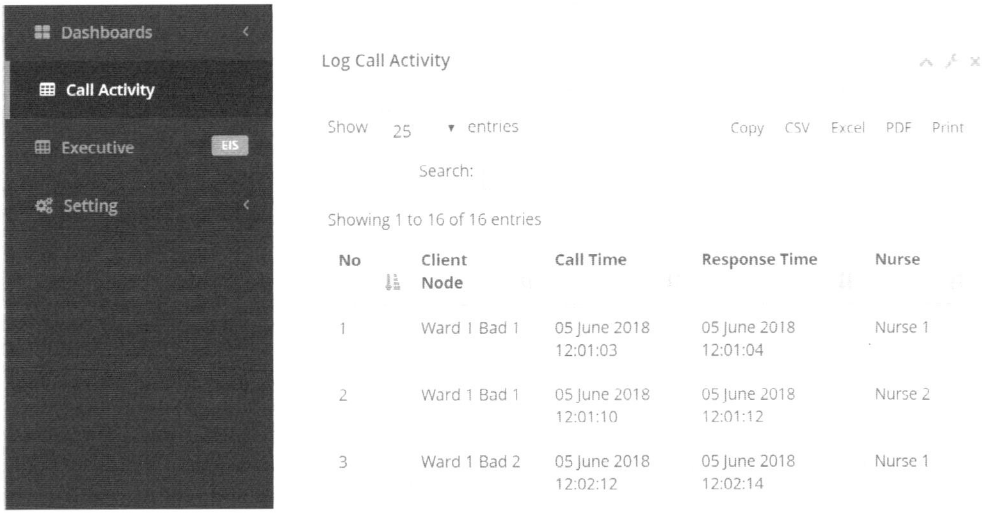

Figure 7. Implementation of server node.

Table 1 presents data on the difference in facilities from the nurse call system with the concept of Internet of things (IoT) which uses an IP address to transfer data from client to server

Table 1. Difference in facilities.

No	Author	Concept	Client	Log Duration	Nurse Log
1	M. A. Majumder [2]	IoT, IP	36 Client	No	No
2	M. S. Mahmud [6]	IoT, IP	Based IP Class	Yes	No
3	This paper	IoT, IP	Based IP Class	Yes	Yes

4. Conclusion

The nurse call system is an important tool for every hospital. The nurse call system is designed not only to have a facility to call the nurse but to record when the patient calls the nurse and who the nurse is responding to each call. og data stored on a node server in the form of MYSQL database application can be used to create an executive information system that serves to analyze how the quality of nurse respond to each patient call. Utilization of IoT concept can answer the difficulty of installation of the system if using cable and can add a significant number of client node connected. The system created designed client node can directly store data to the server without using additional hardware, so the data store performance is not interrupted by the data storage hardware capability. There are many improvements that can be made in this system such as adding human interface interface in the form of a menu that lists the types of calls according to the needs of the patient so that nurse services to patients will be faster because it comes with the preparation according to the needs of patients, add blue code warning, warning if node client lose connection etc.

References

[1] Mojdeh S, Nadi M M, Gordiz K, Khalili G and Sadri A 2013 Especial nurse call system for ICU patient *J. Biomed. Sci. Eng.* **2013** 843–6

[2] Majumder M A 2016 Low Cost Wireless Nurse Call System with Webserver & Pager *Glob. J. Comput. Sci. Technol.* **16**(1)

[3] Sayekti I 2013 Bel Pemanggil Perawat Berbasis Wireless Menggunakan Xbee *J. Tek. Elektro*

Terap. 174
[4] Ardiyanto F Alat pemanggil perawat dengan antarmuka rs-485 1–7
[5] Xia F, Yang L T, Wang L and Vinel A 2012 Internet of Things *Int. J. Commun. Syst.* **25** 1101–2
[6] Mahmud M S, Majumder M A and Tushar A K 2017 Real-Time Feedback-Centric Nurse Calling System with Archive Monitoring using Raspberry Pi *Int. Conf. Networking, Syst. Secur.* **4**
[7] Mardjoko P B and Wulandari M Bel Pemanggil Perawat Rumah Sakit Berbasis Wireless Menggunakan Android, *JUTEI* **1** 31–40
[8] Aswin S, Gopalakrishnan N, Jeyender S, Prasanna R G and Kumar S P 2011 Design development and implementation of wireless nurse call station *Annu. IEEE India Conf*
[9] Sharma C 2015 design development and implementation of wired nurse calling system *Int. Conf. Green Comput. Internet Things* 10–14
[10] Adejoke E J, Samuel I T and Omolewa O T 2017 Eye blink Enabled Nurse Call and Security Alert System for Hospital Patients, **8**(10) 229–34
[11] Rahmatulloh A and Nursuwars F M S 2017 Implementasi Load Balancing, Web Server Menggunakan Haproxy Dan Sinkronisasi File Pada Sistem Informasi Akademik Universitas Siliwangi, *J. Teknol. dan Sist. Inf.* **3**(2) 241–8

Self-adaptation modeling for service evolution on the Internet of Things (IoT)

Aradea[1], I Supriana[2], K Surendro[2], and H Mubarok[1]

[1]Department of Informatics, Faculty of Engineering, Siliwangi University
[2]School of Electrical Engineering and Informatics, Bandung Institute of Technology

Email: aradea@unsil.ac.id; iping@informatika.org; endro@informatika.org; husni.mubarok@unsil.ac.id

Abstract. One of the most important concepts of the internet of things is related to software services, where the system must be able to provide real-time data collected from various environments. So that software services represent a variety of physical and virtual real-world objects that can grow very fast. This condition is related to the ability of self-adaptation software services. However, the existing model of self-adaptation is generally not paying attention to the requirements for service evolution. The objective of this paper is to introduce self-adaptation modeling techniques which consist of, first, domain modeling of the internet of things to represent real-world context; second, developing an inference engine for context inference. As a form of evaluation, this model is applied to the patient monitoring system that will relate to the concept of self-adaptation of various systems, devices, actors, and environment. The case study results show the system's ability to anticipate changes in context and growth needs.

1. Introduction

A software service is required to have smart solutions, such as managing various resources, data and others. The internet of things (IoT) is a concept that can answer the challenge. Currently there are several definitions of IoT, but basically aims to expand the benefits of internet connectivity that can integrate real-world objects as physical and virtual representations on an ongoing basis. The IoT concept offers many advantages, such as efficiency and effectiveness of application, access speed, service accuracy, and more. The various advantages offered will not have a significant impact if only temporary. This is related to how far the concept offered can meet the dynamism and diversity of its environment.

Self-adaptation model has been present and can be a solution to issues related to complexity of the system, such as demands for autonomy, flexibility, scalability, reliability, speed, etc. [1]. Self-adaptation is a system that automatically can take appropriate action based on its knowledge of what is happening in the system [2], the system can modify its behavior in response to changes in the system itself or its environmental changes [3]. However, the study of self-adaptation related to the requirements of system evolution still requires further study [4], because a number of current researches are mostly aimed at requirements specification, monitoring, and verification, while studies on evolution for self-adaptation systems still lack of research [5-6].

The self-adaptation system has gained much attention from researchers, such as Morandini [7] and Dalpiaz [8] developing an adaptive system by taking the case of a patient monitoring system. The proposed model highlights the importance of modeling variability and adaptability at run-time. With the same case study, we modified it according to the proposed model, and added system evolution capabilities not yet covered in both works. In addition, researchers in this field have exploited many autonomic computing concepts through various approaches, such as Gabriel [9] focusing on quality of service specifications through context-awareness models, Dhaminda [10] adopting a natural language approach to goal pattern model, Arciani [11] utilizing abstract state machines to achieve context awareness and self-awareness, Knauss [12] uses machine learning approaches to introduce contextual requirements, as well as Paz [13] through formal methods of proposing dynamic adaptation mechanism planning. The concept adopted by these researchers was adopted in the proposed model. However, we add a domain problem representation through goal modeling and evolutionary functions, so that the model relates to the requirements engineering area in dealing with the variation of operational changes and allows it to accommodate its growth. Based on this fact, this paper proposes a self-adaptation modeling technique in the IoT concept. Section 2 of this paper describes the proposed model, section 3 application of the proposed model into a case study, section 4 discusses related work, and section 5 contains conclusions and future work.

2. Self-adaptation model

All elements and activities in the system environment require modeling and control that can represent their behavior in various conditions in the real world [14]. Self-adaptation model is proposed to realize a software service that is capable of reflecting every element within its environment and has the adaptability to its changes. The model consists of (a) domain model, is a basic function for the IoT context and application logic, (b) control model, is an inference engine to manage domain model through adaptation logic.

The approach used in this model is an extension of previous research conducted by Supriana (2016) and Aradea (2017) [15] [16]. Domain models were developed by adopting a goal-based approach [17] through model extensions [4], mapped to the concept of autonomic computing [18]. The model generally has the main activity, ie monitors, analyzers, planners, executors, and knowledge (MAPE-K). Adaptation patterns are controlled using an autonomic manager. The autonomic manager contains the rule engine in the form of event-condition-action (ECA) rules, and in our version, the concept is extended by using the rule editor model, where the rule changes can be done directly on the knowledge base.

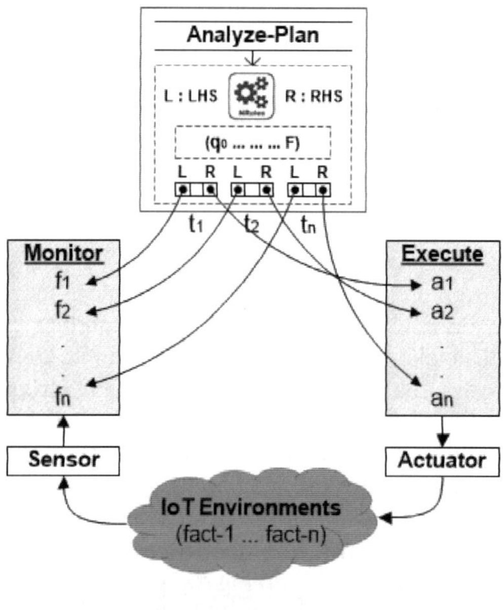

Figure 1. Self-adaptation model.

Figure 1 shows the self-adaptation model developed. Each computational process is transformed into the model domain based on the mapping rules [19] through a component-based software approach [20], as well as utilizing several design patterns [21]. Each root goal in the model domain with AND-decomposition is defined as a component of the analyze & plan (AP), while each sub-goal can be defined as a monitor component (M) and/or an execute component (E), Knowledge (K) is represented as rule-based systems with the following basic structures:

WHEN<*event*>; *state transition*
IF<*condition*>; *conditions to trigger action*
THEN<*action*>; *action when the event takes place*
VALID-TIME<*time_period*>; *a period of time during adaptation needs to be applied*

The sensor monitors the context information that becomes the input set as <*event*> and is captured by the context processor, where Σ: context information is a set of facts (f_1. f_n) captured from an IoT environment event and can determine state changes. The controller evaluates the <*condition*>, referring to a certain <*event*> that occurs in the state set, and this controller will trigger the performer action component to do the <*action*> adaptation through the transition function to the <*condition*> that applies at a certain <*time_period*>. Q: the action (a_1. a_n) behavior is a set of state system behavior, from the initial state (q_0) to the state that can be the target (F). δ: the transition (t_1. t_n) function as the time of the adaptation action based on an evaluation of the condition to change the initial state (q_0) to the target state (F). The quality of the inference engine depends on the selection of one state, where the state selected as the adaptation action in the right hand side (RHS) class is determined by the expression in the class left hand side (LHS) which is the compatibility between the rules and facts. The strategy developed for this mechanism is the forward strategy, which is the reuse of existing components and matches the required ones.

3. Case study: patient monitoring systems

In order to evaluate the model, we developed a patient monitoring system (PMS) case study. The PMS services have made significant advances and are supported by medical technologyas to generate various devices for software services. For example smart-home scenario that can help patients in carrying out their daily activities. This system will connect with various devices, such as cameras, meal sensors, medicine sensors, thermometers, oxymeters, smart shirts and other devices. In addition, this system will connect with other systems such as hospitals, nutrition consultancy services, clinics, etc., and other important things will involve various professions as system actors, such as doctors, nurses, social workers and other relevant actors. This fact shows that the PMS services has a variety of interrelated elements, and eventually will cause various problems related to software services.

Case study of PMS is a case problem have been discussed by fellow researchers [7, 8]. In this discussion, case descriptions are modified in accordance with the proposed model. The goal modeling for patient monitoring system can be seen in figure 2. The system has three sub-goals to be achieved, ie feeding schedule, medication schedule, and patient service. This system will connect with several sensors, including meal sensors associated with real-time monitoring plan, food ordering, and activation alarm-1, other sensors are medicine sensors associated with the plan to monitor patient prescriptions and activation of alarm-2. Coordination of all plans will be regulated through the achievement of patient service goals.

Figure 2. Patient monitoring system modeling.

The adaptation mechanisms required for this case consist of (a) adaptation to changes and growth of context information related to patient feeding activities, (b) adaptation to changes and growth of context information related to the activity of taking medicine for the patient. The system requirement of this case can be illustrated as follows:

- There are a number of properties to monitor: (a) external properties: a number of sensors, system actors, and other related systems within the IoT environment, (b) internal properties: functional and non-functional systems, represented as goals and soft goals in figure 2.
- The system state at runtime is a combination of both values of the property. Violations will be analyzed based on the symptom of each detected event.
- There are several rules required in this case, namely (a) a set of rules for organizing components, when new events arise and the unavailability or incompatibility of context information, (b) a set of rules for handling the addition and/or changes of new functions.

Based on the modeling of figure 2, the three sub-goals "follow medical instructions" are AND-decomposition, meaning that the variables in each goal are dependent on each other. This can be represented by setting up detectable events or symptoms, such as mealtime events, deliver food events, medicine supply events, and more. Each event can be set through the rule engine, as follows:

- Mealtime events are generated when changes are detected or unavailable of context information, thus having to perform system reconfiguration.
- Deliver food events are generated when the patient's food unavailability is detected, thus having to order food from outside service providers.
- Medicine supply events are generated when an unsuitable supply of drugs is detected, thus calling an assistant or nurse

The setting of mealtime events is manifested based on the "eating time" component function as a monitor component (M), which determines the function of the execute (E) "deliver food" and "eat meals" components. The decomposition of this sub-goal is OR-decomposition, thus indicating the variability of the context information (*eating_time*), which consists of breakfast, lunch and dinner. While delivering food events will be related to order availability for patients (*order_food*), whether available or not. In addition, the system components are also associated with the resources of the meal sensor, and it may be possible unexpected events or errors (*event_error*), such as sensor damage, lost connections, etc. Based on the description, then obtained three system variables, namely: "eating_time", "order_food" and "event_error", so the plan can be set as follows: *plan (eating_time, order_food, event_error)*. The settings of the system behavior relate to the rule specification, in this case for example set as follows:

Rule-1:

if (eating_time = time.breakfast or time.lunch or time.dinner) and (order_food = true) and (event_error = null) then plan = activate alarm

Rule-2:

if (eating_time = time.breakfast or time.lunch or time.dinner) and (order_food = false) and (event_error = null) then plan = create new service_action

Rule-3:

if (eating_time = null) and (event_error = null) then plan = change service_action

Rule-4:

if (event_error = not null) then plan = change service_action

Rule-5:

if not [criteria] then plan = change service_action

This plan is realized to overcome variability in fulfilling goal and softgoal. For example, goal achievement for feeding schedule and sheduled to take medicine will use the "patient service" component function as a component of analyze and plan (AP), as it contributes positively (++) to "availability" and "relevance" softgoals, compared to analyze and plan (AP) each that only contribute positively (+), will even be affected negative contribution (-) when detected change. The rule specifications can be mapped into ECA rules (table 1).

Table 1. ECA-feeding schedule.

Event (E)	Condition (C)	Action (A)
mealtime_events	*(eating_time = time.breakfast or time.lunch or time.dinner) ; (order_food = true) ; (event_error = null) ;*	*P_1 = activate_alarm*
mealtime_events deliver_food_events	*(eating_time = time.breakfast or time.lunch or time.dinner) ; (order_food = false) ; (event_error = null) ;*	*P_2 = create new service_action*
mealtime_events	*(eating_time = null) ; (event_error = not null) ; not [criteria];*	*P_3 = change service_action*

Adaptation action in table 1 is an adaptation for changes in context information, for example:

- *P_1 = activate_alarm*, set alarm messages based on prevailing conditions. In this case the system will connect with the repository, sensor, etc.

As for fulfilling the evolution of the system and as a follow-up of adaptation, the system can handle its growth. For example:

- *P_2 = create new service_action*, adding a new function to order patient food from a catering service provider based on information/ advice on nutrition consultancy services. In this case the system will connect with other systems within the IoT environment.
- *P_3 = change service_action*, function to call the assistant/ nurse when the patient's feeding schedule is not available, or contact the service desk when interruption occurs. In this case the system will connect with other system actors within the IoT environment

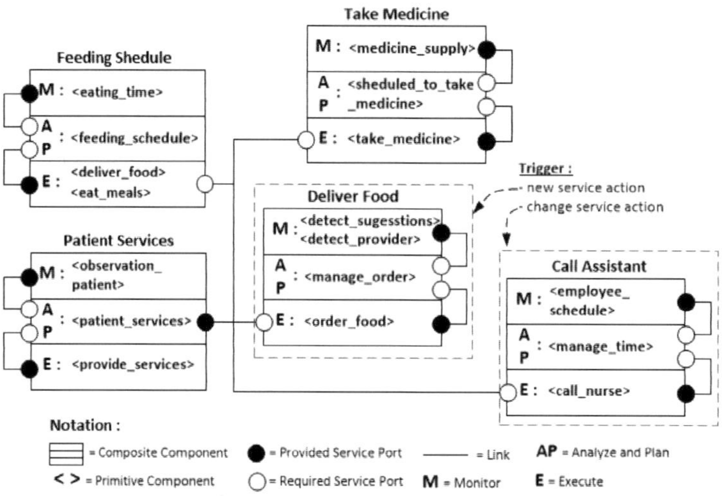

Figure 3. The addition of the new functionality of PMS components.

As another example, when the medicine supply events arise due to an unexpected supply of drugs, the system will call an assistant or nurse. These actions show the adaptability of PMS when it comes to connecting with various devices, actors, and other related systems within the IoT concept. The dashed lines in figure 3 illustrate the addition of new functions to existing system components

4. Conclusion

The self-adaptation model proposed in this paper focuses on the requirements for service evolution within an IoT concept, where systems are required to have the ability to adapt flexibly with various devices, system actors, and other related systems. The proposed approach emphasizes the importance of understanding the IoT environment through the development of model domains, as well as the ability of context inference in anticipating changes to the system and its growth needs.

The results of the case studies show the proposed model allows to anticipate changes in the range of environmental entities. In addition, the model also takes into account its evolutionary needs, so the system can be expanded easily and as needed. We believe this model can contribute to the IoT domain, where the system must be viewed based on its life cycle that will continue to grow. As future work, we plan to extend the context inference strategy to formulate conflict resolution and apply it to more complex cases.

References

[1] Supriana I and Aradea 2015 Automatically relation modeling on spatial relationship as self-adaptation ability *International Conference on Advanced Informatics: Concept, Theory and Application (ICAICTA)* vol. **2** (Chonburi-Thailand: IEEE Xplore) p 157–162

[2] Ganek A G and Corbi T A 2003 The dawning of the autonomic computing era *IBM Systems Journal* **42**(1) 5–18

[3] Cheng B H C, Eder K I, Gogolla M, Grunske L, Litoiu M, Muller H A, Pelliccione P, Perini A, Qureshi N, Rumpe B, Schneider D, Trollmann F and Villegas N 2014 Using models at runtime to address assurance for self-adaptive *LNCS Springer* 101–136

[4] Aradea, Supriana I and Surendro K 2018 Self-adaptive software modeling based on contextual requirements *Telecommunication, Computing, Electronics and Control (TELKOMNIKA)* **16**(3) 1276–88

[5] Sucipto S and Wahono R S 2015 A Systematic Literature Review of Requirements Engineering for Self-Adaptive Systems *Journal of Software Engineering* **1**(1) 17–27

[6] Knauss A 2015 *The Capture and Evolution of Contextual Requirements: The Case of Adaptive Systems* Dissertation (Hannover: University of Victoria)

[7] Morandini M, Penserini L, Perini A and Marchetto A 2017 Engineering requirements for adaptive systems *Requirements Engineering* **22**(1) 77–103
[8] Dalpiaz F, Giorgini P and Mylopoulos J 2013 Adaptive socio-technical systems: a requirements-based approach *Requirements Engineering* **18**(1) 1–24
[9] Tamura G, Villegas N, Muller H, Duchien L and Seinturier L 2013 Improving context-awareness in self-adaptation using the DYNAMICO reference model *8th Int. Syp. Software Engineering for Adaptive and Self-Managing Systems* 153–62
[10] Abeywickrama B D, Bicocchi N and Zambonelli F 2012 SOTA: towards a general model for self-adaptive systems *Proc. 2012 IEEE 21st International WETICE* 48–53
[11] Arcaini P, Riccobene E, Scandurra P 2015 Modeling and Analyzing MAPE-K Feedback Loops for Self-adaptation *SEAMS 15' Proc. 10th Int. Sym. Software Eng. For Adaptive and Self Managing Systems* (USA: IEEE Press) 13–23
[12] Knauss A, Damian D, Franch X, Rook A, Müller HA and Thomo A 2015 ACon: A learning-based approach to deal with uncertainty in contextual requirements at runtime *Journal of Information and Software Technology* **70** 85–99
[13] Paz A and Arboleda H 2016 A model to guide dynamic adaptation planning in self-adaptive systems *Electronic Notes in Theoretical Computer Science* **321** 67–88
[14] Aradea, Supriana I and Surendro K 2014 An overview of multi agent system approach in knowledge management model *International Conference on Information Technology Systems and Innovation (ICITSI 2014) - Proceeding (*Bandung) 62–69
[15] Supriana I, Surendro K, Aradea and Ramadhan E 2016 Self-adaptive cyber city system *International Conference on Advanced Informatics: Concepts, Theory and Applications (ICAICTA)* IEEE Malaysia
[16] Aradea, Supriana I, Surendro K and Darmawan I 2017 Integration of self-adaptation approach on requirements modeling *Recent advances on soft computing and data mining. SCDM. Advances in Intelligent Systems and Computing* vol. 549 ed T Herawan, R Ghazali, N Nawi, M Deris (Cham, Switzerland: Springer) p 233–43
[17] Bresciani P, Perini A, Giorgini P, Giunchiglia F and Mylopoulos J 2004 Tropos: An agent-oriented software development methodology *Journal of Autonomous Agents and Multi-Agent Systems* **8**(3) 203–36
[18] Kephart JO and Chess DM 2003 The vision of autonomic computing *Computer* **36**(1) 41–50
[19] Nakagawa H, Ohsuga A and Honiden S 2012 Towards dynamic evolution of self-adaptive systems based on dynamic updating of control loops *Sixth International Conference on Self-Adaptive and Self-Organizing Systems (SASO),* (IEEE Press) pp 59–68
[20] Hirsch D, Kramer J, Magee J and Uchitel S 2006 Modes for software architectures *In EWSA* pages 113–26
[21] Abuseta Y and Swesi K 2015 Design patterns for self adaptive systems engineering *International Journal of Software Engineering & Applications (IJSEA)* **6**(4)

IOP Conf. Series: Materials Science and Engineering **550** (2019) 012027 doi:10.1088/1757-899X/550/1/012027

Visualization of a two-dimensional tree modeling using fractal based on L-system

E W Hidayat[1], I K G D Putra[2], I A D Giriantari[2] and M Sudarma[2]

[1]Siliwangi University, Siliwangi Street, Tasikmalaya 46115, Indonesia
[2]Udayana University, Panglima Besar Sudirman Street, Denpasar 80234, Indonesia

Email:ekawahyu@unsil.ac.id

Abstract. The purpose of this paper is to present a visualization of two-dimensional trees using fractals and L-Systems. Grammar in L-System consisting of letters, axioms, and production rules can be used to construct trees in various forms according to the production rules used. With the formation of grammar from the tree pattern in a fractional manner, the grammar can be reused to construct other tree patterns at a higher level. The iterations made on certain production rules visually affect the tree model. Experiments were carried out on the L-System generator to see the results of tree visualization by testing various grammars. Testing was done by trial and error in production rules to get a realistic visualization of a two-dimensional tree model. From the experimental results produced a variety of grammar, where modifications to the production rules produced different tree models. Visually, from the iteration results, the higher the iteration value, the two-dimensional tree was not formed properly. When the iteration value or n > 5 the two-dimensional tree model results from an iteration it lost detail from the branch.

1. Introduction

The form of self-similarity naturally exists in nature, according to Mandelbrot called as Fractal. The fractal object has infinite details and has a similar structure of self at different magnification levels. This property means that the more fractal objects enlarged the more detailed objects will be obtained. Details of the fractal object are not visible directly but will appear gradually in the event of magnification. There are three levels of self-similarity in fractals [1], namely Exact self-similarity, Approximate self-similarity and Statistical self-similarity. Approximate self-similarity is a natural form that matches the paradigm of self-similarity in nature, such as trees, clamshells, clouds, lightning, river tissue, and blood vessels. Fractals have been applied to various aspects, for example, applied to various forms of mathematical computations that are to produce tree branch patterns [2], leaf pattern [3], and fibrous roots pattern [4]. Tree Modeling is the part of the fractal with Approximate self-similarity. Trees have very complex structures but are well defined. One of the major factors that organize tree structure is the concept of self-similarity as characterized by the nature of fractals. Fractal geometry can be generated using a number of methods namely Lindenmayer System (L-System) and Iterated Function System (IFS).

Imitation of natural forms as a basis seed has been done to imitate the fractal geometry-based butterfly pattern. Objects of fibrous roots can also be produced using L-System and combined with existing batik patterns [4]. Other studies offer methods and algorithms for constructing butterfly patterns by adopting two-dimensional IFS models [5]. While other studies were carried out to construct a line graph on two-dimensional fractals with the L-System approach. This study shows that the fractal dimensions of the graphs obtained in all the cases analyzed are the same as the original graph, both for the original graph and for the line graph by identifying the node class that reflects the graph symmetry. Fractals are applied to the construction of tree branch patterns [2]. Other researchers have developed L-System to become an Adaptive L-System so that it can be used to produce geometric branching structures in various forms and to illustrate the shape of leaves and tree branches with finer nodes [3]. Other studies propose an approach in studying the modeling of Zinnia plant growth using the L-System method using Mathematica software [6]. The study conducted by Lim (2017) compared L-System applications for plant modeling, music rendering, generating musical scores [7]. In the study several applications such as TreeSketch, L-Studio, and L-Py.

The research roadmap in this study is as a prefix in visualizing tree models from two-dimensional shapes to three-dimensional shapes which can later be used for three-dimensional generated trees for application development in the fields of Games, Augmented Reality and Virtual Reality. Research challenge in this research is how to visualize trees using L-System which is limited by Angle, Axiom, and Production Rules. This paper will explore the use of fractal geometry as a basis for simulating tree shapes using the L-System approach and visualize computational two-dimensional tree construction from the L-System grammar made up of four iterations.

2. Fractal based on l-system

Fractals are roughly fragmented geometric shapes, can be divided into several parts, and each part is a clone of size equal to or greater than the original shape of the whole. Fractals are defined as branches of geometry that show that a set of repeated geometric patterns with scaled scales can be used to describe geometric patterns contained in nature or to describe the complexities of mathematical models [1, 5]. Fractals show that the forms in nature are not dimensionless integers but dimensionless fractions, such as the branches of a tree will be seen as miniature of the tree as a whole, not exactly similar but similar [8]. A fractal can be generated by repeating a pattern, usually in a recursive or iterative process. Mandelbrot uses the word fractal to describe objects that do not have clear dimensions but have self-similiarity.

The use of fractals in modeling a two-dimensional tree using L-System means that the tree has a self-similarity characteristic, in which each part of the tree is an imitation of size equal to or greater than the original shape as a whole which is repeated many times on a minimized scale. The repetition can be formed through grammar in the L-System. The following figure shows that in each branch of a tree in the constellation using L-System it is a repetition of the seed shape.

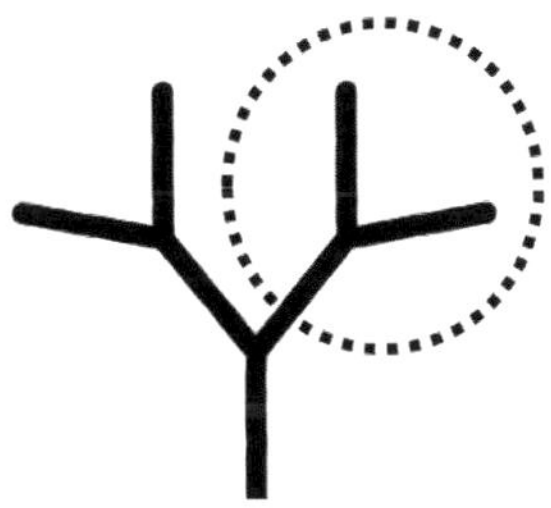

Figure 1. Fractal tree using L-System with $\delta = 25^0$, Axiom P: F, ω: F [+F] [-F], The tree branch image in the circle line shows the result of the loop from the first iteration. The branches of the tree are self-similar to the seeds. So it can be said that the tree is recursive and has fractal properties.

3. L-system

L-System or Lindenmayer System is a parallel rewrite system and is included in a formal grammatical type. This system was introduced by Aristid Lindenmayer (1925-1989) a Hungarian theoretical biologist and botanist from Utrech University in 1968 to model the pattern and process of the growth of various types of algae. L-System has been successfully used to produce models of many natural forms, including higher plants.

L-System is a context-free grammar in which each production rule applies only to one symbol on a set. Other symbols are not affected by the production rule. The basic idea of L-System is to form an object starting from the initial structure by swapping or replacing some parts of the grammar notation (rewriting rules) on a rule by sequentially looping mechanisms that refer to a self-similarity. The loop is performed according to the number of desired iterations [9].

The L-System was initially in form to provide a formal description of the development of simple multicellular organisms and illustrates the relationship between plant cells. Then, the system is expanded to describe higher plant species and complex branch structures. Recursive behavior in L-System leads to self-similarity so that natural forms such as fractals can be easily drawn using L-System. The forms of plants and other forms of organisms are quite easy to define because by increasing the rate of repetition, this form will grow and become more complicated [9].

3.1. L-System grammar

The grammar of L-system is very similar to the grammar in the Chomsky hierarchy. The L-System grammar is defined according to Prusinkiewicz (1990) and Mishra (2007) as a pair of 3 tuples $G = (V, \omega, P)$ where V (alphabet) or letter is a set up to V and formal symbols, ω (start, axiom, initiator) or axiom is a string ω of the V symbol that defines the initial state of the system [9, 10]. The string set of V is denoted as V*, and P is a set of production rule a symbol mapping $a \in V$ to *string* $w \in V^*$ is written as $p: a \rightarrow w$. A variable can be replaced with a combination of constants and other variables. Production consists of two strings namely predecessor and successor. For all symbols a if $a \in V$ has no production rules, it is assumed that the symbol is mapped to itself so that a becomes the constant or terminal of L-System.

3.2. Rewriting production rule

The L-System fractal consists of a symbolic alphabet that can be used to create strings, a collection of production rules that extend each symbol to some larger symbol strings, the initial "axiom" string to begin construction, and a mechanism for translating the string generated into geometric structures. The principle of rewriting rules in L-System grammar is exemplified by L-System notation used by Lindenmayer to model algae growth.

For example, given two symbols A and B. Each symbol declared as a production rule. A produced AB and B produced A, then every iteration of each symbol is rewritten according to its production rules. The product of L-System is defined as a sequence $\{g_n\}$ with $n = 1, 2, 3, 4, \dots, k$ where the first iteration g_0 is axiom w. The process can be seen in table 1 showing the L-System writing process for Algae growth with 7 iteration from *g0* to *g6*.

Table 1. The writing of L-system algae growth [9].

Grammar		Result
Variabel	: A B	g0: A
Axiom	: A	g1: AB
Production Rule	: A → AB	g2: ABA
	B → A	g3: ABAAB
		g4: ABAABABA
		g5: ABAABABAABAAB
		g6: ABAABABAABAABABAABABA

3.3. Visualization of L-system

In L-System there are symbols that can be interpreted graphically. If it is assumed to be a unit of length h and angular rotation δ, then the commands of the symbols in L-System used for visualization of trees in this study are as follows:

- F : drawing forward one unit along a h. (draw line forward).
- + : spin clockwise with angle δ (rotate right).
- - : spin counter clockwise by angle δ (rotate left).

IOP Conf. Series: Materials Science and Engineering **550** (2019) 012027 doi:10.1088/1757-899X/550/1/012027

- [: save current position and move according to the next command (begin new branch).
-] : back to the original position stored by the symbol "[" (end branch).

3.4. L-system algorithm
Solving the problem of how to interpret the grammar above can be described in the form of systematically arranged steps and then to interpret grammar into a graphic can be aided by the general algorithm of fractal object interpretation on L-System with the following steps:

- Enter value of generation (k), branch inclination angle (δ) and line segment length (h).
- Determine the starting angle (a_0) and enter the a_0 value to get the starting point (F_0) ; then enter F_0 on the production (P) so as to produce P_0.
- Perform the iteration process i for $i = 1, 2, 3, \dots, k$ so get the next angle value (a_1) and forwarded to get the next point value (F_i) ; then enter F_i on production rule P_{i-1} to get P_i.
- Plotline segments generated from axioms and production rules.

4. Results and discussion
The turtle graph grammar reconstruction into a tree graph was implemented using a generator built using JavaScript and visualization of the reconstruction results viewed using the internet browser. The following are four experimental results from several production rules used can be seen in the figure 2 to figure 6.

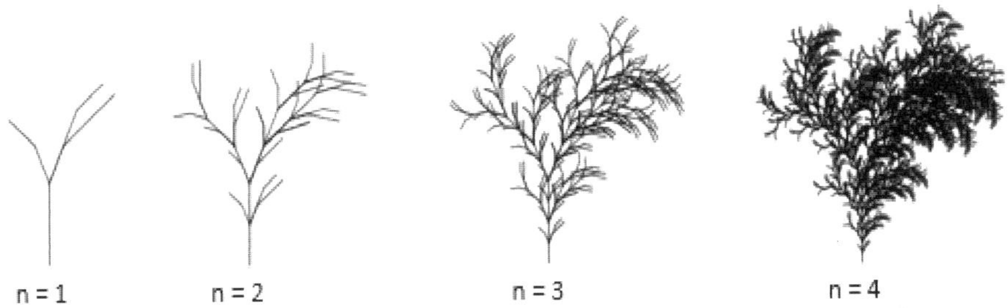

Figure 2. Fractal tree using L-system with $\delta = 25^0$, Axiom P: F, ω: FF [+F+FF] [-F-F] [+FF+F].

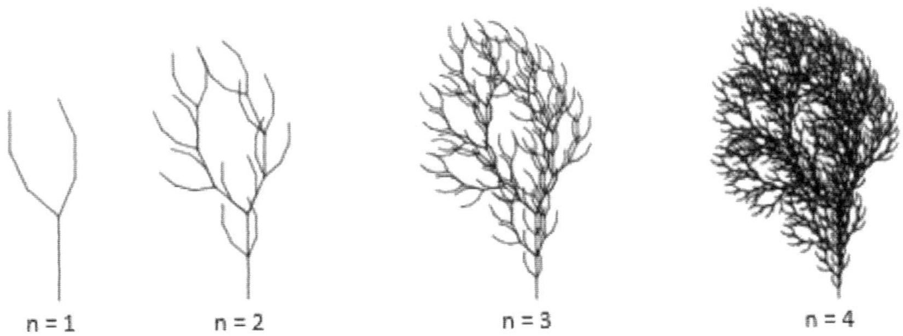

Figure 3. Fractal tree using L-system with $\delta = 25^0$, Axiom P: F, ω: FF-[-F+F+F]+[+F-F-F].

Figure 4. Fractal tree using L-system with $\delta = 25^0$, Axiom P: F, ω: FF+[+F-F+F]-[-F+F-F].

Figure 5. Fractal tree using L-system with $\delta = 25^0$, Axiom P: F, ω: FF [+F] [-FF] [-F+F].

Figure 6. Fractal tree using L-system with $\delta = 25^0$, Axiom P: F, ω: FF+FF [+F-F] [-F+F].

Figure 2 through figure 6 show the experimental results for generated model trees. It can be seen that production rules in the construction will produce different models. Visually, figure 2 to figure 5 approaches the visualization of trees, but figure 2 and figure 3 are visually more realistic. While figure 6 successfully constructed grammar but fails to visualize the tree model.

5. Conclusion and future research

In this paper, a variety of L-System grammar had been tested to reconstruct fractal trees with the same branch slope for all experiments, namely $\delta = 25^0$. The axiom used was F. The deterministic grammar was used as an approach. The experimental obtained several grammars forming two-dimensional trees visually to see which iteration that were good for modeling trees. some trials and errors must be done in the production rules. From the iteration results, it can be seen that the higher the iteration value, the two-dimensional tree was not well-formed visually. When the iteration value or $n > 5$ iteration result trees lose detail from the branch. The results of the trial show that different production rules will produce different fractal trees. For further research, modeling trees in three dimensions using L-System must be tested first using the L-System grammar for a two-dimensional tree can form a better visual tree. Because L-System grammar consists of two types, namely Deterministic and Non-Deterministic, it is necessary to conduct further studies to see the results of the reconstruction in a non-deterministic manner.

References

[1] Mandelbrot B B 1982 *The Fractal Geometry of Nature* (New York: W H Freeman and Company)
[2] Warchalowski W and Krawczyk M J 2017 *Line Graphs for Fractals* (Commun Nonlinear Sci Numer Simulat: Elsevier) pp 506–512
[3] El-Latif Y M A 2010 A New Conception in Constructive Branching Structures and Leaves Using L-System *JCSE* **4** 240–52
[4] Kusuma P D 2017 Fibrous Root Model in Batik Pattern Generation *JATIT* **95**(14) 3260–9
[5] Barnsley M F 1988 *Fractals Everywhere* (Sandiego: Academic Press)
[6] Suhartono 2013 Pemodelan Pertumbuhan Tanaman Zinnia Menggunakan Lindenmayer System Dengan Mathematica *Cauchy: Jurnal Matematika Murni dan Aplikasi* **3**(1) 33–7
[7] Lim C K Tan K L and Suppramaniam V 2017 Comparison of L-System Applications Toward Plant Modelling Music Rendering and Score Generation Using Visual Language Programming *Int. Conf. Appl. Sci. Tech.* 020086:1–6
[8] Buchanan T 2017 *Power of Pattern Fractal* (Huntington: Teacher Created Materials)
[9] Prusinkiewicz P and Lindenmayer A 1990 *The Algorithmic Beauty of Plants* (New York: Springer-Verlag)
[10] Mishra J and Mishra S 2007 *L-System Fractal* (Netherland: Elsevier)

Monitoring of AH-11A fused-silica drift in RCM-LIPI and its implementation

L Khairiyati[a)], N F Yayienda[b)], and M Azzumar [c)]

Research Center for Metrology
Indonesian Institute of Sciences
Indonesia

Email: [a)]lukluk.khairiyati@gmail.com; [b)]nibrasfitrah@gmail.com; [c)]muhammad.azzumar@lipi.go.id

Abstract. We have been monitoring our standard capacitors of AH-1100 fused-silica drift for 8 years to maintain and obtain the actual values of them. One of the advantages of the standard capacitors is to be used as a reference standard for ensuring the performance of an AC high voltage capacitive divider, that is commonly used to assess a voltage level on a power distribution from plants to end-users. This paper describes the monitoring of AH-1100 and its validation performed by comparing to the calibration results of KRISS, the national metrology institute of The Republic of Korea. The AH-1100 drifts is found to be less than 0.0052μF/F/year, and the comparison results show a good agreement, making the monitoring results have been validated. Moreover, the implementation has been performed with expanded uncertainties of 1.7 fF for 10 pF and about 22 fF for 100 pF.

1. Introduction

A voltage level on a power distribution shall be checked to know that an electrical quality from power stations (plants) to industrial or residential customers (end-users) is effective and also to ensure the continuity of distribution is running well. To assess it, having an AC high voltage value, engineers use an AC high voltage capacitive divider that had been calibrated. A work principle of the capacitive divider is that consider the two capacitors connected in series across an AC voltage. As the two capacitors are in series, the charge Q on them is the same, but voltage across them will be different and related to their capacitance values, as V=Q/C [1]. Therefore, to ensure the AC high voltage capacitive divider has good performance, it must be calibrated and has a metrological traceability to SI through standard capacitors as shown in figure 1.

In Research Center for Metrology LIPI (RCM-LIPI) as National Metrology Institute of Indonesia, has Fused-silica of Andeen-Hagerling 11A (AH-11A) mounted in AH-1100, consisting nominal values of a 10 pF and two 100 pF (namely 100-C and 100-D), as primary standard of capacitance value [2]. One of the big uncertainty contributions of the capacitance standard using AH-1100/AH-11A is drift, which is evaluated by using manufacturer specification and a year duration since the last calibration date.

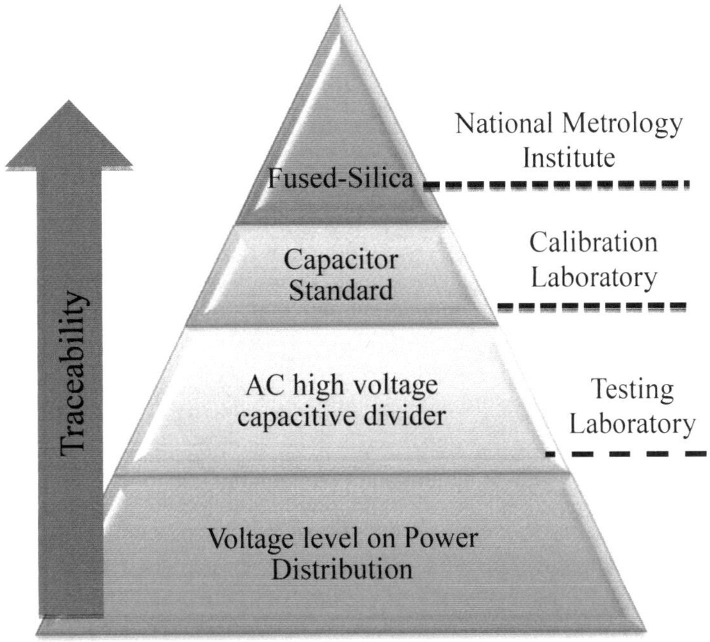

Figure 1. Metrological traceability [1].

To improve the uncertainty of the drift, we have been performed monitoring of AH-11A for more than 8 years. The monitoring is performed by using Andeen-Hagerling 2700 (AH-2700) capacitance bridge based on the direct measurement method. The monitoring results are used as a drift correction to estimate each value of AH-11A as well as the each standard error of the estimation is also taken into account to uncertainty evaluation. To validate these monitoring results, we do a comparison with the 2017 calibration values of KRISS-South Korea. Then, we implement AH-11A based on the monitoring results to calibrate capacitance parameter of LCR meter. To validate this implementation, we also do a comparison with the calibration values of SCL-Hong Kong.

2. Monitoring of AH-11A and uncertainty

2.1. Method of monitoring

Monitoring measurements are performed at a test voltage of 15 Vrms and frequency of 1 kHz, which is commonly used for measurement of standard capacitors in the metrology field as well as international comparisons of capacitance are also carried out around this frequency, by direct measurement method using AH-2700.AH-2700 employs a three-terminal measurement configuration that permits high precision measurement and avoids stray capacitance which is very affected on low capacitance measurements [3, 4].

The environmental temperature of laboratory is kept at 23 ± 2°C. Also, the internal temperature of AH-11A fused-silica is controlled by using a temperature-controlled oven that is powered by via AH-1100 frame.

2.2. Monitoring results and its evaluation

The monitoring of drift is taken from November 2009 to January 2018. In figure 2, symbol 'x' represents reading values of 10 pF.

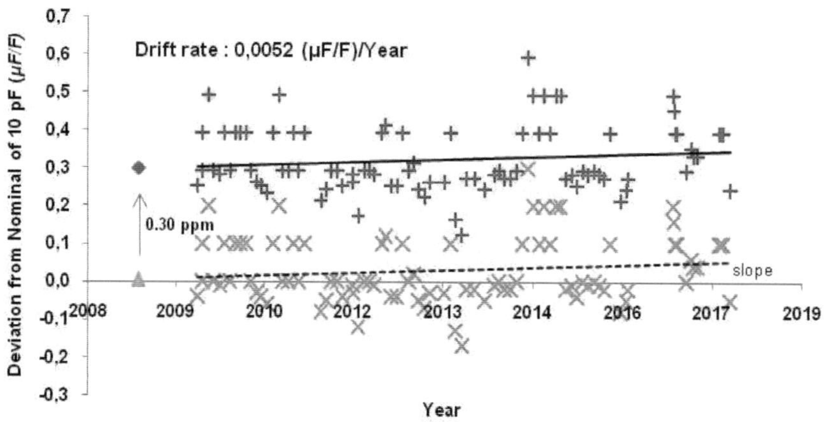

Figure 2. Monitoring results of 10 pF AH-11A.

In this monitoring, we need to evaluate the performance of AH-2700 to know its correction, even if on December 2008 the AH-2700 had been verified only in KRISS but no calibration so we did not have a correction for it. In the direct measurement for measuring a source, the known correction of a calibrated meter shall be necessary. Therefore, to get the correction of AH-2700, we do an extrapolation of reading value data for estimating a reading value in a time when AH-11A was calibrated, and then we obtain the estimated correction of AH-2700 by subtracting the calibration value of KRISS (with symbol '♦' in figure 2) with the estimated reading value (with symbol '▲' in figure 2). After the correction of AH-2700 has been known, we implement this correction to all reading values of AH-11A as the corrected monitoring data,which can be used to obtain the estimated value of AH-11A, and it can be seen in figure 2 with symbol '+'. Also, from this evaluation, we can obtain the drift of 10 pF, that is 0.0052 μF/F/year within the fitting error of -0.086 μF/F as the error of estimate in uncertainty evaluation.

This monitoring is also performed for both 100-C and 100-D, and the obtained drifts are respectively -0.035 μF/F/year and -0.037 μF/F/year within each the fitting error of 0.018 μF/F and 0.025 μF/F.

2.3. Uncertainty of AH-11A

The uncertainty of AH-11Ain each nominal has been updated since the last calibration issued by KRISS. The updated uncertainty contribution of AH-11A consists of its calibration uncertainty of KRISS, extrapolation to mean measurement date, instability of temperature, voltage measurement, and accuracy of AH-2700. The reported expanded uncertainties are stated at a confidence level of 95% with coverage factor $k = 2$, i.e. 6.1 μF/F for 10 pF, 5.9 μF/F for 100-C, and 5.9 μF/F for 100-D, which improves our uncertainty contribution of AH-11A to be around 1 μF/F.

2.4. Validation of monitoring result

Before these results are implemented, we need to check its validity. These are validated by calculating the normalized error E_N value of the monitoring value to the 2017 calibration results issued by KRISS which are considered as the reference values. Also, E_N use expanded uncertainty with a confidence level of 95%.The obtained E_N are tabulated in table 1.

Table 1. The E_N values based on monitoring results.

Nominal	KRISS		RCM-LIPI		E_N
	MV	Exp Unc	EV	Exp Unc	
10	10,00000	0,00005	10,00000	0,00006	-0,04
100	99,9999	0,0005	100,0000	0,0006	-0,12
100	99,9999	0,0005	100,0000	0,0006	-0,11

Although the updated uncertainty is less 1 μF/F than the current uncertainty contribution of AH-11A that uses 10 times manufacturer specification as a drift contribution, the updated uncertainties have good agreement with KRISS measurements. It means that its monitoring results have been validated.

3. Implementation to LCR meter measurement

3.1. Method of measurement

LCR meter is set at test voltage of 1V, frequency of 1 kHz, and zero DC BIAS voltage (OFF) by direct measurement using capacitors of AH-11A as reference. Each measurement was taken by 5 repeated measurements.

3.2. Measurement results and its evaluation

The measurements of LCR meter reading are taken from February 2017 to January 2018.

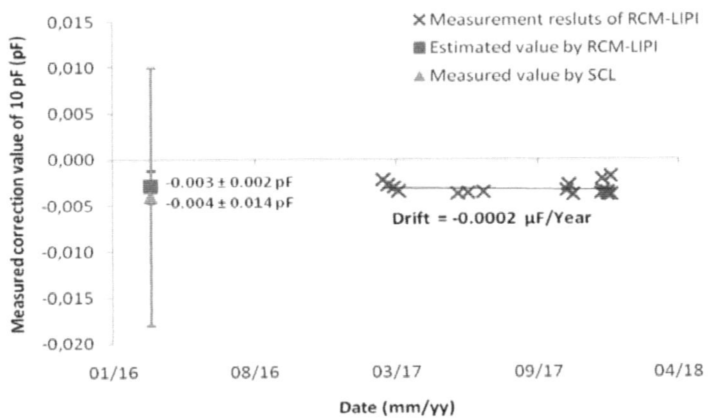

Figure 3. LCR reading measurement for 10 pF.

In figure 3, each symbol 'x' represents an average correction value of LCR meter for 10 pF AH-11A in each measurement every month. This correction value was obtained by subtracting the estimated value of AH-11A in each measurement date with the meter reading value of LCR meter.

For the need of comparison, we need to estimate LCR meter correction to the same time when LCR meter was calibrated by SCL. It aims to minimize the different value of correction measurement between RCM-LIPI and SCL due to time-dependent change of LCR meter. Therefore, to get the estimated correction of LCR meter in the same time with SCL, we do an extrapolation using linear fitting of those correction values, with the estimated drifts of LCR obtained were -0.0002 μF/year for 10 pF, 0.005 μF/year for 100-C, and 0.005 μF/year for 100-D within each its fitting error to be included to uncertainty evaluation of LCR as error of its estimated correction.

The difference of tes voltage, for monitoring and calibrating LCR, on AH-11A is not included as correction of AH-11A values, since the actual voltage coefficients are not known. So we include an uncertainty based on the manufacturer specification.

Lead/fixture effect is estimated by measuring the relative change of LCR reading with re-zeroing in every doing a measurement in different time but still same day.

3.3. Validation

Results of multiple measurements by RCM-LIPI or by SCL are compared by calculating the normalized error E_N value where the calibration results of SCL are considered as the reference values in these comparisons. Also, E_N use expanded uncertainty with a confidence level of 95%. The obtained E_N are 0.08, 0.9, and 0.9. These results show that RCM-LIPI's measurements have good agreement with SCL's measurements.

4. Conclusions

The monitoring of AH-11A 10 pF and two 100 pF in RCM-LIPI have been doing for 8 years. The obtained drifts are less than -0.0052 μF/F/year within the fitting error of about 0.02 μF/F and the updated expanded uncertainties are less than 6.1 μF/F which it improves the uncertainty contribution of AH-11A to be about 1 μF/F. These have been validated by KRISS measurement. Also, implementation of monitoring results to calibrate LCR meter has been performed with expanded uncertainties of 1.7 fF for 10 pF and about 22 fF for 100 pF, and it has been validated by SCL measurement.

The LCR meter can be used for calibrate the standard capacitors that are to be used as a reference standard for ensuring the performance of an AC high voltage capacitive divider, that is commonly used to assess a voltage level on a power distribution from plants to end-users.

References

[1] Kurupakorn C and Pattarakijkul D 2016 Development of 100 kVAC High Voltage Measurement System (Thailand: Electrical Metrology Department National Institute of Metrology (NIMT))
[2] Andeen-Hagerling Inc 2003 AH 1100 Capacitance Standard Frame AH 11A Fused-Silica Capacitance Standard Ohio USA
[3] Jonhson L *et al* 2008 Final Report : APMP Supplementary Comparison of Capacitance at 100 pF *Metrologia* **45** 1–52
[4] Andeen-Hagerling Inc. 2003 Operation and Maintenance Manual Ultra Precision Capacitance AH 2700A 50 Hz-20 kHz Ultar-Precision Capacitance Bridge Ohio USA

Acknowledgment

The authors would like to express our thanks to the management team at the Research Center for Metrology—Indonesian Institute of Sciences (RCM-LIPI), which supported the research in the form of facilities and infrastructure. We are also indebted to our friends, who helped in ensuring that the study ran smoothly, whether directly or indirectly.

Enhanced biogas production by mesophilic and thermophilic anaerobic co-digestion of palm oil mill effluent with empty fruit bunches in expanded granular sludge bed reactor

R Fitrah, A Ahmad and I Amri

Department of Chemical Engineering, University of Riau,
HR Subrantas Rd. km 12.5 Pekanbaru 28293

E-mail: adri@unri.ac.id

Abstract. This study aimed to utilize Palm Oil Mill Effluent (POME) and preheated Empty Fruit Bunches (EFB) as a source for enhanced biogas production in Expanded Granular Sludge Bed (EGSB) Reactor. The reactor was operated continuously for 48 days with 2 days hydraulic retention time (HRT), POME: EFB mixing ratios of 1: 0 ; 0.10 ; 0.15 ; 0.20 ; temperature of 35 ± 2^{0}C and 55 ± 2^{0}C. The best improved was achieved from co-digestion of POME with EFB at a mesophilic temperature of 35 ± 2^{0}C and mixing ratios of 1: 0.10, highest total biogas volume achieved was 7,400 ml (0.1359 – 0.1619 m^3 CH_4/kg COD removed), Volatile Fatty Acid (VFA) /Total Alkalinity (TA) ratios 0.1418 – 0.1582, and COD removal 61.97 – 78.87 %. The Methane content was 46.10 %. Under a Thermophilic temperature of 55 ± 2^{0}C and mixing ratios of 1: 0, 10, the highest total biogas volume achieved was 1.350 ml (0.0859 – 0.1358 m^3 CH_4/kg COD removed) VFA/TA ratios increased to 0.4498 – 0.6685 with COD removal 61.97 – 74.65 %.

1. Introduction

The growth of Indonesia's population would encourage the increase in electricity consumption. In the year 2018, United Nations (UN) stated that Indonesia's population had reached 266.79 million. The number will continue to increase and reached its peak in the year 2062 with 324.76 million inhabitants [1, 2].

The demand for electricity would increase from 283,000 MWh in 2018 to 819,000 MWh in 2030. The Ministry of Energy and Mineral Resources try to build the new power plants with the aim of fulfilling electrification ratio of Indonesia 100% in 2025 with the target power plant sourced from renewable energy for 23% [3]. Riau Province has potential renewable energy resources to meet this 23 % target by utilizing Palm Oil industrial waste into electricity. By 2016, 20,942,857 m^3 of Palm Oil Mill Effluent (POME) and 8,028,095 tons of Empty Fruit Bunches (EFB) is generated from 7.33 Million tonnes of CPO produced in Riau Province. It is the largest among another province in Indonesia.

Biogas utilization from POME was environmentally friendly because it can reduce Greenhouse Gas emissions. EFB had been used by researchers as a co-substrate with POME to enhance the biogas production [4-7].

O-Thong *et al.* (2012) reported the effect of pretreatment method EFB in co-digestion with POME in biogas production rates as well as against the ability of biodegradable [4]. A second substrate is mixed in a 320 ml serum bottle in batch conditions. The maximum Biomethane production rates from this study were 82.7 m^3 CH_4/ton mixture (ratio of 6.8:1), Biomethane yield of 392 ml CH_4/g VS at the temperature of 55⁰C and HRT 45 days. O-Thong (2012) also stated that EFB and POME blends have the potential of great biomethane [4]. Kim *et al.* (2017) reported the co-digestion POME with EFB in mesophilic temperature [5]. A second substrate was mixed in a 150 ml serum bottle in batch condition. The maximum Biomethane production rates from this study were 50 ml CH_4/g VSS. The Biomethane generated was 350 ml CH_4/g COD with substrate ratio 1:0.31, the temperature of 35⁰C, HRT of 16 days and pH 7.6.]. Kim *et al.* (2017) reported that co-digestion EFB and POME is one of the appropriate methods of controlling pollution and recover energy in the oil palm industry [5]. Ohimain and Izah (2014) reviewed the performance of various types of anaerobic biodigester in terms of biogas production using Nigeria's POME historical data from 2004 to 2013 [2]. From these data, it is known the Expanded Granular Sludge Bed (EGSB) reactor generated the highest Biomethane in the shortest HRT when compared to other types of the anaerobic biodigester.

This research aimed to utilize POME and preheated EFB as a source for enhanced biogas production in (EGSB) reactor as well as determine the best variations of temperature and substrate ratio to get the highest volume of biogas production.

2. Materials and methods

2.1. Materials and experimental setup

The materials used in this study were POME obtained from Sei Galuh Palm Oil Mill and EFB from Sei Rokan Palm Oil Mill of PT. Perkebunan Nusantara V, granular sludge from PT. Insansandang Internusa's UASB reactor, Nitrogen and 10% NaCl solution.

Other materials included $K_2Cr_2O_7$ 0.05 M solution; Ag_2SO_4 0.05 M; FAS; H_2SO_4 0.02 N; NaOH 0.1 N; Ferroin, PP and MO indicators. The total volume of EGSB reactor used in this study was 30,570 ml with the effective volume of 19,820 ml. The following is a schematic setup of the EGSB reactor.

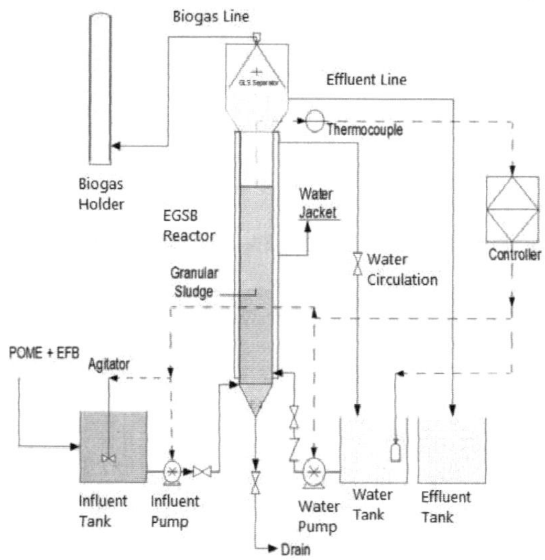

Figure 1. EGSB reactor schematic setup.

Research variables

Dependent variables

1. Size of the EFB < 1 mm [5]
2. The effective volume of the reactor was 19,820 ml
3. HRT of 2 days

Independent variables

1. POME: EFB Ratio: 1 ml: 0; 0.10; 0.15 and 0.20 gr [7]
2. Reactor temperature 35 ± 2^0 C [5, 8] and 55 ± 2^0 C [4]

2.2. Research procedures

Materials preparation

1. POME was obtained from bottom outlet Fat Pit of Sei Galuh Palm Oil Mill.
2. EFB was taken from Sei Rokan Palm Oil Mill, minced and dried in an oven for 6 hours with the temperature of 105^{0}C. The dried EFB was sieved using 20 mesh filter (to obtain the average diameter of EFB powder ± 0.85 mm or below 1 mm)
3. Granular Sludge was obtained from UASB reactor treating textile waste in PT. Insansandang Internusa. The microbial consortia in granular sludge were *Micrococcus, sp; Nitrosomonas, sp; Nitrobacter, sp; and Methanobacter sp.*

Analysis of POME characteristics

Analysis of POME characteristics was held at the laboratory of Bina Marga Department in Riau Province.

Operation of EGSB reactor

The reactor was operated with variation in temperature 35 ± 2^{0}C (mesophilic) and 55 ± 2^0 C (thermophilic), POME to EFB variation ratio 1 ml: 0; 0.10; 0.15 and 0.20 gr and 2 days HRT. The study was divided into 8 runs presented in table 1.

Table 1. Operational stage of EGSB reactor.

Run	Temperature (^{0}C)	Co-digestion ratios (POME: EFB)
1st	35 ± 2	1: 0.00
2nd		1: 0.10
3rd		1: 0.15
4th		1: 0.20
5th	55 ± 2	1: 0.00
6th		1: 0.10
7th		1: 0.15
8th		1: 0.20

3. Results and discussion

Characterization of POME

POME used in this study was taken directly from Sei Galuh Palm Oil Mill's Fat Pit outlet then analyzed in the Material Testing Lab of Dinas Bina Marga, Riau Province. This analysis was needed as a comparative data for influents' COD analysis, especially for POME and EFB mixture. POME characterization analysis result can be seen in table 2.

Table 2. POME characteristics obtained from sei galuh palm oil mill.

Parameters	Value	Threshold*
pH	4.58	6.0 – 9.0
BOD_5 (mg/l)	-	100
COD (mg/l)	55.000	350
TSS (mg/l)	-	250
Total Alkalinity (mg/l)	1.220	N/A
Volatile Fatty Acid (mg/l)	180	N/A
Total Nitrogen (mg/l)	-	50
Max. Debit (m^3/ton CPO)	-	2.5

**Indonesia's Ministry of Environment Regulation No. 5,2014*

Influent's COD

The analysis of influent's COD was used as the basis for calculation of EGSB reactor's COD removal. The value was obtained from the influent's COD analysis per run with three repetitions. The results presented in table 3.

Table 3. Influents' COD analysis on every run.

Run		POME: EFB	Influent's COD (mg/l)			
Mesophilic	Thermophilic		Sample 1	Sample 2	Sample 3	Average
1	5	1: 0.00	55.000	50.000	55.000	53.333
2	6	1: 0.10	60.000	57.500	60.000	59.167
3	7	1: 0.15	67.500	67.500	67.500	66.667
4	8	1: 0.20	77.500	77.500	87.500	83.333

3.1. COD removal

1st Run (T = 35 ± 2⁰ c; Co-digestion ratio 1:0)

Compared to the early startup stage in day 8 to 12 (78.44 to 81.25%), the COD removal in first run decline, within the range of 48.44 – 67.19%. But the reactor still in a stable condition. The production of biogas generated in the first run was better compared to early steady state (day 8 to 12). The COD removal in the first run presented in figure 2.

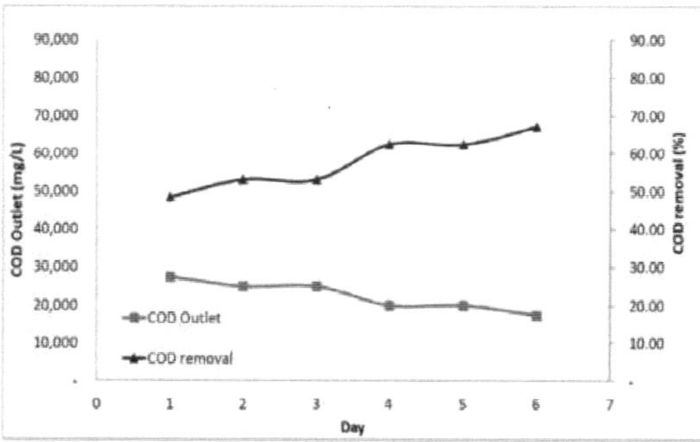

Figure 2. COD removal on 1st run.

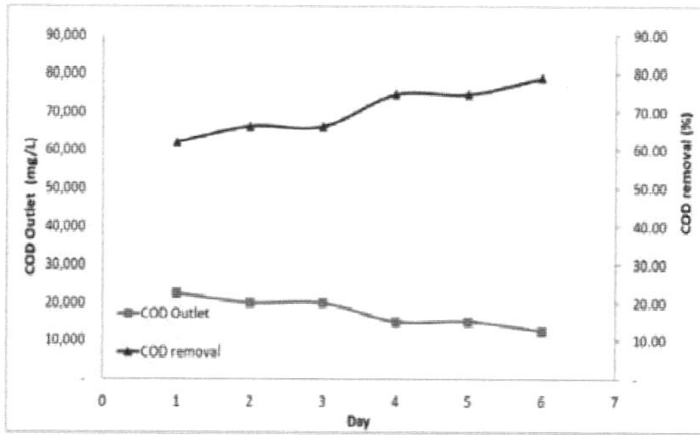

Figure 3. COD removal on the 2nd run.

2nd Run (T = 35 ± 2⁰C ; Co-digestion ratio 1: 0.10)

The COD removal on the 2nd run was increased when compared to 1st run within the range of 61.97 – 78.87%. The influent's COD increased due to the addition of EFB with co-digestion ratio 1:0.10. The average influents' COD were 59,167 mg/l.

The effluents' COD on the 2nd run was better when compared to 1st run, i.e. within the range of 22,500 – 12,500 mg/l. From this data, it can be stated that the addition EFB as a co-substrate, increased the organic material settle-ability at the bottom of the reactor, so the effluents' COD on the 2nd run was relatively low when compared to the previous run.

During 6 days continuous operation on 2nd run, EGSB reactor exhibited good performance. The COD removal on the 2nd run is presented in figure 3.

3rd Run (T = 35 ± 2⁰C ; Co-digestion ratio 1: 0.15)

The COD removal on the 3rd run was also increased when compared to 2nd run. The COD removal vary within the range of 70.00 to 81.25%. The average influents' COD on the 3rd run was 66,667 mg/l. Compared to the previous run, the effluents' COD was better, i.e. within the range of 20,000 – 12,500 mg/l.

The Co-digestion ratio of 1: 0, 15 increased the organic material settle-ability. Thus effluents' COD on 3rd run better than 2nd run. The COD removal on the 3rd run is presented in figure 4:

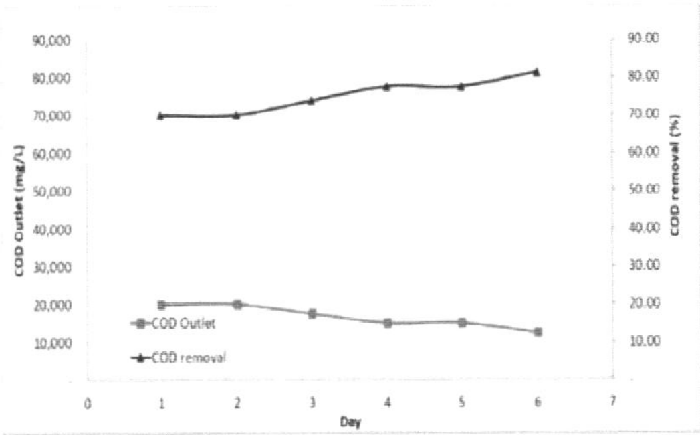

Figure 4. COD removal on the 3rd run.

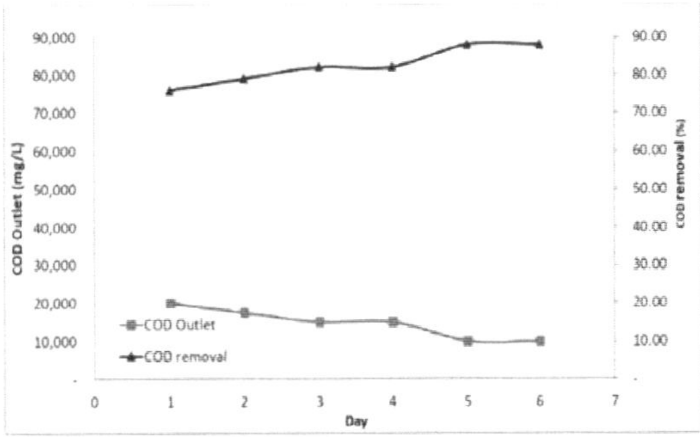

Figure 5. COD removal on the 4th run.

4th Run (T = 35 ± 2^{0}C ; Co-digestion ratio 1: 0.20)

The highest COD removal during the operational stage; 88% was achieved in the 4th run on day 5 and 6. The COD removal on 4th run varies within 76-82%. Influents' COD on this run were 83,333 mg/l. The effluents' COD achieved on the 4th run was the best compared to 3 previous runs. The value vary within 10,000-20,000 mg/l. The effluents' clarity were increased compared to the previous run. The COD removal on 4th run presented in figure 5:

5th Run (T = 55 ± 2^{0}C ; Co-digestion ratio 1:0)

On this 5th run, the EGSB reactor operated in Thermophilic condition (Temperature of 55 ± 2^0 C) and there is no addition of EFB as co-substrates. The influents' COD value on 5th to 8th run was equal to the 1st to 4th run because the POME to EFB ratios was identical.

The effluents' COD were increased slightly compared to the Mesophilic condition. From the observations during the 5th run, thermophilic condition on EGSB reactor was more unfavorable for Methanogens bacteria in the granular sludge bed. The conversion of organic acid into biogas by Methanogens bacteria became obstructed. The effluents coming out of the reactor looked like a bit murkier compared to the effluents on 1st run.

The COD removal at this stage varies within 48.44 – 62.50%. The COD removal on 5th run presented in figure 6.

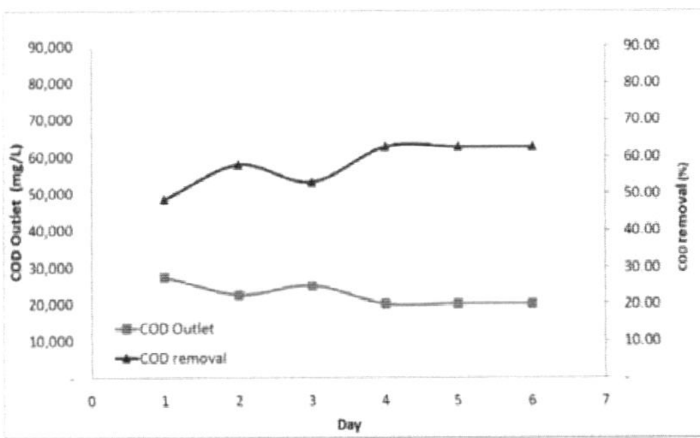

Figure 6. COD removal on *the* 5th run.

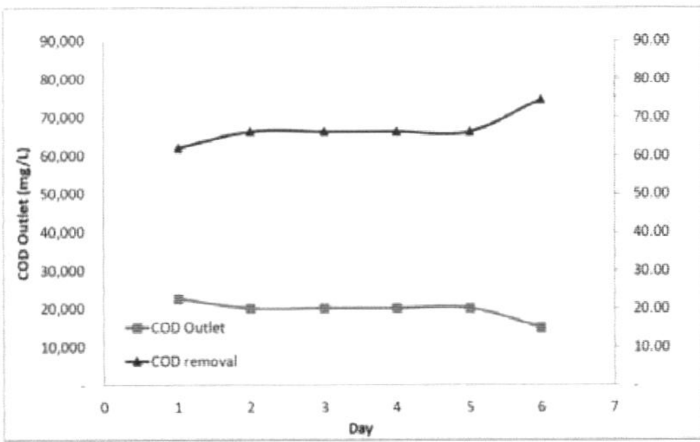

Figure 7. COD removal on the 6th run.

6th Run (T = 55 ± 2^{0}C ; Co-digestion ratio 1: 0.10)

Co-digestion ratios on 6th to 8th run were the same as on 2nd to 4th run. The difference was in the temperature conditions used. The condition of the reactor on 6th to 8th run was Thermophilic (Temperature 55 ± 2^0 C).

From the observations made during the 6th run, it is known that the thermophilic condition causes negative effects in terms of biogas production. The organic acid conversion by Methanogens bacteria was limited, thus increased the VFA concentration in the reactor. The volume of biogas generated in this run was far below the 2nd run.

There was no major difference in terms of effluents' COD value compared to the 5th run. The effluents' COD on 6th run vary within 15,000-22,500 mg/l. On 2nd to 5th operational day, the effluents' COD was stable at 20,000 mg/l.

The COD removal vary within 61.97 – 74.65% as presented in figure 7.

7th Run (T = 55 ± 2^{0}C ; Co-digestion ratio 1: 0.15)

The EFB addition to the ratio of 1:0.15 on this run was not increase the biogas production said reactor. During the 7th run, it was observed that the thermophilic condition of the reactor help increased the substrate decomposition rate. More organic substrate converted to organic acid caused increased the VFA concentration and pH, hence the reactor became unstable.

The average influents' COD on the 7th run was 66,667 mg/l, effluents' COD increased slightly compared to the previous run, i.e., in the range of 17,500-25,000 mg/l. The COD removal vary within 62.50-73.75% as presented in figure 8.

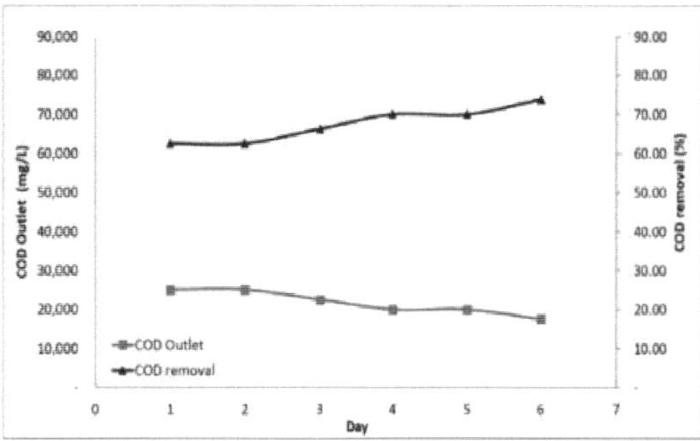

Figure 8. COD removal on the 7th run.

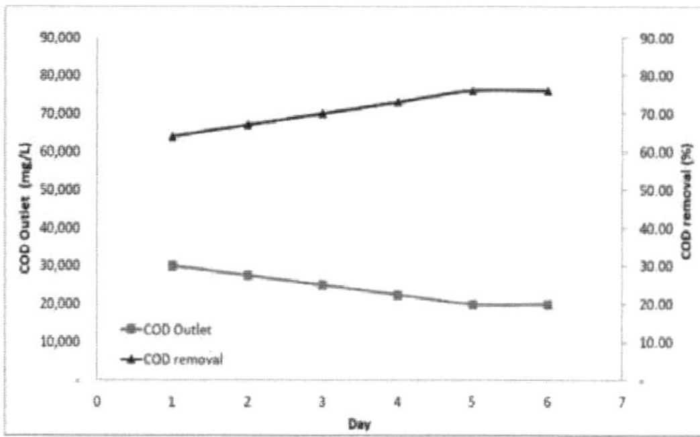

Figure 9. COD removal on the 8th run.

8th Run (T = 55 ± 2^{0}C ; Co-digestion ratio 1: 0.20)

The effluents' COD on the 8th run was much higher compared to any other previous, i.e., in the range of 20,000 – 30,000 mg/l. The physical appearance of the effluents on the 8th run was pale yellow colored liquid with the strong sour smelling. The COD removal on this run was 64.00 – 76.00% as presented in figure 9.

3.2. *Volatile fatty acid to total alkalinity ratios*

The volatile fatty acids to total alkalinity ratios in the EGSB reactor reflects how high the value of organic acids which has been converted by bacterial fermentation of organic matter. When linked with the COD removal and biogas volume generated, this ratio indicates the equilibrium of organic materials/fatty acids/methane gas that is formed [9, 10].

EGSB reactor's VFA/TA ratios performance-based detailed as follows:

1. Under 0.15; The reactor was stable, most of the organic material was successfully converted into fatty acids and Biomethane.
2. Between 0.15 to 0.20; reactor needs some attention and approaching the state of overloading.
3. Between 0.20 to 0.25; the reactor was on the verge of overload, there were limited bacteria present in the reactor that could convert the organic material to biogas.
4. Above 0.25; the reactor has been acidified, an organic material that goes largely converted into fatty acids, and only a fraction are converted into methane gas [9].

Details of VFA/TA ratios on every run in EGSB reactor presented in table 4.

Table 4. VFA/TA ratios on every run.

Temperature	*Run*	POME: EFB ratios	VFA: TA ratios					
			Day 1	Day 2	Day 3	Day 4	Day 5	Day 6
35 ± 2^{0}C (Mesophilic)	1st	1: 0.00	0.1477	0.1489	0.1515	0.1505	0.1545	0.1396
	2nd	1: 0.10	0.1538	0.1427	0.1419	0.1419	0.1566	0.1488
	3rd	1: 0.15	0.1663	0.1548	0.1442	0.1442	0.1492	0.1575
	4th	1: 0.20	0.1575	0.1457	0.1501	0.1501	0.1094	0.1281
55 ± 2^{0}C (Thermophilic)	5th	1: 0.00	0.5758	0.5256	0.4647	0.4647	0.5700	0.6053
	6th	1: 0.10	0.5385	0.4498	0.6683	0.6683	0.5531	0.6386
	7th	1: 0.15	0.5122	0.5190	0.4794	0.4794	0.4184	0.4678
	8th	1: 0.20	0.5135	0.5395	0.5705	0.5705	0.5836	0.5977

The reactor was likely exhibited stable performance in mesophilic condition although it has received a fairly high organic load, whereas in the thermophilic condition reactor tend to be more acidic (high concentration of VFA).

3.3. Biogas volume

The addition of EFB as co-substrate with the ratio of 1:0.10 enhanced the biogas production in mesophilic. It can increase the production of biogas by 3.06% compared to a single substrate (POME). EGSB reactor was at stable condition during mesophilic operational temperature, although it had actually approached the overload condition.

Inversely proportional to the mesophilic condition, the thermophilic condition with various POME to EFB ratios have no effect against the increase of biogas volume generated. The addition of co-substrate only increases the settle-ability of influents at the bottom of the EGSB reactor.

Mesophilic condition

The reactor exhibited good performance in converting organic substrate to organic acid but limited to the biogas conversion. The highest biogas accumulation occurred on 2nd run, where POME to EFB co-digestion ratio was 1:0.10 and temperature at 35 ± 2^0 C. During mesophilic condition, biogas production observed was not always constant.

During the operational stage, it was observed that the unstable production rate of biogas closely related to influent pump operations, expansion of granular bed and influent upflow patterns resulted by nozzle outlet. A large number of Biogas bubbles trapped among the granular sludge bed were presumed as the main cause of this incident. The volume of production of Biogas in conditions of Mesophilic is presented in figure 10.

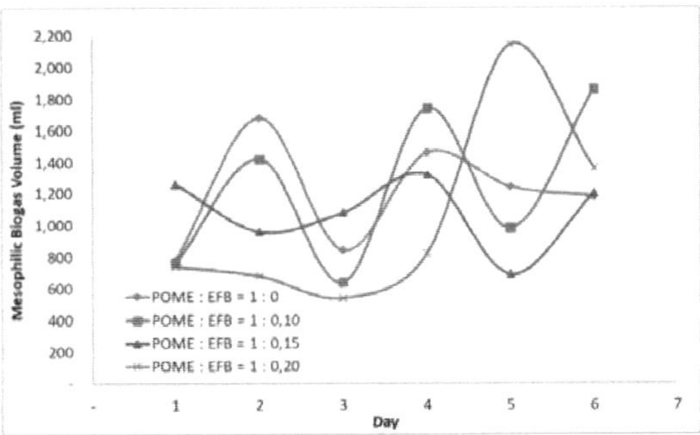

Figure 10. Biogas volume on mesophilic condition.

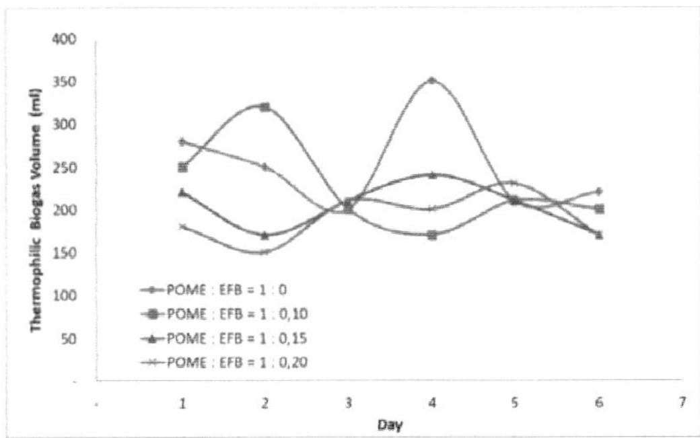

Figure 11. Biogas volume on thermophilic condition

Thermophilic condition

The biogas generated during thermophilic condition were less compared to the mesophilic, i.e., in the range of 150 – 350 ml. The various co-digestion ratio did not have any positive correlation in terms of biogas production by EGSB reactor. The biogas volume in thermophilic condition presented in figure 11.

3.4. Study comparison

POME and EFB co-digestion process in EGSB reactor at Mesophilic (temperature of 35 ± 2^0 C) and the ratio of 1:0.10 would enhance the biogas production. The study showed that the yield of Biomethane was 0.1359 - 0.1619 m^3 CH_4/kg COD, slightly higher when compared to previous research by Yejian *et al.* (2008) (0.112 m^3 CH_4/kg COD) [11] and Yejian *et al.* (2015) (0.108 m^3 CH_4/kg COD) [8].

The higher Biomethane yield in this study was generated by the addition of EFB as co-substrate, which benefits the bacterial consortia in granular sludge as for their source of carbon-rich nutrients. The initial drying at the temperature of 110^{0}C for 3 hours would help the bacterial consortia in doing EFB's lignocellulose hydrolysis and acidogenesis [4]. Breaking the lignocellulose would enhance the carbon content of the substrate mixtures, hence increased the opportunities of acetates conversion into biomethane by bacteria consortia.

The Biogas volume achieved in thermophilic (temperature of 55 ± 2^{0}C) with the same co-digestion ratio of 1: 0.10 were less compared to the mesophilic. The Biomethane yield was 0.0859 - 0.1358 m^3 CH_4/kg COD. They were far below the Biomethane yield achieved on previous studies, i.e. [4]; 0.338 m^3 CH_4/kg COD. O-Thong (2012) also used EFB as co-substrate with a ratio of 1: 6.8, in thermophilic (temperature of 55^0 C), HRT 45 days.

O-Thong also has 2 (two) of POME biogas research in 2011 with UASB and EGSB reactor, HRT 5 days without using co-digestion processes [12]. The results of the study showed a lower biomethane production when compared to this research, i.e. 0.0053 m3 CH_4/kg COD for UASB reactor and 0.0036 m^3 CH_4/kg COD for EGSB reactor.

Basically, the substrates decomposition rate were much higher in thermophilic conditions, because it helps the microbial consortia to grow rapidly [13]. Thus, the lignocellulose hydrolysis and acidogenesis were more likely to be done in this condition compared to the mesophilic [4]. So, it was lowered the reactor's pH which was 4.52 to 4.72 in thermophilic and 6.13 to 6.81 in mesophilic.

The low production of biomethane in thermophilic occurred due to the long chain fatty acids contained in undiluted POME. Long-chain fatty acids such as Palmitic acid (over 50 mg/g of dry weight) and oleic acid (above 200 mg/l) would inhibit the decomposition process by microbial consortia and Biomethane production in the reactor [4, 14, 15].

4. Conclusion

The biogas volumes achieved in mesophilic were much higher compared to the thermophilic condition. The largest biogas volume obtained at mesophilic (temperature of 35 ± 2^0 C), and POME to EFB ratio = 1: 0.10, was 7,400 ml for 6 days of operation (0.1359 – 0.1619 m^3 CH_4/kg COD). The COD removal was 61.97 – 78.87%. The reactor operation was stable, as related to the ratio of VFA/TA 0.1418 – 0.1582. The Biogas content analysis showed that CH_4 generated was 46.10%; 26.70% of CO_2; 0.80% of O_2 and 1.119 ppm of H_2S. The biogas volume achieved in thermophilic (temperature of 55 ± 2^0 C) and POME to EFB ratio of 1: 0.10, was 1,350 ml for six days of operation (0.0859 – 0.1358 m^3 CH_4/kg COD). The COD removal was 61.97 – 74.65%. The reactor was already unstable, and pH varies from 4.52 to 4.72. The VFA/TA ratios were also increased rapidly, i.e., 0.4498 – 0.6683 (> 0.25; reactor was acidified, most of the organic material were converted into fatty acids but only some fractions were converted into methane gas). The Biogas content analysis showed that CH_4 obtained was 41.25%; 29.80% of CO_2; 0.76% of O_2 and 1.126 ppm of H_2S.

References

[1] APHA (American Public Health Association) 2005 *Standard Methods for the Examination of*

Water and Wastewater 20th Edition 20th Edition, ed Lenore S et al (Washington DC: American Public Health Association)
[2] Ohimain E I and Izah S C 2014 Potential of biogas production from palm oil mill effluent in Nigeria *Sky Journal of Soil Science and Environmental Management* **3** 50 – 58
[3] National Energy Board 2015 Indonesia Energy Outlook, Ministry of Energy and Mineral Resource
[4] O-Thong S, Boe K, and Angelidaki I 2012 Thermophilic anaerobic co-digestion of oil palm empty fruit bunches with palm oil mill effluent for efficient biogas production *Applied Energy* **93** 648–54
[5] Kim H S, Choi S, Ju H and Jung J 2013 Mesophilic co-digestion of palm oil mill effluent and empty fruit bunches, *Environmental Technology* **34**(13-14) 2163 -70
[6] Ahmad A, Shah, Othman, and Abdullah 2014 Aerobic and anaerobic co-cultivation of Nannochloropsis oculata with oil palm empty fruit bunch for enhanced biomethane production and palm oil mill effluent treatment *Desalination and Water Treatment* **56** 2055-65
[7] Ahmad A, Shah, Othman and Abdullah 2014 Enhanced palm oil mill effluent treatment and biomethane production by co-digestion of oil palm empty fruit bunches with Chlorella sp. *Can J Chem Eng* **92** 1636–42
[8] Yejian Z, Hairen Y, Xiangyong Z, Zhenjia Z, and Li Y 2011 High-rate mesophilic anaerobic digestion of palm oil mill effluent (POME) in Expanded Granular Sludge Bed (EGSB) reactor *Int. Confer. Agricul. Natural Res. Engine. Adva. Biomed. Engine.* **3-5** 214-19
[9] Malina J F and Pohland F G 1992 *Design of anaerobic processes for the treatment of industrial and municipal wastes, Water Quality Management Library* vol. 7 ed Malina JF and Pohland F G (Florida, USA: CRC Press)
[10] Hasanah A S, Kurnani B A and Joni I M 2016 Bacteria isolate growth examination from cow feces for production engineering of methane gas with subbituminous coal media *E-Jurnal Unpad* Universitas Padjadjaran
[11] Yejian Z, Li Y, Lina C, Xiuhua L, Zhijian M and Zhenjia Z 2008 Startup and operation of anaerobic EGSB reactor treating palm oil mill effluent *J. Env. Sci.* **20** 658-63
[12] O-Thong S, Boe K and Angelidaki I 2011 Comparison of UASB and EGSB reactors performance, for treatment of raw and deoiled palm oil mill effluent (POME), *J. Hazard. Mater.* **189** 229–34
[13] Najafpour, Hasnain, Isa, Mohamed, Nasrollahzadeh and Zinatizadeh 2006 High-rate anaerobic digestion of palm oil mill effluent in an up-flow anaerobic sludge-fixed film bioreactor, *Process Biochem.* **41** 370–79
[14] Speece R E 1996 *Anaerobic biotechnology for industrial wastewaters* (Nashville, Tenn: Archae Press)
[15] Yu HQ, Mu Y and Fang HHP 2004 Themodynamic analysis of product formation in mesophilic acidogenesis of lactose *Biotechnol Bioeng.* **87**(7) 813–22

Acknowledgment
This study was supported by PT. Perkebunan Nusantara V.

Self-synthesize and flexural strength test porcelain from Indonesian natural sand

J Gunawan, D Taufik, V Takarini and Z Hasratinigsih

Department of Dental Materials, Faculty od Dentistry, Universitas Padjadjaran, Jl. Raya Bandung-Sumedang, KM 21, Jatinangor, Sumedang, Jawa Barat,45363, Indonesia.
E-mail: ocephjoseph@gmail.com

Abstract. Dental porcelain is usually used for aesthetic indirect restoration. Porcelain powder is still imported, although Indonesia is rich with natural sand as porcelain raw material. Flexural strength is a mechanical property which is related to porcelain's ability to resist occlusal force. The objective of this study is to synthesize porcelain from the mixture of Sumateran and Javanese natural sand and evaluate its flexural strength. This experiment used explorative descriptive method. Feldspar Pangaribuan (65% wt), silica Belitung (25%wt), kaolinite Sukabumi (5%wt), and potassium salt (5%wt) was mixed with wet-ball mill and dissolved by fritting to produce porcelain powder. Five samples were made by liquid- phase method in size 3mm x 2cm x 7cm. XRD test was used to analyse crystalline formed and flexural strength of these samples was measured by Netzch Universal Testing Machine with 7.5kg initial load and calculated by flexural strength formula. The result was porcelain powder which has been sintered on 1150 ^{0}C contain leucite (potassium aluminium silicate) crystalline on XRF test with average flexural strength 26.678 MPa. Leucite crystalline is the main indicator for synthesizing dental porcelain because leucite is not usually found on other ceramic. Leucite crystalline is known for increasing strength, preventing crack propagation, and resisting pyroplastic flow. Ten percent of leucite on X-Ray Fluoroscence shows the percentage is less than other dental porcelain. The conclusion of this research is that porcelain succeeded to be self-synthesized from the mixture of Sumateran and Javanese natural sand and having flexural strength 26.678 MPa.

1. Introduction

Porcelain is a type of ceramic which contains glass matrix phase and at least one crystalline phase. Porcelain was used as one of indirect restoration material in dentistry [1, 2]. For its esthetic and mechanical properties, porcelain is fit to restore massive damaged teeth structure. Mechanical property, flexural strength, is really important on this biomaterial as this is related to porcelain's ability to resist occlusal force [1, 3].

On the other hand, production price of dental porcelain is expensive. This might be caused by dental porcelain materials still imported from abroad. In fact, all raw materials of dental porcelain can be obtained in Indoneisa. Feldspar, as the main constituent, can be found at its best at Aceh and North Sumatera. Silica with least impurities can be found at Belitong and kaolinite can be found at any region in Indonesia. [4, 5]

This study aims to synthesize porcelain from the mixture of Sumateran and Javanese natural sand and evaluate its flexural strength. Hypothesis of this experiment is dental porcelain can be synthesized from the mixture of Sumateran and Javanese natural sand (indicator: leucite crystalline on XRD & XRF analysis) and have average flexural strength similar with marketed porcelain's.

2. Materials and methods

This experiment was done in several steps: mixing and fritting raw materials, sintering, X-Ray Fluorescence testing, and flexural strength test.

On this experiment, porcelain powder was made with composition of 65%wt Pangaribuan feldspar, 25%wt Belitung silica, 5%wt Sukabumi kaolinite, and 5%wt potassium salt. Initially, all the raw materials was milled by wet ball mill with 97% ethanol for 4 hours. Milling output was dried with drying cabinet on 50°C for 24 hours. Then, dry mixtures were molded to aluminous crucible for fritting. Fritting was done on the furnace with firing graph as below (figure 1) [6, 7].

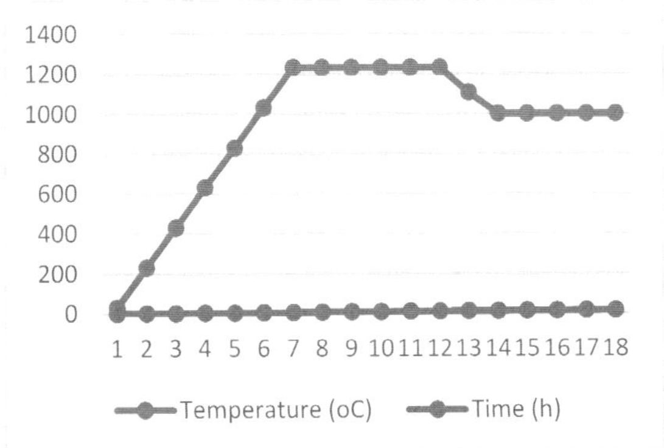

Figure 1. Fritting graph

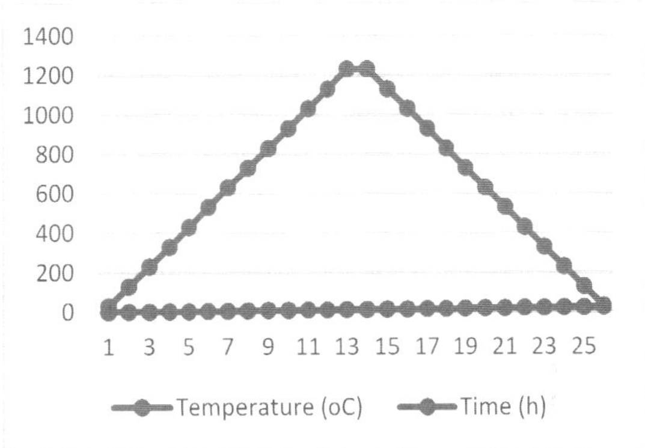

Figure 2. Sintering graph.

On the temperature of 1000°C, crucible was taken out and quenched in iced water until water's temperature is equal with crucible's. Fritting product was then grounded until 200 mesh sized. This powder, called as porcelain powder, manipulated with little aquades for making bar samples as flexural strength test needed. Five samples with size of 70mm x 20mm x 3 mm was sintered (figure 2) on phosphate-bonded gypsum impression [7, 8].

Five samples were tested by Netzsch universal testing machine with 50mm jig range, 1, 6mm lever diameter, and 7.5 kg smallest load. This machine went on the intercalation load of 0.25kg/second. Calculation of flexural strength then performed on the data obtained by the following equation (1) [5, 7].

$$FS = \frac{1.5FL}{bt^2} \tag{1}$$

Where FS: flexural strength; F:biggest load before sample fractures; L: sample length; b: sample width; t: sample thickness.

Broken samples after flexural strength testing were then prepared by milling to 200 mesh sized and tested by Philips Analytical X-Ray BV Lelyweg 1 type 9430 01830401 to wit the crystalline produced in the porcelain.

3. Results and discussion

Dental porcelain on this experiment was synthesized by using 65% wt Pangaribuan feldspar, 25% wt Belitung silica, 5% wt Sukabumi kaolinite, and 5% wt potassium salt. Feldspar was used as glass matrix former, silica as filler and framework, kaolinite as adhesive, and potassium salt increase potassium component. Sintered porcelain shown in white colour and had fairly good translucency (figure 1). [1, 8] Color and translucency are related to particle size and porcelain maturity. Suitable particle size will provide good transparency along with the mechanical properties. The smaller particle size, the closer and stronger bond between particles which means less translucency but greater mechanical properties. While less firing temperature can reduce translucency and mechanical properties [9, 10].

Figure 3. Sintered porcelain.

XRD test results show 2θ most intense are on 27.13; 25.97; and 26.59. The analysis was then performed in accordance with ICDD 15-0047: 3 highest peak of crystalline leucite was at 27.34; 25.93; and 16.52. This indicates the presence of quartz and crystalline leucite in porcelain composition.

X-Ray Fluorescence presented approximately 10% crystalline leucite was on this porcelain. Approximate 10% leucite crystalline found from stoichiometry of the percentage of SiO2 (58.60%), Al2O3 (20.52%), K2O (10.38%) (table 1). This shows that dental porcelain has been succeeded to be synthesized. On the other hand, leucite crystalline concentration on this dental porcelain is lower than the marketed dental porcelain. This crystalline was formed from the mixing of materials, especially silicate, alumunium oxide, and potassium oxide, which greatly affect the formation of triaxial pottasium alumino silicate. Ratio between pottasium oxide and sodium oxide on the raw materials alse affect the leucite production, the more potassium the more leucite will be formed. Leucite crystalline need to be more considerable as it improves mechanical properties, including strength and hardness [2, 3, 10].

Table 1. The X-Ray fluorescence data of self-synthesized porcelain.

Element	Percentage (%)	Element	Percentage (%)
SiO_2	58.60	Al_2O_3	20.52
K_2O	10.38	Na_2O	7.58
CaO	1.16	PbO	0.59
MgO	<0.001		

The result of flexural strength test on self-synthesized dental porcelain was 26.678MPa (table 1), while marketed dental porcelain's is 65-120MPa. Differences on testing standard (ANSI/ADA 69) were occured on sample size and testing machine requirement. Bigger and thicker sample size on this experiment induced smaller flexural strength. This occured as thicker materials has less strain and easier to crack on flexural strength test [7, 11]. Testing machine requirement on ANSI/ADA 69 is smallest load 5.4kg and intercalation load speed 0.25kg/second. On other hand, testing machine used here has heavier initial load and faster intercalation load speed. This mattered on the more easily samples fractured and the calculation become less accurate. The force loaded on samples becomes dinamic force that strength tested not only flexural but also impact strength [5, 7, 12].

Table 2. Flexural strength test data of self-synthesized porcelain.

Sample Number	F (kg)	L (cm)	B (cm)	T (cm)	Flexural Strength Test (kg/cm^2)
1	11.25	5.0	1.628	0.424	288.289
2	12.50	5.0	1.653	0.432	303.900
3	12.50	5.0	1.621	0.497	234.140
4	12.00	5.0	1.644	0.451	269.146
5	11.00	5.0	1.704	0.424	269.310
				AVERAGE	272.957

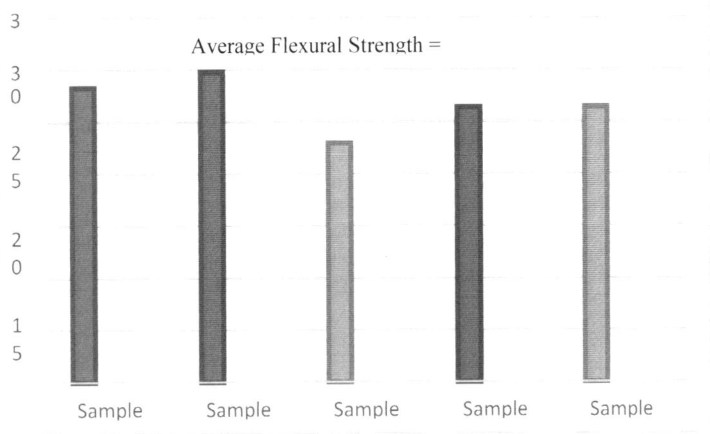

Figure 4. Flexural strength data (MPa).

On the other hand, suitable sintering temperature and time are taking effect on flexural strength. On pre-experiment procedure, that was found that sintering temperature and time is 1150^{0}C in an hour. However, on this experiment with larger size of sample, sintering temperature and time should be higher because of the unmatured porcelain. The calculation of sintering temperature should be made correctly as this affect the maturity and pores production. The unmature and overheating porcelain that will lead to pores production are decreasing dental porcelain flexural strength. [10, 13]

4. Conclusion

From this study, it is shown that the synthesized porcelain contains leucite crystalline which means this porcelain can be used as dental porcelain. The flexural strength of this porcelain is 26.678 MPa and this should be tested further by the method that is suitable with dental porcelain standard (ANSI/ADA) so that can be compared with marketed dental porcelain. Visualization of synthesized porcelain which is probably opaque can indicate this porcelain as body layer, however there is a probability to produce translucent porcelain by manipulate the composition.

References

[1] Babu P, Alla R, Alluri V, Datla S R, and Konakanchi A 2015 *AJMSE* **3** 13–18
[2] Helvey G A 2013 Classification of dental ceramics *Inside Continuing Education* **13** 62–68
[3] Raghavan R N 2012 *Ceramics in Dentistry, Sintering of Ceramics - New Emerging Techniques*, Dr. Arunachalam Lakshmanan (Ed.) (Croatia: InTech), Available from: http://www.intechopen.com/books/sintering-of-ceramics-new-emerging-techniques/ceramics-in-dentistry
[4] Adrian 2001 Potensi batuan pegmatif Pangaribuan sebagai sumber alumina dan alkali untuk pembuatan gelas dan kaca *Makalah Direktorat Jenderal Geologi dan Sumber Daya Mineral* **3** 1–20
[5] Direktorat Jenderal Agro Industri & Departemen Perindustrian 2009 Roadmap industri keramik *Jurnal Dirjen Agroindustri* **21** 1–10
[6] Katz S 1987 Translucent dental porcelain composition, its preparation and a restoration made thereof. *European Patent Application* **9** 1–15
[7] Panzera C, Mead B, and Kalser LM 1999 Dental Porcelain Composition *United States Patent Application* **17** 1–6
[8] Shenoy A and Shenoy N 2010 Dental ceramics: An update *J. Conserv. Dent.* **4** 195– 03
[9] Rahaman M N 2007 *Ceramic Processing* (Parkway: Taylor & Francis Group) **2** 365–81
[10] Ranogajec J, Djuric M, Radeka M, Jovanic P 1999 Influence of particle size and furnace atmosphere on the sintering of powder for tiles production *Ceramics - Silikaty* **2** 71–77

[11] Carter CB and Norton MG 2007 *Ceramic Materials Science and Engineering* (New York: Springer Science) **1** 365–80
[12] Denry I and Holloway J A 2010 Ceramics for dental applications: A review *Materials (Basel)* **1** 351– 68
[13] Gozneli R, Kazazgolu E and Ozkan Y 2013 Flexural properties of leucite and lithium disilicate ceramic materials after repeated firings *JDS* **9** 144–50

Acknowledgment
The authors thank the Centre for Ceramics, Ministry of Industry, Indonesia. The deepest gratitude was addressed to Universitas Padjajaran for financial support through the HIU (hibah Intenral Unpad).

Numerical investigation of droplet evaporation phenomena in the spray methods

F F Florena [1], F Faizal [2,3], D W Maulana [2], S T Satrio [2], C Panatarani [2,3], I M Joni [2,3] and S Viridi [1]

[1] Department of Physics, Faculty of Mathematics and Natural Sciences, Bandung Institute of Technology, Jl. Ganesha No 10, 40132, West Java, Indonesia
[2] Nanotechnology and Graphene Research Center, Universitas Padjadjaran, Jl. Raya Bandung-Sumedang KM 21, 45363, West Java, Indonesia;
[3] Department of Physics, Faculty of Mathematics and Natural Sciences, Universitas Padjadjaran, Jl. Raya Bandung-Sumedang KM 21, 45363, West Java, Indonesia;

E-mail: c.panatarani@phys.unpad.ac.id

Abstract. Physical phenomena of the droplet evaporation occurred in the spray pyrolysis reactor tube has been studied. This model simulation was expected to predict the process occurred in the reactor tube and predicted the size and morphology of particle. The dynamical model of particles influenced by macroscopic parameters such as reactor wall temperature, flow rate carrier gas and concentration of precursor have been successfully simulated. The size of the particle, the number of aqueous vapors around droplet, air temperature around the droplet, temperature of droplet and solution concentration profile of particle in the reactor tube were simulated. The desired final particle can be achieved by controlling these factors.

1. Introduction

The spray method is one of the nanoparticle synthesis methods that are now being developed by researchers. The spray is the generation of small droplets originating from the liquid phase medium. The method commonly used in the synthesis of nanostructured particle solids is spray drying and spray pyrolysis. The method is initiated from a solution of metal nitrate or other atomized solution, and then after going through a heating system, the solid particles will be formed. Spray methods have many advantages, including; the resulting particles are in the form of spherical, the diameter distribution is uniform and can be controlled from micrometers to sub-micrometers; the purity of the particles produced is very high, the process takes place continuously [1–5].

Simulation is a bridge between theory and experiment that has been accepted as one of the methods of material research and development. An experiment material can be preceded by computer simulation to determine the conditions needed or predict the results. Computer simulations can function to complement experiments. The advantage of computer simulations is easy to perform. The results obtained from computer simulations can be a benchmark for the experimental methods carried out.

A computational approach was used to investigate the spray drying process. In spray drying experiments, there are several parameters such as carrier gas flow (Q), reactor wall temperature (T_w), initial concentration of droplet solution (C_o), the number of concentrations of droplets entering the reactor (N_o) and initial droplet (dp_o) diameter size that affects the reactor length needed for droplets to be dry particles, the size of the final droplet, the amount of water vapor concentration on the droplet surface, the amount of water vapor concentration around the droplet, the temperature on the droplet surface and the air around the droplet. Therefore computer simulations can be varied so that the parameters can predict the results obtained and can be used as a reference for experiments spray drying.

In the current study, the models presented by Jayanthi [6] and Kodas [7] were used. This research aims to simulate physical phenomena (particle dynamics) of droplets in a tube reactor using the spray drying methods and to get a deep understanding of mechanisms involved in the particle formation to obtain desired final particles. Relationships between the operating conditions and final properties of particles are investigated.

2. Experimental procedure

Study on droplet evaporation and particle formation in the reactor are depicted in figure 1. The model describes changes in droplet diameter and continuous particle morphology based on multiple evaporations of droplets along the tubular aerosol reactor and together with the distribution of solution concentrations in the droplet [1]. The droplet (monodispersed droplets) assumed has diameter d_p, droplet concentration N_o and solution concentration C_o. The droplet is transported into the reactor using the average carrier gas flow Q, relative humidity RH^o and initial temperature T_o. The reactor has length L and diameter $2R$ and temperature T_W. As the solvent evaporates, the size of the droplet decreases and the concentration of the solution inside increases while moving inside the reactor [1]. The concentration of the solution in the droplet is in the critical area during the nucleation of the solution, precipitation, decomposition and solid processes that occur in the particles.

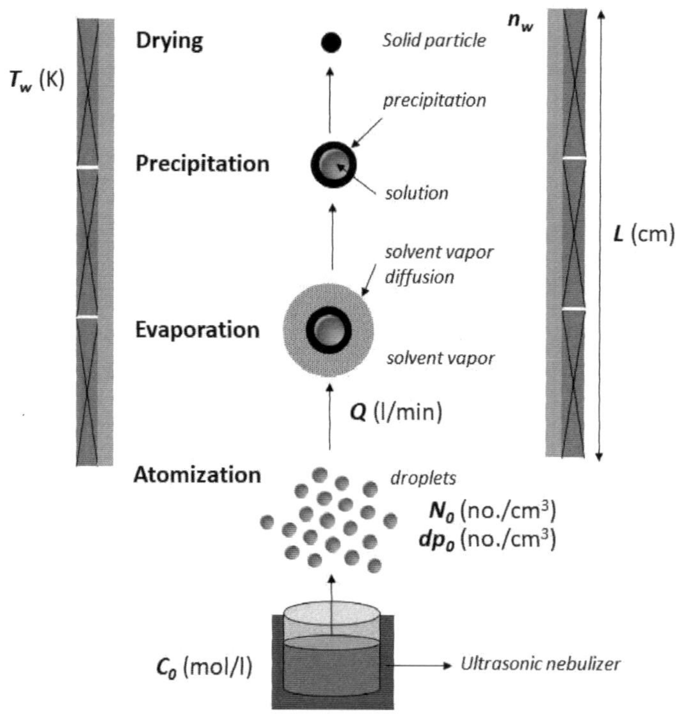

Figure 1. Illustration of simulation model.

Evaporation of solvents causes the concentration of a solution to increase on the droplet surface and form a concentration gradient which is the result of diffusion of the solution. At critical supersaturation (CSS), if the concentration of the solution at the center of the droplet ($\frac{r}{R} = 0$) is higher than the equilibrium concentration (ES) [1,8,9], the solid phase will nucleate and produce a solid particle. Meanwhile, if the concentration of the solution in the center of the droplet is smaller than the equilibrium concentration (ES), hollow particles will be formed [1].

The modeling is equated with limiting the problem. Changes in diameter d_p (residence time) along the reactor can be stated as follows:

$$\frac{dd_p}{dt} = \frac{4D_v m_l}{\rho_p dp N_A}(n - n_s) \tag{1}$$

with D_v (cm^2 /s) is the vapor diffusion coefficient in air at T, m_l (g/mol) is the mass of the water molecule, n is the concentration of vapor (molecule / cm^3) in the air at T, n_s is the concentration of vapor (molecule / cm^3) at the droplet surface at T_s and ρ_p are droplet density, which is a function of particle size.

Changes in droplet surface temperature (T_s) over the residence time along the reactor is given by latent heat of evaporation of water and heat transfer from the air surrounding it [1].

$$\frac{dTs}{dt} = \frac{1}{C_p d_p}\left[3H_L \frac{dd_p}{dt} + \frac{6h_s}{\rho_p}(T - T_s)\right] \tag{2}$$

With C_p is droplet heat capacity, H_L is latent heat from evaporation of water and h_s is heat transfer coefficient at the droplet. Changes in the concentration of vapor with respect to residence time are given by:

$$\frac{dn}{dt} = \left[-2\pi d_p D_v N_o(n - n_s) - \frac{2K_m(n - n_w)}{R}\right] \tag{3}$$

N_o (no / cm^3) is the number of droplet concentrations, n_w a is the concentration of vapor at the reactor wall, K_m is vapor mass transfer coefficient for laminar flow, R is the reactor tube radius. The first term of equation (3) states the concentration of water vapor entering the air due to evaporation of droplets. The second term states the vapor lost due to the diffusion of water vapor into the reactor wall. Air temperature along the residence time due to the transfer of heat from the air to the droplet and from the reactor wall [10]:

$$\frac{dT}{dt} = \frac{1}{FC_{pa}}\left[-\pi^2 R^2 d_p{}^2 N_o h_s(T - T_s) + 2\pi R h_w(T_w - T)\right]\left(\frac{dz}{dt}\right) \tag{4}$$

with F is the molar flow rate of air, C_{pa} is the heat capacity of surrounding gas and h_w is the heat transfer coefficient at the reactor wall. By using the ideal gas law, aerosol residence times along the reactor can be stated as follows:

$$\frac{dt}{dz} = \frac{0.06\pi R^2}{Q}\left(\frac{To}{T}\right)\frac{(1 - yw)}{(1 - yw_o)} \tag{5}$$

t is residence time (s), z (cm) is the coordinate distance of the reactor, Q is the carrier gas flow rate (cm^3/s), T is the air temperature (K), y_w is the mole fraction of vapor in surrounding gas, y_{wo} is the mole fraction of vapor in surrounding gas at z = 0. The correction factor $\frac{(1-yw)}{(1-yw_o)}$ is calculated for changes vapor contained in the air flow as a result of evaporation of droplets. Assuming n_w = 0, the fraction of moles of water vapor in the air can be expressed as:

$$yw = \frac{n}{n + n_s} \tag{6}$$

whereas the mole fraction of water vapor in air at z = 0 can be expressed as:

$$yw_o = \frac{n_o}{n_o + n_s} \tag{7}$$

Equations (1) to (5) are coupled differential equations. The solutions to these equations are done simultaneously using the 4th order Runge-Kutta method. Order 4 of the Runge-Kutta method is one method for finding solutions to the first-order differential equation. The constants used in the model are obtained from Reid *et al.* [11] and some are calculated. The boundary condition of the mass and energy transfer on the reactor wall are determined by the reactor system. In the case of diffusion drying, the reactor tube is wrapped in a drying agent, so the wall behaves like a sink that is perfect for steam (n_w) and constant wall temperatures.

The initial conditions at the reactor inlet (z = 0) are: t = 0, $d_p = d_{po}$, $n = n_o$, and $T_s = T = T_o$. The initial vapor concentration with relative humidity in surrounding gas (RH^o)) is:

$$n_o = \frac{RH^o p_s{}^o}{kT_o} \tag{8}$$

with $p_s{}^o$ is the equilibrium vapor pressure to the temperature T_o and k is the Boltzmann constant. The vapor concentration at the droplet surface is given by:

$$n_s = \frac{p_s}{kT_s} \tag{9}$$

p_s is the droplet surface vapor pressure at temperature T_s. The presence of dissolved substances in water, resulting in the lowest balance of vapor pressure through pure solution rather than through pure solvent $p_s{}^o(T_s)$ is:

$$p_s{}^o(T_s) = a p_s{}^o(T_s) \tag{10}$$

with a is water activity in droplet solution which is influenced by the molality of the solution m and the osmotic coefficient of water φ, with the equation:

$$a = exp(-0.036m\varphi) \tag{11}$$

with φ is a function that alternates from m obtained from Pitzer and Mayorga [12]. By assuming the ideal mixed solution of dissolved and water, the droplet density is given by:

$$\rho_p = \rho_L + \left(\frac{d_{ps}}{d_p}\right)^3 (\rho_S - \rho_L) \tag{12}$$

with ρ_L is the density of pure water, ρ_s is the dissolved density and d_{ps} is the equivalent diameter of a complete drying droplet. The average molecular droplet is given:

$$b = \frac{1000\rho_s}{M_s\rho_L\left[\left(d_p/d_{ps}\right) - 1\right]} \tag{13}$$

with M_s is the soluble molecular weight (NaCl).

In relation to the higher transport resistance for diffusion of the solution in the droplet compared to the evaporation of solvents on the droplet surface, the concentration profile spreads in the droplet with the highest solution concentration on the surface. Assuming $d_p\left(dd_p/dt\right)$ is constant, then the concentration of solute in the radial direction r of the droplet center is [7]:

$$C = C_o\,exp\left[\left(\frac{K}{4D_L}\right)\left(\frac{r}{r_p}\right)^2\right] \tag{14}$$

with C_o is the solute concentration at the center of the droplet, $r_p = d_p/2$ is the radius of the droplet, D_L is the coefficient of diffusion of the solution while K is the mass balance on the surface of the droplet:

$$K = \frac{2D_v m_l(n_s - n)}{\rho_v N_A} \tag{15}$$

3. Results and discussion

The results obtained refer to some variations in the value of flow rate *(Q)*, variations in the value of reactor wall temperature (T_w) and variations in the value of the initial concentration (C_o) of NaCl solution. The three variations carried out in the program simulation, and in the experimental process, then the size of the droplet simulation results and experimental results are compared

3.1. Variation in Q (flow rate)

In this simulation, the flow rate was varied in three values: 1 l/min, 3 l/min, and 5 l/min. The effect of the average gas flow on the evolution of the size of droplets along the reactor can be seen in figure 2.

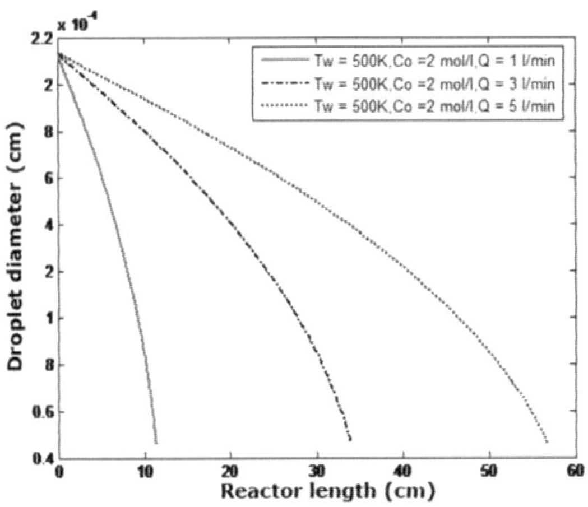

Figure 2. Effect of flow rate on the evolution of the size of NaCl droplets throughout the reactor.

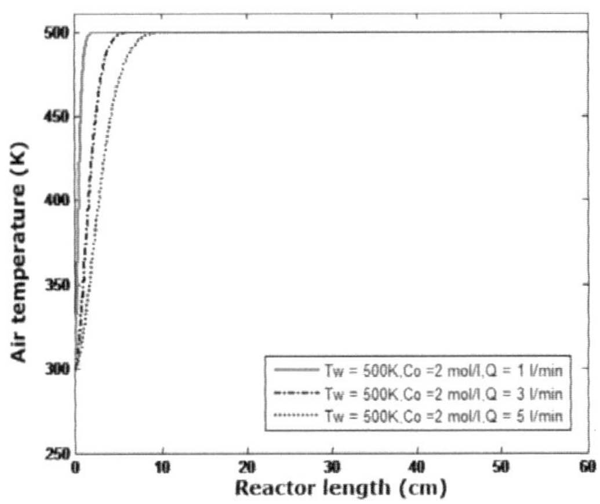

Figure 3. The effect of the average gas flow on air temperature.

Based on figure 2, the droplet diameter size evolved from 2.14×10^{-4} cm (the initial size of the droplet calculated on the ultrasonic nebulizer frequency at 1.7 MHz) up to 4.60×10^{-5} cm. The result shows that the size of the final droplet diameter is not influenced by variations in Q (flow rate) because the final droplet size remains the same (4.60×10^{-5} cm). However, the achievement of the last droplet size on the length of the reactor is different. The reactor length needed to reach dry particles (perfect evaporation process) is 11.3563 cm for $Q = 1$ l/min, while for $Q = 3$ l/min is 34.069 cm and for $Q = 5$ l/min is 56.7816 cm. The difference in reactor length needed to reach the final size of the droplet is due to the difference in the carrier gas flow rate carrying the droplet. The greater the Q value, the faster the carrier gas flow so that with the same evaporation time, the reactor length needed to reach the final size is different. The evaporation time needed by droplets for one evaporation process is 0.002 seconds. Therefore with the same evaporation time and different flow rates, the required reactor length is different. Therefore, the greater the Q value, the longer the reactor length for the droplet to completely evaporate the solvent and become a dry particle. The heat required by the droplet originating from the furnace that flows radiatively to the reactor wall, the heat reactor wall then flows through the air. The profile of the air temperature around the droplet can be seen in figure 3.

In figure 3, it can be seen that the value of air temperature rises from 299 K (initial air temperature, obtained from the measurement results in the experiment) up to 500 K. The difference in temperature transition rate towards stable values occurs for variation in Q. For $Q = 1$ l/min the air temperature reaches a stable value around the reactor length of 2.5 cm so that the air temperature has the same value with the reactor wall temperature. For $Q = 3$ l/min the reactor length needed to reach a stable temperature is around 5 cm while at $Q = 5$ l/min the length of the reactor required to achieve a stable value is about 10 cm. The difference in the transition temperature rate is due to differences in the velocity of the carrier gas. The smaller the Q value, the slower the speed of the carrier gas, the more time it takes for heat to flow from the reactor wall to the air, which will result in faster temperature transition speeds. In the other side, if the value of Q increased, the rate of the carrier gas will become faster so that the time needed for the heat to flow from the reactor wall into the air will reduce the speed of the transition temperature. The heat contained in the air would be flowed by gas particles to the droplet surface so that the heat is used by droplets to vaporize the solvents contained in the droplets. Heat can flow because of the temperature difference between gas and droplets. Temperature profiles on droplets can be seen in figure 4.

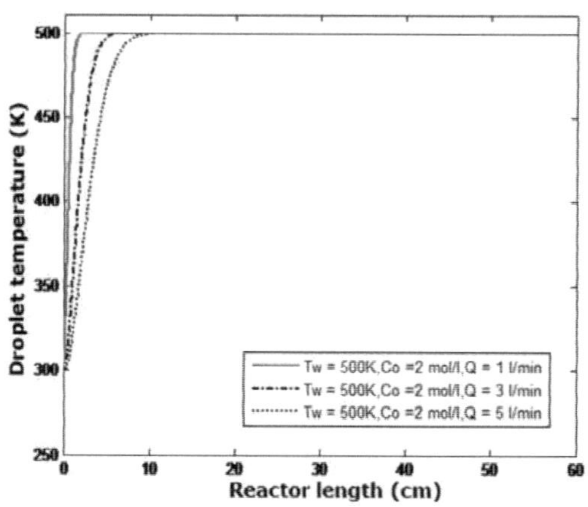

Figure 4. Effect of average gas flow on droplet temperature.

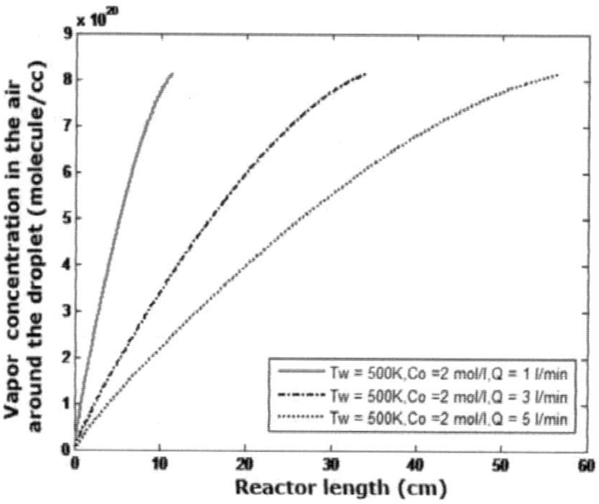

Figure 5. The effect of the average gas flow on the concentration of water vapor in the air around the droplet.

The result shows that the droplet temperature values increase from 299 K (initial air temperature, obtained from the measurement results) to 500 K. At the final condition (at the time of evaporation) the droplet temperature reaches 499.9212 K. This value is smaller than the air temperature (499.9595 K). The droplet temperature is quite enough to evaporate the solvent in the droplet because in this case water is used as the solvent and it requires temperatures of around 100° C or 373 K to evaporate completely. The droplet temperature transition rate is the same as the air temperature profile. Because the solvent in the droplet evaporates, the concentration of water vapor in the carrier gas increases. The profile of water vapor concentration in the gas around the droplet is shown in figure 5.

In figure 5, the increase in the concentration of water vapor rate is different for each value of Q. The smaller the value of Q, the faster the concentration of water vapor increases. This is because the flow velocity of the carrier gas was slow enough so that the time needed for heat to flow to the droplet is getting longer, as a result, the concentration of water vapor rate is getting faster. However, the amount of water vapor in the carrier gas during the evaporation time is the same. Meanwhile, if the value of Q is getting bigger, then the concentration of water vapor rate will increase. During the evaporation time the droplet concentration of water vapor in the carrier gas increases. The solvent in the droplet that evaporates into water vapor not only diffuses into the surrounding gas but also distributes to the droplet surface. The water vapor concentration profile on the droplet surface can be seen in figure 6.

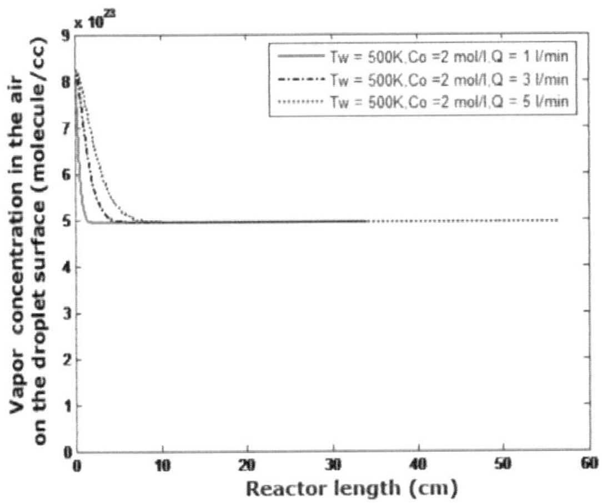

Figure 6. Effects of the average gas flow on the concentration of water vapor on the surface of the droplet.

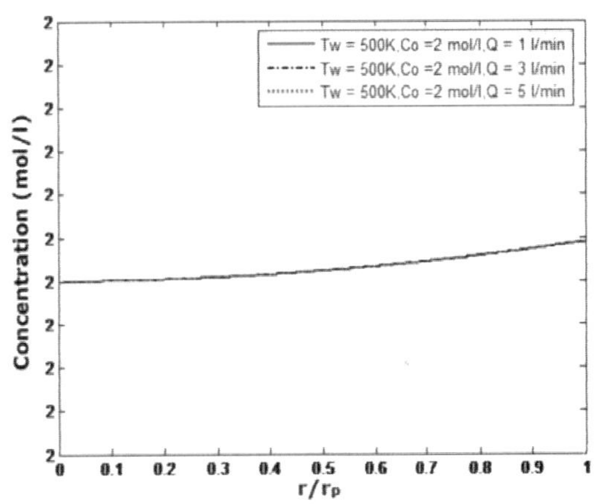

Figure 7. Effect of average gas flow on solution concentration on droplets.

Figure 6 shows that the concentration of water vapor on the droplet surface decreases with respect to the length of the reactor. This is because the heat on the droplet surface is used by water vapor particles to increase its kinetic energy and moved from the droplet surface to the air with the carrier gas. During the spray drying process, the concentration of the solution is changed gradually. The concentration profile of the solution on droplets could be seen in figure 7. It can be seen that the concentration of the solution on the droplet is higher when it gets closer to the surface. This is because when the solvent in the droplet underwent an evaporation process, the solvent carries dissolved particles on the droplet to the surface through diffusion which causes the concentration of the solution on the droplet surface to become higher than the solution concentration at the center of the droplet. However, the difference in the Q does not affect the magnitude of the concentration distribution in the droplet as the difference in Q only affects the gas flow rate and the droplet heating time. However, because the temperature used is constant, the difference in Q does not change the concentration distribution in the droplet, because the concentration distribution in the droplet is influenced by the temperature on the droplet and the initial concentration of the solution before the anatomized solution becomes droplets. The simulation results show that the last droplet size obtained was 4.60×10^{-5} cm or 460 nm.

3.2. Variation in Tw (wall temperature)

Value of reactor wall temperature is also an essential element in physical phenomena that occur in the reactor. The value of temperature was varied at 400 K, 500 K, and 600 K. The profile of the droplet size of the reactor temperature variation can be seen in figure 8. The result showed that the size of the droplet diameter evolved from 2.14×10^{-5} cm (the initial size of the droplet obtained from the measurement of frequency (at 1.7 MHz) from the ultrasonic nebulizer) to a certain size which varies in each reactor wall temperature variation. In the temperature range of 400-600 K, the linearly increasing temperature profile will result in a longer reactor length needed to reach the final size of the droplet. This is because of the higher the reactor wall temperature, the higher the heat given to the air around the reactor. Therefore, the kinetic energy of gas particles increases so that the gas velocity of the droplet particle carrier is getting fast and makes the required reactor length become longer. This will cause less heating time so that the droplet particles dry more entirely and cause the droplet size to be higher as the temperature increases. At 400 K the last droplet size obtained was $22. \times 10^{-5}$ cm, at 500 K the final droplet size obtained was 4.60×10-^{5} cm, while at 600 K the last droplet size obtained was 4.93×10^{-5} cm. The effect of the reactor wall temperature on the air temperature around the droplet can be seen in figure 9.

Figure 9 shows that the air temperature was transitioned from 299 K to 400 K, 500 K and 600 K according to the variation in the reactor wall temperature. The higher the temperature of the reactor wall, the higher the air temperature around the droplet. The air temperature profile will affect the droplet temperature profile. The droplet temperature profile can be seen in figure 10. As with air temperature, the higher the reactor wall temperature, the higher the droplet surface temperature in its steady state. The temperature transition rate is the same since the Q value in this simulation is the same.

The profile of water vapor concentration in the air around the droplet can be seen in figure 11. The result shows that the smaller the reactor wall temperature, the more water vapor is produced, result in a faster water vapor concentration rate. At 400 K the concentration of the final water vapor produced is 8.2235×10^{20} molecules / cm^3, at 500 K the concentration of the final water vapor produced is 8.13142 $\times 10^{20}$ molecules / cm^3 and at 600 K the final water vapor concentration produced is 8.04199×10^{20} molecules / cm^3.

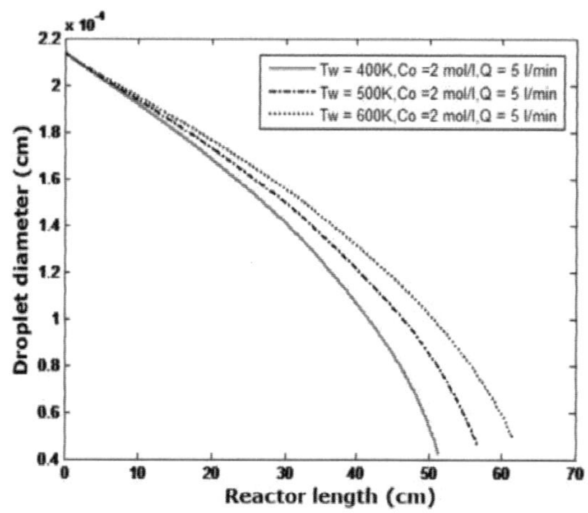

Figure 8. Effect of reactor wall temperature on the evolution of the size of NaCl droplets throughout the reactor.

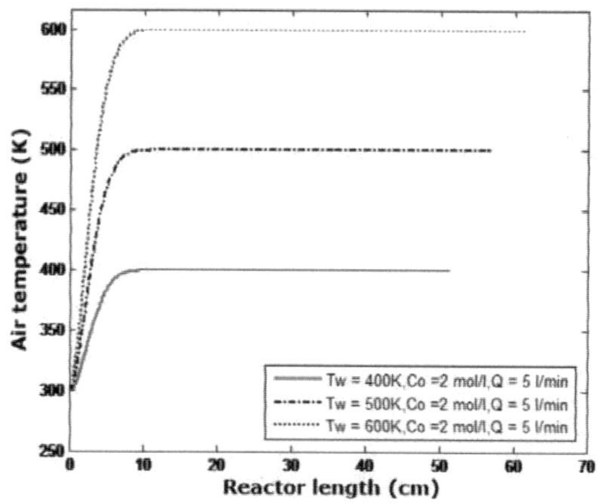

Figure 9. Effect of reactor wall temperature on the air temperature around the droplet.

IOP Conf. Series: Materials Science and Engineering **550** (2019) 012031 doi:10.1088/1757-899X/550/1/012031

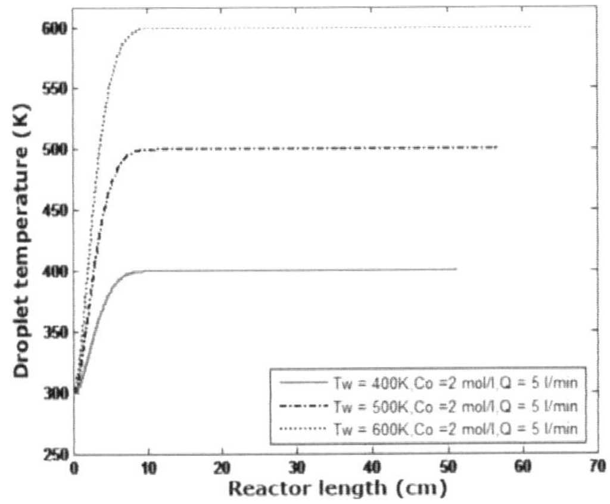

Figure 10. Effect of reactor wall temperature on the temperature of the droplet.

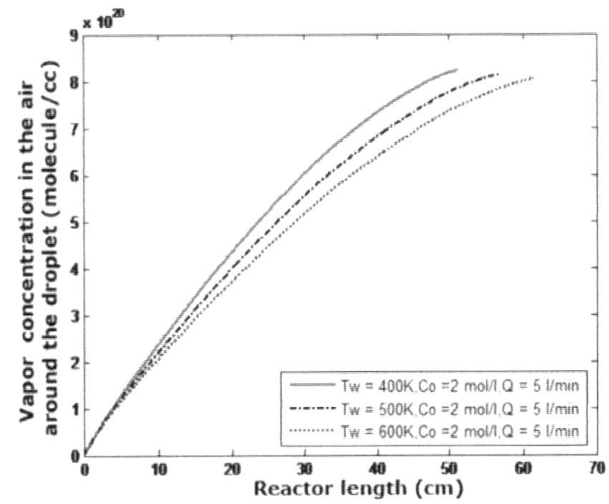

Figure 11. Effect of reactor wall temperature on the concentration of water vapor around the droplet.

The water vapor concentration profile on the droplet surface can be seen in figure 12. The result shows that the higher the reactor wall temperature is given, the less the amount of water vapor concentration attached to the droplet surface. This is because as the temperature increases, the kinetic energy of water vapor attached to the droplet surface become higher so that the water vapor can diffuse toward the surface into the air. The concentration profile of the solution on droplets can be seen in figure 13. The result shows that the concentration on the surface of the droplet is higher as the reactor wall temperature increases since the higher the reactor wall temperature, the higher the droplet temperature. As the droplet temperature increases, the diffusion coefficient between the solvent (water) and the gas around the droplet becomes increase, causing more solvents to become water vapor particles and causing the concentration of the solution on the droplet surface increases.

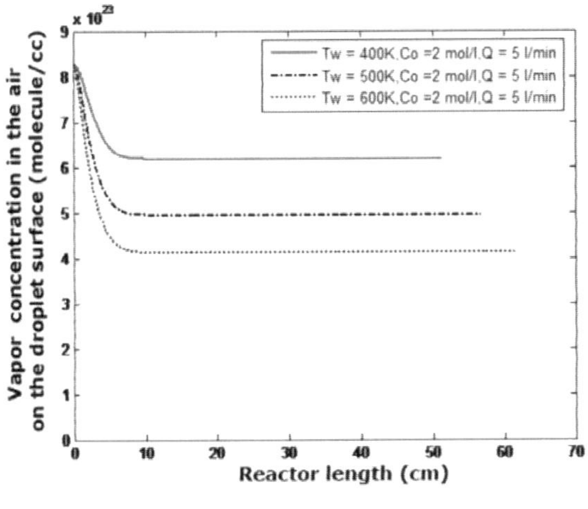

Figure 12. Effect of reactor wall temperature on water vapor concentration on the droplet surface.

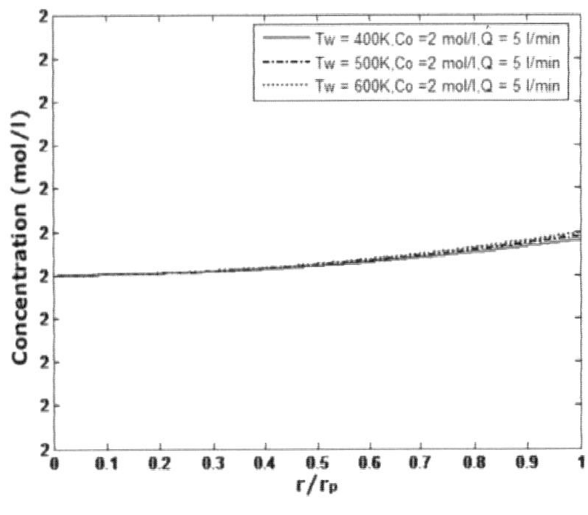

Figure 13. Effect of reactor wall temperature on solution concentration in droplets.

3.3. Variation in Co (initial concentration)

The initial concentration of the solution are varied by 0.1 mol / l, 0.5 mol / l, 1 mol / l, and 2 mol / l. The simulation results of this different solution concentration can be seen in figure 14. In the figure, it can be seen that the concentration of the solution in the droplet is different for each condition as the initial concentration were varied.

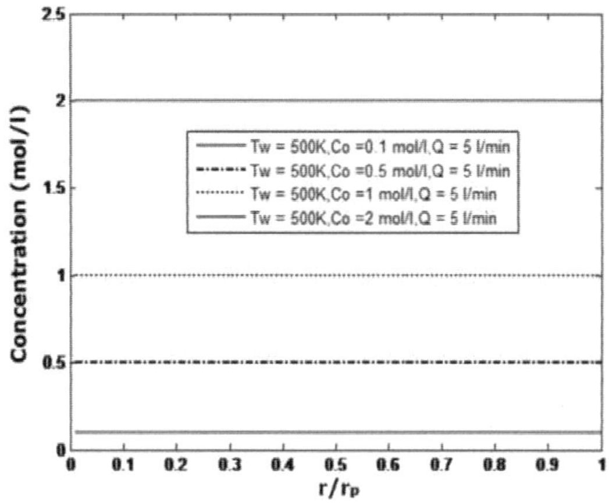

Figure 14. Effect of the initial concentration of the solution on the concentration of the solution on the droplet.

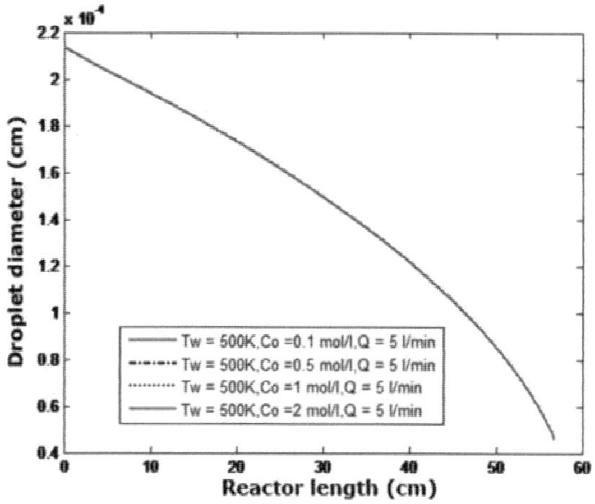

Figure 15. Effect of initial concentration of solution on the evolution of the size of NaCl droplets throughout the reactor.

In figure 14, changes in the solution concentration at the droplet center or at the surface of the droplet are not visible. However, this is not the case since from the center of the droplet until the surface of the droplet distribution the concentration of the solution changes very little. While the effect of variations in solution concentration on droplet size can be seen in figure 15. In the picture, it can be seen that each variation in the value of the initial solution concentration does not affect the final particle size and does not affect the length of the reactor required by the droplet to reach its final size.

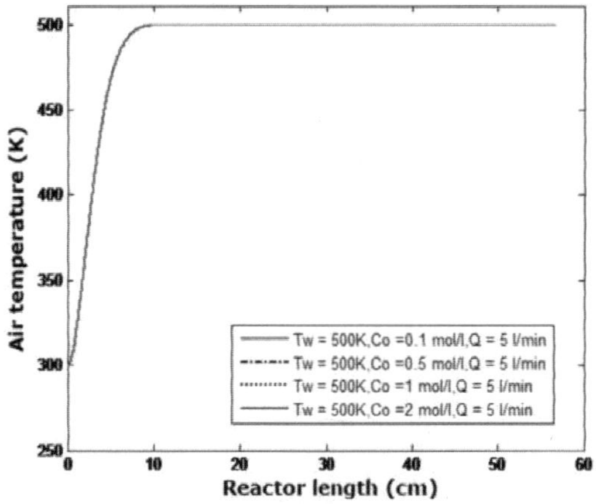

Figure 16. Effect of the initial concentration of the solution on the air temperature in the reactor tube.

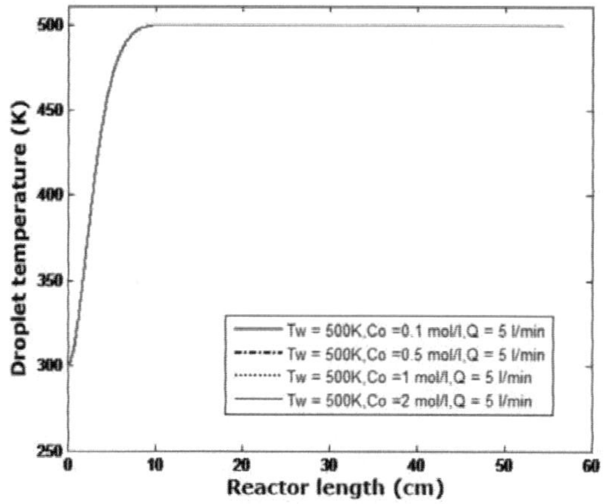

Figure 17. Effect of the initial concentration of the solution on the temperature on the droplet.

Similarly, the air temperature, droplet temperature, water vapor concentration in the air around the droplet, and the concentration of water vapor on the droplet surface, all of which do not change with variations in the concentration of the solution. The profile can be seen in figure 16-19.

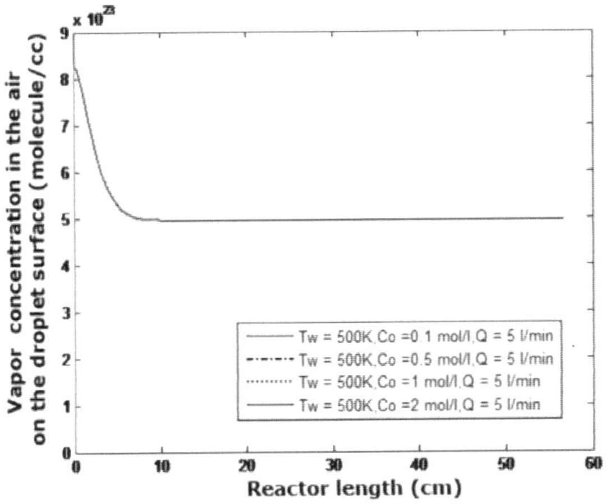

Figure 18. Effect of the initial concentration of the solution on the concentration of water vapor on the surface of the droplet.

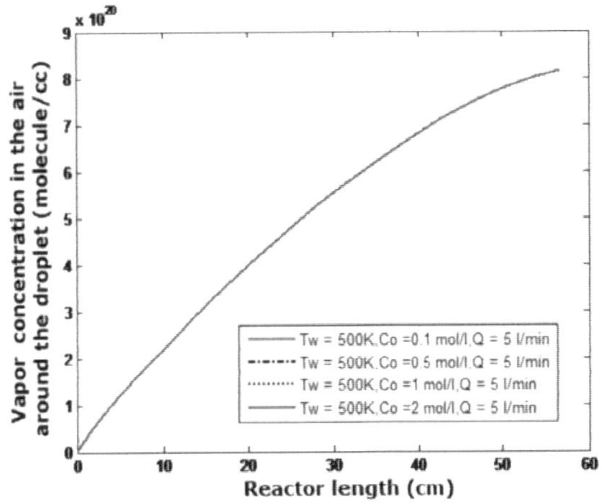

Figure 19. Effect of the initial concentration of the solution on the concentration of water vapor in the air around the droplet.

4. Conclusion

The dynamics of the droplets have been successfully simulated by the influenced of reactor wall temperature (T_w), the flow rate of carrier gas (Q), and initial concentration (C_0). Variation Q affects the flow rate of the carrier gas carrying the droplet and the length required for the droplets to reach their final shape, but that variation does not affect the final size of the particle. T_w variations affect the flow rate of carrier gas carrying droplets and the size of droplets. C_o variation does not affect the flow rate of the carrier gas transporting the droplet so that the length of the reactor required by droplet to reach its final size is fixed for each variation. In this modeling *Co* variation seems to have no effect on the droplet size.

Nomenclature

b	average molecular droplet (mol/g)	d_{ps}	equivalent diameter of a complete drying droplet (cm)
C_0	initial solution concentration or solute concentration at the center of the droplet (mol/ l)	D_L	solute diffusion coefficient (cm/s)
C_{pa}	heat capacity of surrounding gas (J/mol K)	D_v	solvent vapor diffusion coefficients (cm/s)
C_p	heat capacity of droplet (J/ mol K)	ES	equilibrium saturation concentration (mol/ l)
CSS	critical supersaturation concentration (mol/ l)	F	average molar flow of air (mol/s)
C	solute concentration at a radius r (mol/l)	H_L	latent heat of vaporization (J/g)
d_{p0}	initial droplet diameter (cm)	h_s	heat-transfer coefficient at the droplet surface (J/cm^2 s K)

h_w	heat-transfer coefficient at the reactor wall (J/cm^2 s K)	RH^0	relative humidity in surrounding gas (-)
K	constant, cm^2/s	r_p	droplet radius (cm)
K_m	vapor mass transfer coefficient for laminar flow(cm/s)	R_{p0}	initial droplet radius (cm)
L	reactor length (cm)	t	time (s)
m_l	mass of water molecule (g/mol)	T_0	initial temperature of reactor inlet (K)
M_s	solute molecular weight (g/mol)	T	temperature of surrounding gas (K)
N_0	vapor concentration at the droplet surface (no./cm^3)	T_s	temperature at the droplet surface (K)
N_A	Avogadro constant (no./mol)	T_w	temperature at the reactor wall (K)
n	vapor concentration in surrounding gas (molecules/cm^3)	y_{w0}	vapor mole fraction in surrounding gas (-)
n_s	vapor concentration at the droplet surface (molecules/cm^3)	y_w	vapor mole fraction in surrounding gas at the reactor inlet (-)
p_s	droplet surface vapor pressure (dyne/cm^2)	z	Reactor coordinate distance (cm)
$p_s{}^o$	equilibrium vapor pressure (dyne/cm^2)	ρ_L	density of the pure water (g/cm)
Q	carrier gas flow rate (l/min)	ρ_s	dissolved density (g/cm)
R	reactor tube radius (cm)	ρ_v	density of the droplet (g/cm)
r	radial coordinate (cm)		

References

[1] Lenggoro I W, Hata T, Iskandar F, Lunden M M and Okuyama K 2000 An experimental and modeling investigation of particle production by spray pyrolysis using a laminar flow aerosol reactor *J. Mater. Res.* **15** 733–43

[2] Florena F F, Faizal F, Maulana D W and Wibawa B M 2018 Ultraviolet and Violet-Blue Emission of Zn doped MgO / Mg (OH) 2 Particles Prepared by One Step Spray Pyrolysis Method **1** 57–66

[3] Panatarani C, Muharam D G, Wibawa B M and Joni I M 2013 Blue Luminescent of ZnO:Zn Nanocrystal Prepared by One Step Spray Pyrolysis Method *Mater. Sci. Forum* **737** 20–7

[4] Messing G L, Zhang S and Jayanthi G V 1993 Ceramic powder synthesis by spray pyrolysis *J. Am. Ceram. Soc.* **76** 2707–26

[5] Nandiyanto A B D and Okuyama K 2011 Progress in developing spray-drying methods for the production of controlled morphology particles: From the nanometer to submicrometer size ranges *Adv. Powder Technol.* **22** 1–19

[6] Jayanthi G V., Zhang S C and Messing G L 1993 Modeling of solid particle formation during solution aerosol thermolysis: The evaporation stage *Aerosol Sci. Technol.* **19** 478–90

[7] Kodas T T and Hampden-Smith M J 1999 *Aerosol processing of materials* vol 33 (Wiley-Vch New York)

[8] Eslamian M, Ahmed M and Ashgriz N 2009 Modeling of solution droplet evaporation and particle evolution in droplet-to-particle spray methods *Dry. Technol.* **27** 3–13

[9] Bandyopadhyay A A, Pawar A A, Venkataraman C and Mehra A 2015 Modelling size and structure of nanoparticles formed from drying of submicron solution aerosols *J. Nanoparticle Res.* **17** 1–14

[10] Bergman T L, Incropera F P, DeWitt D P and Lavine A S 2011 *Fundamentals of heat and mass transfer* (John Wiley & Sons)

[11] Reid R C, Prausnitz J M and Poling B E 1987 The properties of gases and liquids

[12] Pitzer K S and Mayorga G 1973 Thermodynamics of electrolytes. II. Activity and osmotic coefficients for strong electrolytes with one or both ions univalent *J. Phys. Chem.* **77** 2300–8

Acknowledgements
The authors thank to Universitas Padjadjaran for the financial support under Academic Leadership Grant (ALG) 2018.

Electromagnetic-wave sintering of alumina ceramics from nano-sized particles: possible material for high-pressure cell for millimeter-wave electron spin resonance

T Nawate[1], Y Yamamoto[1], Y Kanie[1], S Mitsudo[1], Y Fujii[1] and T Sakurai[2]

[1] Research Center for Development of Far-Infrared Region, University of Fukui, Bunkyo 3-9-1, Fukui 910-8507, Japan
[2]Center for Supports to Research and Education Activities, Kobe University, 1-1 Rokkodai, Nada, Kobe 657-8501, Japan

Email: tanuping@gmail.com

Abstract. Electromagnetic-wave sintering of alumina ceramics using 28 GHz gyrotron has been performed aiming for a high fracture toughness value in order for use as pistons for pressure cells of high-frequency electron spin resonance (ESR) measurements. We have tried to improve the fracture toughness by using nano-sized alumina powder (140 nm in average particle size) having smaller particle size than our previous work (400 nm in average particle size). Rapid densification was observed around the sintering temperature of 1200 °C. We obtained the relative density over 99 % above 1400 °C. It was found that alumina ceramics made in this work at sintering temperatures have smaller grain size and higher density simultaneously as compared to the previous work. These results suggest that alumina ceramics made from powder of smaller particle size sintered by electromagnetic-wave sintering possibly have high fracture toughness which can be used as materials for the pressure cell for ESR.

1. Introduction

Pressure is one of important parameters in measuring physical properties. Electron spin resonance (ESR) is one of the most reliable methods to detect electron spins and obtain information about magnetic properties of materials from the microscopic point of view. Recently, Ohta and co-workers developed high-pressure, high-field and multifrequency electron spin resonance system in the millimeter wave (MMW) / sub-millimeter wave (SMMW) region by using a clamped-type piston-cylinder pressure cell [1]. In order for high-sensitive ESR measurements, the ceramic material for the piston of the pressure cell is required to have two properties simultaneously: 1) Less attenuation of electromagnetic waves and 2) hardness and toughness value withstanding the high pressure. They examined several commercially available products of ceramics and found that zirconia-based ceramics have high fracture toughness and rather low electromagnetic wave transmittance which decreases significantly with increasing proportion of zirconia, and that alumina has relatively low fracture toughness and high electromagnetic wave transmittance [1].

In the conventional sintering method of ceramics which is heating from outside of the sample with an electric furnace or even fire etc., the temperature inside the sample changes by thermal conduction from the outside. Since this method requires thermal equilibrium condition (or close) inside the sample, it takes several hours to several tens of hours to complete sintering. On the other hand, electromagnetic-wave sintering is an internal heating method based on absorption of the electromagnetic wave by the sample, which makes possible to heat uniformly the sample and in a shorter sintering time.

The continuous MMW/SMMW with an order of ten kilowatts can be produced only by the gyrotron at present. Progress in the high-frequency and high-power radiation source technology stimulates the electromagnetic wave sintering technology. We have been studying MMW/SMMW sintering of ceramics such as alumina, zirconia, silica xerogel and so on by using gyrotorns [2,3]. In our previous work on MMW sintering of alumina [2], the sintered alumina had higher density than those of conventional method with any sintering temperature. In addition, microstructure evaluation demonstrated that grain growth of MMW annealed alumina was faster than in conventional annealing. It indicates that the applied MMW enhances mass transport and solid-state reaction rates during sintering. Such empirical observations of microwave enhancements have been broadly known as microwave (or non-thermal) effect. Therefore, electromagnetic-wave sintering has a potential to make materials with unprecedented physical properties or functions.

It is known that a grain growth in the sintering process causes decreasing of fracture toughness value [4]. Since the average grain size of the MMW sintered alumina was smaller than that of conventional alumina at high density region [2], one can expect to make the MMW sintered alumina with an improved fracture toughness property which can be used for pressure cell for ESR measurements.

We have been sintered alumina ceramics by electromagnetic wave sintering using 28 GHz gyrotron. In this paper, we report the results with newly adopted nano-sized alumina particles as a starting material which is smaller than what is used in our previous work [2] in order to prevent grain growth. After the experimental backgrounds and sintering procedures in the following sections 2 and 3 respectively, we will describe the properties of sintered alumina including analyses of scanning electron microscope (SEM) images in section 4. We will also discuss on a comparison with our previous work.

2. Experimental

In all sintering process in this study, we have used MMW heating system where a 28 GHz gyrotron is used as the radiation source as is used in our previous work [2].

The schematic diagram of a pressure cell device for ESR system is shown in figure 1. The top/bottom backups and pistons must be tough enough as well as having good transmittance of MMW. The sizes of the pieces are summarized in table 1. We have developed alumina ceramics for top and bottom backups sintered by the MMW heating system with 28 GHz gyrotron by using alumina powders AES - 11C (Sumitomo Chemical Co. Ltd., 99.8 %, 400 nm in average particle size, and 7-8 m^2/g of specific surface area). Since a relatively strong force is applied to the piston, higher strength and toughness is required for the material of pistons. In order to obtain a higher toughness, we used in this study smaller alumina powder TM-DAR (Taimei Chemical Co., Ltd., 99.95%, 140 nm in average particle size, and 13.3 m^2/g of specific surface area). Further, the size of the sintered sample needs to be □6 x 20 mm in order to carve it into a shape of the top piston with the length of 10 mm. Making such a long shape sample is a challenging for us, because it may become difficult to pull out air and water from the long-rod shaped sample in the procedure of making green sample which is described later.

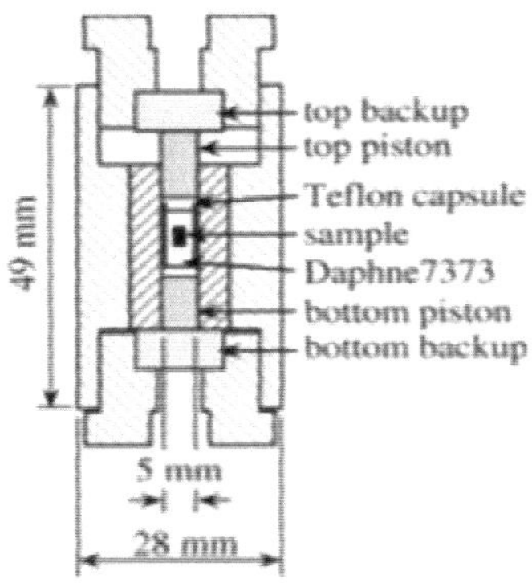

Figure 1. The schematic diagram of a pressure cell device for ESR system [1].

Table 1. The size of the pressure

	Height(mm)	Diameter (mm)
top backup	6	12
bottom backup	6	12
top piston	10	5
back piston	8	5

The fracture surfaces of the obtained sample were observed using a field-emission scanning electron microscope (SEM for short, S-2600H, Hitachi).

3. Procedure of sintering

We basically followed the previously reported procedure to make the green sample [5]. First, 200 g of alumina powder (TM-DAR), 50.0 g of distilled water, 0.06 g of magnesium oxide as an additive and 2.0 g of dispersant Aron A 6114 (ammonium carboxylate) were charged into a stirring vessel with 150 g of alumina ball for stirring and stirred for 18 hours with a dedicated machine. Next, the resulting muddy was degassed by using a vacuum pump in order to remove the air mixed in stirring. Then the muddy was poured into a silicon mold placed on the porous resin connected to a rotary pump. The air and the moisture were taken out from the muddy through the pores of the resin by pumping, then we obtained green samples. Note that for such a long-rod shape sample about 20 mm, viscosity of the muddy should be optimized before the degassing procedure to obtain good shape of the green sample. Otherwise we may obtain a sample with a hole inside the rod after the pumping, because the surface of the muddy is first dried during the pumping procedure. In our case, we solved it by increasing water to reduce the viscosity and by using a stronger pump.

The obtained compact is dried at 120 ° C in a drier for 2 hours to remove water sufficiently, then pre-sintered at 600 ° C for 2 hours in an electric furnace to remove water and dispersant completely.

The green sample after the pre-sintering was covered by a thermal insulator Denka Arsen Board (BD-1700 LNH, Denki Kagaku Kogyo Co, Ltd.) during the sintering procedure. It is composed of 100% alumina with the maximum operating temperature 1700 °C and with a low absorption of electromagnetic waves. The thermocouple, heat insulating material, and sample were placed as shown in figure 2.

Figure 3 shows a sintering curve used in the electromagnetic wave sintering procedure for this study. Each sample was heated from the room temperature to the sintering temperature at the heating rate of 45 °C/min. After the holding time of 20 min, the irradiation of the electromagnetic wave was stopped and cooled naturally. The sintering temperature was changed from 1000 °C to 1500 °C by 100°C.

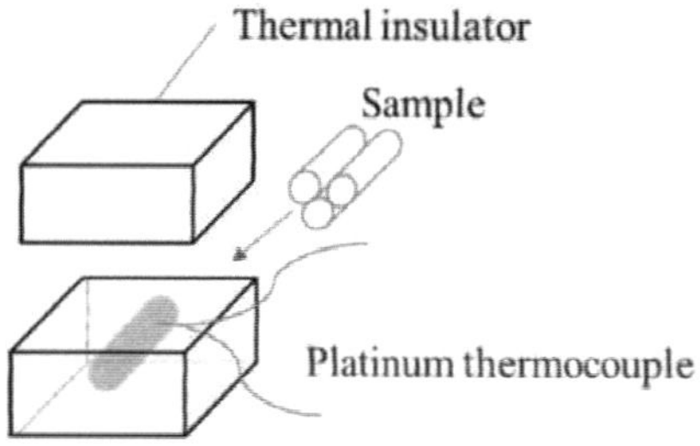

Figure 2. The placement of the sample, the thermocouple and the thermal insulator.

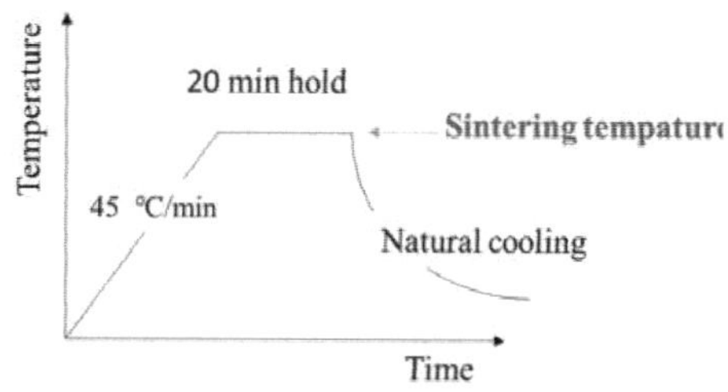

Figure 3. The typical sintering curve showing time change of the temperature.

4. Experimental results and discussions

The density of the sintered sample obtained by electromagnetic wave sintering was calculated by the Archimedes method. We took data of the relative density by averaging over at least three samples. The obtained relative density as a function of the sintering temperature is shown in figure 4.

The relative density at the sintering temperature of 600 °C was obtained from the pre-sintered green sample. A rapid densification around the sintering temperature of 1200 °C was observed. The relative density increased to above 98 % at a sintering temperature of 1300 °C and even higher relative densities (over 99 %) were obtained for sintering temperatures of 1400 °C and 1500 °C.

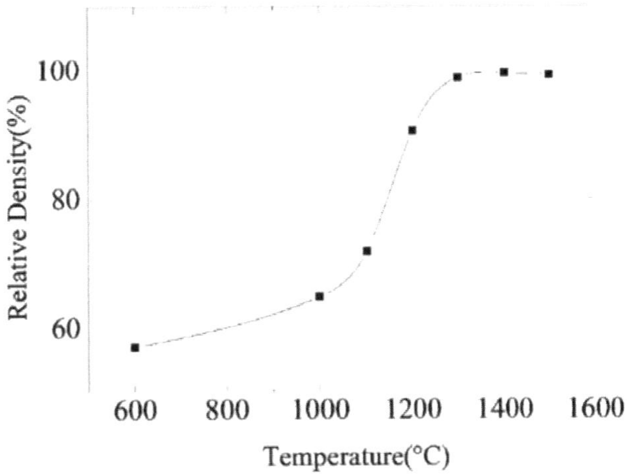

Figure 4. Relative density of sintered alumina sample as a function of the sintering temperature. The line is drawn for guide for eyes.

The sintered alumina ceramics for sintering temperatures above 1300 °C are expected to have high fracture toughness values. In order to examine the grain size, we observed SEM images. The sample for each sintering temperature was selected to have the relative density being closest to the averaged value for the sintering temperature. The captured SEM images are shown in figures 5, 6, and 7 for the sintering temperatures of 1300 °C, 1400 °C and 1500 °C, respectively. It is found that the grain size grows with increasing temperature.

Next, the particle size was measured using "image analysis method". The sintering temperature dependence of the average particle size is shown in figure 8. It is suggested that the grain growth proceeds rapidly around the sintering temperature of 1500 °C. The particle size at 1500 °C became more than twice as large as that at 1400 °C. Considering the relative densities of sintered alumina ceramics at sintering temperatures of 1400 °C and 1500 °C is little different, we can expect that the alumina sintered at 1400 °C possibly has higher fracture toughness value. The toughness value for 1300 °C is possibly also high, though the relative density is slightly smaller (by approximately 1 %) than that for 1400 °C. The measurements of the toughness value must be of interest, which is a future plan.

IOP Conf. Series: Materials Science and Engineering **550** (2019) 012032 doi:10.1088/1757-899X/550/1/012032

Figure 5. SEM image for the sintering temperature 1300 °C.

Figure 6. SEM image for the sintering temperature 1400 °C.

Figure 7. SEM image for the sintering temperature 1500 °C.

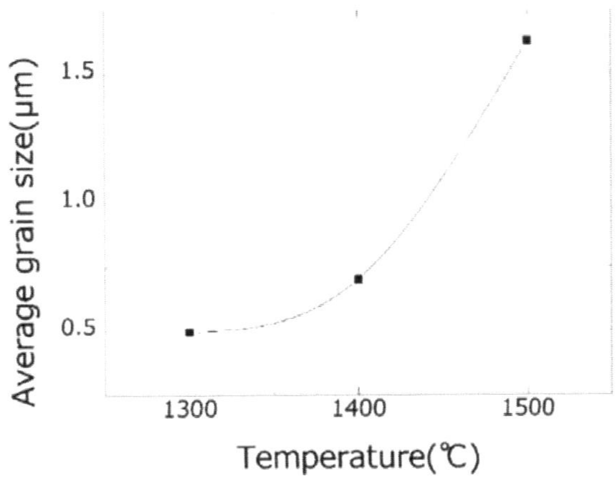

Figure 8. Average particle size as a function of the sintering temperature. The line is drawn for guide for eyes.

The grain size is plotted as a function of the relative density in figure 9, where the data for ceramics made from AES-11C by using the same heating system with 28 GHz gyrotron are also shown for comparison. In this graph, a high fracture toughness value is expected when the data comes to the lower right region. The data for TM-DAR at sintering temperatures of 1300 °C and 1400 °C are in the lower right of the graph than the data for AES-11C. Therefore, we can expect that the alumina ceramics made from TM-DAR for the sintering temperatures of 1300 °C and 1400 °C have higher fracture toughness values than those from AES-11C for any sintering temperatures.

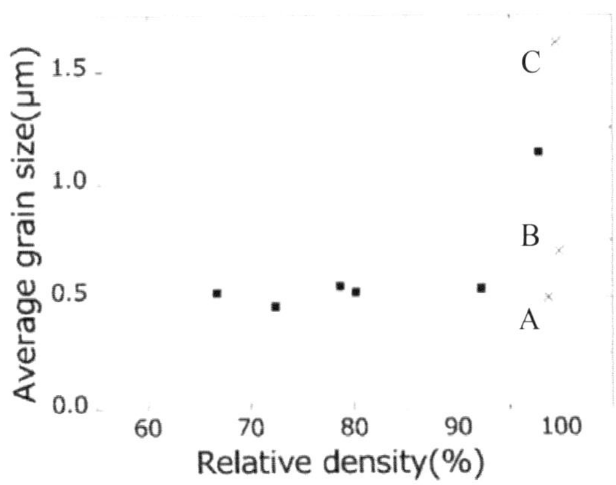

Figure 9. The average particle diameter as a function of the relative density of sintered alumina ceramics made from AES-11C (data from ref. [2], shown as black sold circle) and TM-DAR (this work). The data from TM-DAR is shown as A, B and C for the sintering temperature of 1300, 1400 and 1500 °C, respectively.

5. Summary

We have sintered alumina ceramics by electromagnetic-wave sintering with 28 GHz gyrotron aiming for a high fracture toughness value in order for use as pistons for pressure cells of high-frequency ESR measurements. We have tried to improve the fracture toughness by using nano-sized alumina powder (TM-DAR, 140 nm in average particle size) having smaller particle size than our previous work (AES-11C, 400 nm in average particle size).

We have sintered alumina at sintering temperatures between 1000 °C and 1500 °C. Rapid densification was observed around the sintering temperature of 1200 °C. We obtained the relative density over 98 % at the sintering temperature above 1300 °C and over 99 % above 1400 °C. The grain size of the sintered alumina increased as the sintering temperature was increased. Nevertheless, as compared to the results for AES-11C, it was found that alumina ceramics made from TM-DAR at sintering temperatures of 1300 and 1400 °C have smaller grain size and higher density simultaneously. These results suggest that alumina ceramics made from powder of smaller particle size (140 nm in average) sintered by electromagnetic-wave sintering possibly have high fracture toughness.

In the future, it is planned to pressurize with a pressure cell after shaping the sintered alumina. It is also planned to improve the fracture toughness value of ceramics by adding zirconia. It is expected that the electromagnetic-wave sintering has a potential to produce new physical properties and/or functions which are unattainable by conventional sintering methods.

References

[1] Ohta H *et al.* 2015 *J. Phys. Chem. B* **119** 13755
[2] Aripin H, Mitsudo S, Sudiana I N, Tani S, Sako K, Fujii Y, Saito T, Idehara T and Sabchevski S 2011 *Int. J. Infrared Milli. Waves* **32** 867
[3] Mitsudo S, Inagaki S, Sudiana I N and Kuwayama K 2013 *Adv. Mat. Res.* **789** 279
[4] Hayashi K, Gotoh K and Nishikawa T 1991 *Journal of the Ceramic Society of Japan* **99** 620 (in Japanese)
[5] Sano S, Banno T, Oda K and Shibasaki Y 1996 *Funtai oyobi funmatsu yakin* **868** 43 (in japanese)

IOP Conf. Series: Materials Science and Engineering **550** (2019) 012033 doi:10.1088/1757-899X/550/1/012033

Extraction of essential oils from leaves of the Japanese Lindera umbellata Thunb. by using microwave heating distillation method

Y Yamamoto, T Nawate, and S Mitsudo

Research Center for Development of Far-Infrared Region, University of Fukui, 3-9-1 Bunkyou, Fukui 910-8507, Japan

E-mail: yuuta14@gmail.com

Abstract. In order to investigate effective utilization methods of forest resources and an advanced microwave distillation method, a microwave vacuum distillation system has been developed with commercially available microwave oven. By using this system, essential oil extraction from the Japanese Lindera umbellata Thunb. was carried out under the vacuum condition. The qualitative and quantitative analysis for the components of the essential oil was performed by GC/MS. As the result, the content percentages of nerolidol, selinene and linalool were different from those obtained with the traditional method.

1. Introduction

There are many studies on the essential oils using microwave heating method. The microwave heated distillation method is attracting attention because it enables low cost distillation and it enables different extraction components as compared with the traditional methods [1]. Products market related to aromatherapy is growing 1.2 times over the period from 2011 to 2015 [2]. It is expected that the demand for essential oils will be expected to increase in the forthcoming years. Especially, essential oils from plants, which are inherent in particular countries and regions, are drawing attentions.

On the other hand, in Japan, the efficient use of the forest resources is an important issue in the forest industry. Using forest resources not only as timbers but also as raw materials for essential oils can contribute to solve this problem. Previously, we have studied ceramics sintering by high-power microwave heating [3]. Using these experiences, we have introduced a microwave vacuum distillation system developed by using commercially available microwave oven. The microwave heating has several advantages and interesting features. One of the important features is the direct heating of the core region inside the target. As the high speed heating is possible, the heat treatment time and energy can be reduced. In addition, each component has different absorption rate for the irradiated microwaves. Therefore, it possible that the characteristics of extracted essential oils can be controlled with the microwave irradiation parameters.

The essential oils of Kuromoji which has a very good smell are extracted as the Japanese aroma, and their components are well studied [4]. So in this paper, we compare the results of the essential oil using this method with one using traditional method.

IOP Conf. Series: Materials Science and Engineering **550** (2019) 012033 doi:10.1088/1757-899X/550/1/012033

2. Microwave vacuum distillation system

A microwave vacuum distillation system was constructed by using a commercially available microwave oven. Figure 1 shows a schematic drawing of this system. A still pot made of heat-resistant glass with the capacity of 2.5 little is installed in a microwave oven. The plant which is the raw material of the essential oil is placed in this pot and heated by the microwaves. Microwave power can be set as 200 W, 600 W or 1 kW. The steam which consists of essential oil and water is led to the outside of the microwave oven through the still head and sent into the condenser in which the heated steam is cooled down into the liquid state. Condensed mixture of oil and water is collected into a receiving flask. The essential oil is separated from the water, so as to be a floating on the water surface in the receiving flask. The condenser is the cooled by a water-chiller. An aspirator is used for the system vacuum pumping. A needle valve is used to adjust the pressure. In order to maintain the pressure, the aspirator is cooled with water, and its temperature is controlled during the experiment.

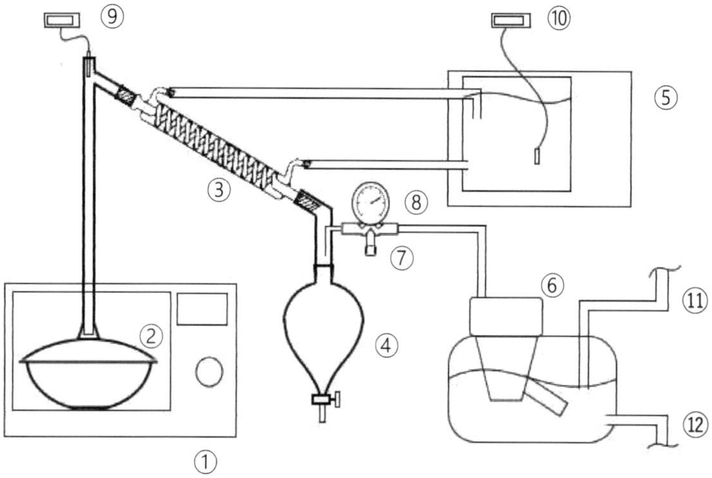

Figure 1. A schematic drawing of microwave vacuum distillation system. 1.Microwave oven, 2.Still pot (Vessel), 3.Condenser, 4.Distillate/receiving flask, 5.Chiller, 6.Aspirator(water vacuum pump), 7.Pressure controller (Needle valve), 8.Manometer, 9.Thermometer(steam temperature), 10.Thermometer(water in chiller), 11.Cooling water inlet, 12.Cooling water outlet.

Figure 2 is a photograph of microwave distillation system.

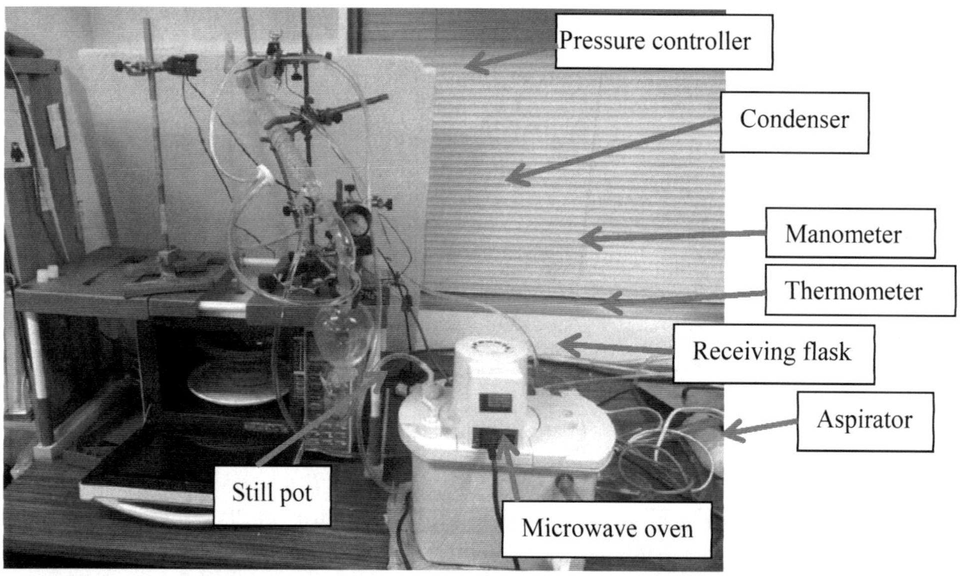

Figure 2. A photograph of our microwave distillation system.

3. Microwave distillation of Kuromoji

In order to investigate the characteristics of microwave heating distillation, the distillation experiments for Kuromoji have been performed. Water of 1 L and leaves of Kuromoji of 300 g were put into the still pot and distilled with the microwave power of 200 W. Temperature of vaper was kept below 80 °C.

The extracted components was analysed using GC/MS(HP6890 Series GC System). Figure 3 shows histogram of the extracted oil. A mass spectrometry (MS) was also performed simultaneously. From this analysis, the top five extracted components from lindera umbellate were identified (see table 1). The indexes indicated in figure 3 correspond to ones in table 1. Nerolidol and selinene are classified as sesquiterpenes. Sesquiterpenes consist of three isoprene units of C_5H_8, and have the molecular formula of $C_{15}H_{24}$. Therefore, Sesquiterpenes are large and heavy molecule components of this essential oil. Generally, the main component of the essential oil of Kuromoji obtained by traditional distillation method is about 50% linalool which is classified as monoterpenes ($C_{10}H_{16}$). This result was different from the fact that the main component of the essential oil of Kuromoji by traditional distillation method was linalool which is classified as monoterpenes. It is considered that monoterpenes, which are lighter molecules than sesquiterpenes, are obtained first in distillation at low temperature. However, in this microwave distillation experiment, sesquiterpenes are mainly extracted. Further studies are needed to clarify whether this is due to special effects such as selective heating by microwave distillation method.

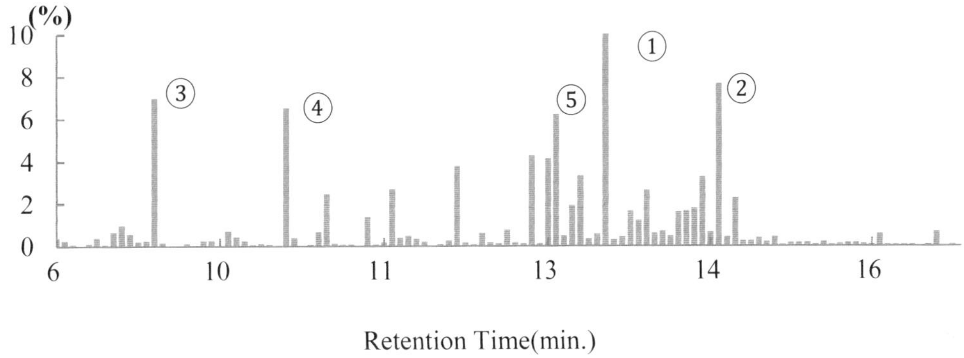

Figure 3. Chromatogram of extracted oil of Kuromoji by microwave distillation.

Table 1. Representative component in essential oil detected by GC/MS.

Peak	Time(min.)	Area (%)	Name	CAS#	Qual
1	13	10.08	1,6,10-Dodecatrien-3-ol, 3,7,11-tr ((6E)-Nerolidol)	040716-66-3	91
2	13.87	7.68	delta.-Selinene	028624-23-9	89
3	8.58	6.98	1,6-Octadien-3-ol, 3,7-dimethyl- (Linalool)	000078-70-6	94
4	10.22	6.55	2,6-Octadien-1-ol, 3,7-dimethyl- (Nerol),	000106-25-2	72
5	12.59	6.2	Naphthalene, decahydro-4a-methyl-1 (β-Selinene)	000515-17-3	86

4. Conclusion

We aimed to develop effective utilization methods of forest resources and advanced research on microwave distillation method. So we've developed microwave vacuum distillation system by using commercially available microwave oven. Essential oil extraction experiment of Japanese essential oil from Kuromoji was carried out under reduced pressure using this system. The qualitative and quantitative analysis of the components in obtained essential oil was performed by GC/MS. As the result, nerolidol and selinene which are classified as sesquiterpenes are mainly extracted. This result was different from the fact that the main component of the essential oil of Kuromoji by traditional distillation method was linalool which is classified as monoterpenes. Usually, it is considered that monoterpenes, which are lighter molecules than sesquiterpenes, are obtained first in distillation at low temperature. However, in this microwave distillation experiment, sesquiterpenes are mainly extracted. Further studies are needed to clarify whether this is due to special effects such as selective heating by microwave distillation method.

References

[1] Maria Goretti de Vasconcelos Silva et al. 2004 *ARKIVOC* **4** 66
[2] Aroma Environment Association of Japan 2016 "Survey report on 2015 aroma market"
[3] S. Mitsudo et al. 2013 *Adv. Mat. Res.* **789** 279
[4] H.Nii et al. 1983 *Nippon Nogeikagaku Kaishi* **57** 151

Acknowledgments

This work was supported by Program for Promoting Regional Revitalization by Universities as Centers of Community: COC+ Program. Kuromoji leaves used for distillation were provided by ShogakiMokuzai Co., Ltd.

IOP Conf. Series: Materials Science and Engineering **550** (2019) 012034 doi:10.1088/1757-899X/550/1/012034

Investigation of NdFeB N52 magnet field as advanced material at air gap of axial electrical generator

N Hiron, Asep Andang and Nundang Busaeri

Electrical Departement of Siliwangi University, Indonesia
Jalan Siliwangi No.24, Kahuripan, Tawang, Kahuripan, Tawang, Tasikmalaya, Jawa Barat 46115

Email: hiron@unsil.ac.id, andhangs@unsil.ac.id, nundangb@unsil.ac.id

Abstract. The aim of this study is to investigate and analyze the NdFeB N52 type magnetic field distribution on a Trapezoidal type coil in an axial type generator. The main problems addressed by this study are how the distribution of magnetic field type NdFeB N52 to the trapezoidal type coil and how the influence of air gap between NdFeB N52 magnets and the coil to the output electric voltage generated from the coil at the speed of rotation. In order to answer the problems, we moved a rectangular type magnet axially towards the trapezoidal type coil with a specified distance, so that we could identify the change of the output voltage. The description of the magnetic field distribution in the stator was simulated by using the Maxwell 3D Design application ran in ANSYS software 16.0. The result of this study shows that the magnet was distributed evenly on the coil. It is also known that based on experiments with 2mm to 6mm of magnet to coils of gaps, the highest output voltage is at 2mm gaps at 750rpm.

1. Introduction

Energy is a major requirement for communication [1-2], transportation and many more, so it requires high efficiency energy sources, one of which is the generation of energy from advanced materials, for example permanent magnets. A neodymium magnet is strong magnet [3] which is also known as NdFeB, NIB or Neo magnet. It is the most widely used type of rare-earth magnet. It is a permanent magnet made from an alloy of neodymium, iron, and boron to form the Nd2Fe14B tetragonal crystalline structure [4]. NdFeB magnets are used in applications such as hard-disk drives and permanent magnet motors [5]. NdFeB magnet made from a combination of an alloy of neodymium, iron and boron, therefore, it is shortened by Nd2Fe14B. Developed in 1982 by General Motors and Sumitomo Special Metals, neodymium magnets are the strongest type of permanent magnet commercially [6] and very useful for energy production need [1, 2, 7].

Even the hybrid bonded magnet, such as NdFeB with ferrites and NdFeB with SmFeN, was reached in many research, but (BH) is not more than 20 MGOe [8]. With the particle size of NdFeB powders 80 – 100 μm, prepared by Hydrogenation Disproportionation Desorption Recombination (HDDR), it was found that the gaps among the particles are 4–10 μm which are mainly filled by an organic binder. The characteristic of NdFeB N52 (figure 1) shows that it has Maximum Energy Product (Bhmax) that is 50-53 MGOe (398-422 MGO (KJ/m3) [9-10]. With this unique nature energy, NdFeB magnets are a favorite in many purposes, from home-based needs to the the healthcare needs. NdFeB permanent magnets contain about 31–32 wt% of rare-earth elements (REEs) [3].

IOP Conf. Series: Materials Science and Engineering **550** (2019) 012034 doi:10.1088/1757-899X/550/1/012034

This paper scope is imulation of generator axial involve of NdFeB class N52 implementation on axial generators using analytical methods using Maxwell simulation from Ansys Application. We test focus on the simulation to find the fact of voltage produced in coils that coused of gaps of NdFeB to coils.

Tests were conducted on simulations with Maxwell and Ansys applications with NdFeB N52 permanent magnet 8 coils and 8 pole.

2. Related works

NdFeB permanent magnet production growth is projected to increase significantly, especially between 2015 and 2020 which will reach more than 120 (kt), especially in China. In addition to the NdFeB type permanent magnet, the ferrite type also has the same role, but based on the economic value of the energy produced, the generator with NdFeB is cheaper than the ferrite type magnet [11]. Figure 1, a and b shows the characteristic of NdFeB N52.

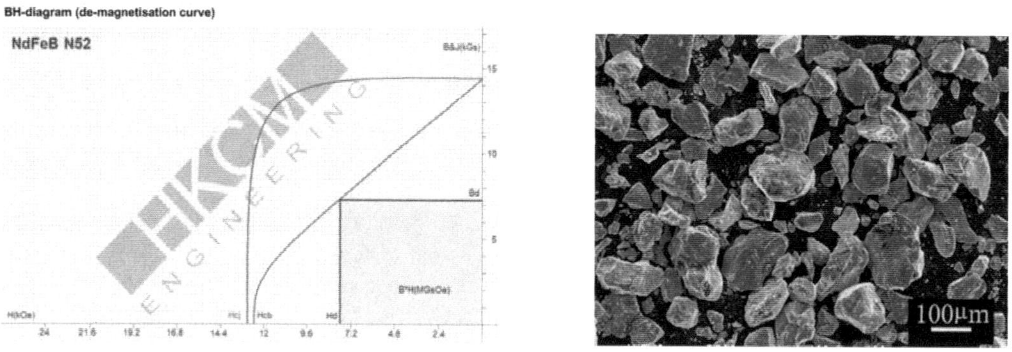

(a) (b)

Figure 1. (a) Characteristic of NdFeB N52 [9], (b) SEM images of the magnetic powder of NdFeB. [8].

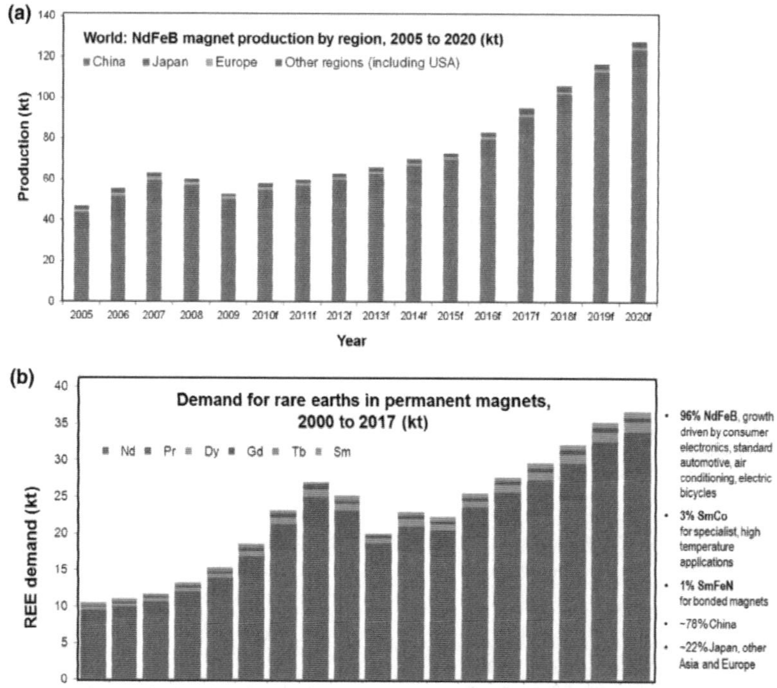

Figure 2. Global production of NdFeB permanent magnets and the demand for the REEs [12], [3]. (a) Total global NdFeB magnet production and prediction: 2005–2020, (b) Total global REE demands for permanent magnets [3].

In industrial scale, China is the largest producer of permanent magnets compared to Japan, Europe and other countries including USA (figure 1). While the world's need for permanent magnets is dominated by the Nd type of permanent magnet (figure 2).

The implementation of NdFeB on generators has been widely used for low speed generators research. This paper is part of the research. N. Georgiev (2017) has succeeded in simulating a 3 phase 16 pole generator using NdFeB magnets [13] and DC generator [14], While axial generators with trapezoidal-shaped coil have been studied in [15]. The differences that come up in those papers, it is not explained the type of class of NdFeB that is used, so that the potential energy value in the magnet cannot be known, Nevertheless NdFeB N52 with MGOe is (398-422 MGO (KJ / m3) is used in the design generator [16]. In N. Georgiev (2017) [13] found the fact that the closer the gap between the magnet and coils, the greater the voltage and power of the generator produced, and this is also affected by the rotor speed.

3. Methodology

In this study, the magnet used was NdFeB N52 type with Energy Product Bhmax 398-422 (KJ/m^3). Figure 3 shows the steps of the simulation process. The ANSYS software version 16.0 was used as the main application to be running the simulation. Then, the process steps consisted of determining the material which will be used which in this case was winding coil made of copper material and formed into trapezoidal (figure 4).

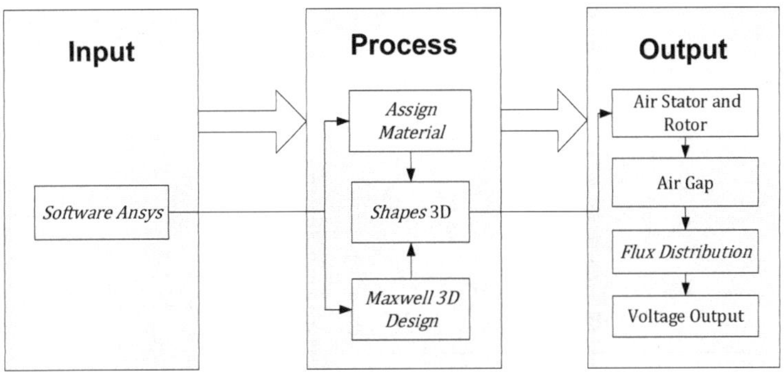

Figure 3. Process of simulation.

The model of winding coil design was trapezoidal (figure 4.a). The winding coil then was placed on the stator and arranged in a row. The number of winding coil, arranged in parallel on the casing of the stator, was total 12 units (figure 4.b).

The chasing of the winding coil was designed in the form of cylinder (figure 5) with the length (ps) was 140 mm, the thickness (ks) was 25mm, the distance between winding coil (g) was 2mm, the outer width (wso) was 57,5 mm, inner width (wsi) was 16mm, stator foot width (wc) was 2mm, the length of the coil (pk) was 80mm, the coil thickness (tk) was 10mm. Coil wire density was 6.36 A /mm^2, diameter 1mm, with turn of coil was 38 turns.

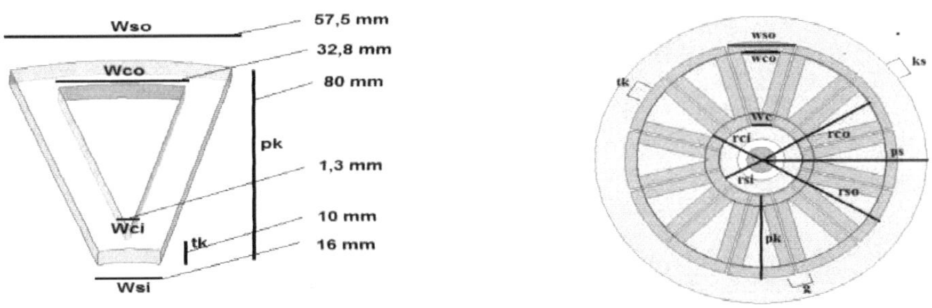

(a) (b)

Figure 4. (a) Design of winding coil with trapeziodal shape, (b) design of winding stator casing.

IOP Conf. Series: Materials Science and Engineering **550** (2019) 012034 doi:10.1088/1757-899X/550/1/012034

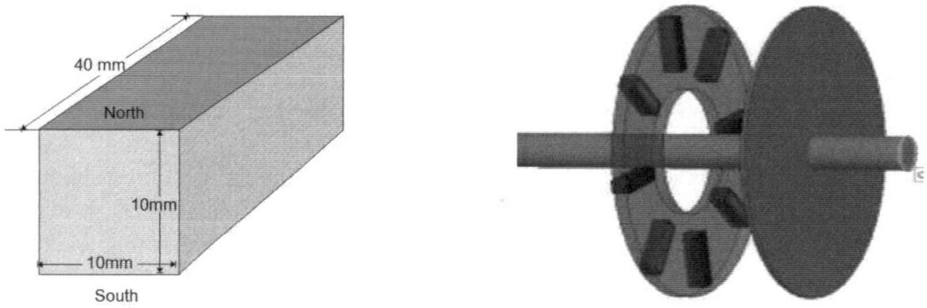

(a) (b)

Figure 5. (a) Dimension of permanent magnet, (b) placing the magnets on the rotor casing.

The permanent magnet used was in rectangular form with a length of 40mm, 10mm wide and 10mm thick (figure 5.a). Nine magnets were arranged in rows on the rotor casing with crossed positions, ie the north pole of the magnets was adjacent to the south pole of the magnet (figure 5.b).

Finally, the stator and rotor designs had been successfully designed based on the design. The stator consists of 12 winding coils, while the rotor consists of 9 permanent magnets. The combination of rotor and stator was arranged axially. The position of the rotor was independent of the stator, so that the rotor could rotate axially while the stator remained stationary (figure 6.).

The fact that the NdFeB N52 magnet was rotated axially resulted in the outer electric voltage of the winding coil which was proportional to the distance between the stator and the rotor.

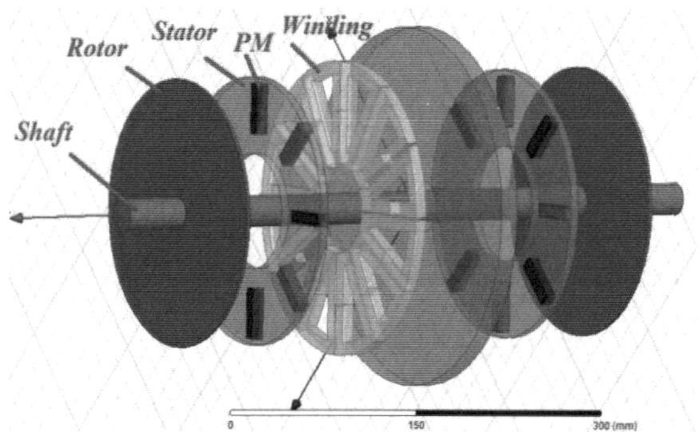

Figure 6. The result of combination of the stator and the rotor on the shaft by software Ansys.

4. Result and discussion

The simulation carried out by using Maxwell 3D Design ran on ANSYS software showed that the distribution of magnetic flux density (B) of NdFeB N52 magnet on winding coil was evenly distributed with the middle of winding coil had the highest magnetic flux density (B) that is 1.2 Tesla, while the lowest magnetic flux density was the outer part, ie 3.26e-5 Tesla. Further result shows that the distribution of magnetic flux density showed arose in all winding coils on the stator casing (figure 7).

Figure 7 shows the value variation of the magnetic flux density (B) on winding coil. There was an air gap between the stator and the rotor with the highest value was 1.23 Tesla at a distance of 2mm and the lowest magnetic flux density was 0.7 Tesla. In the 6mm air gap (figure 8). It proves that the farther the air gap, the smaller the magnetic flux density occurs.

IOP Conf. Series: Materials Science and Engineering **550** (2019) 012034 doi:10.1088/1757-899X/550/1/012034

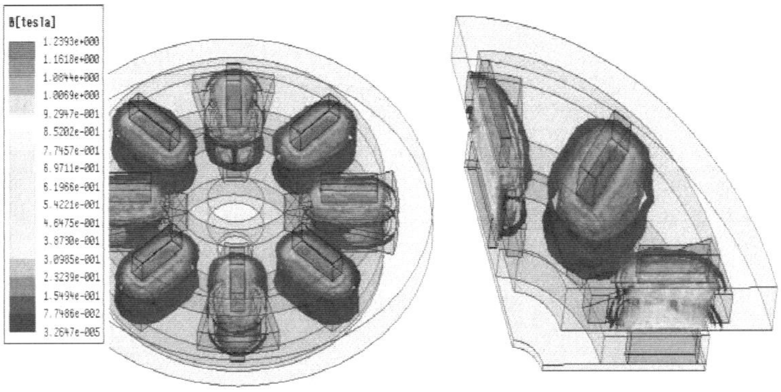

Figure 7. The magnetic flux distribution (B) of the stator to the winding coil.

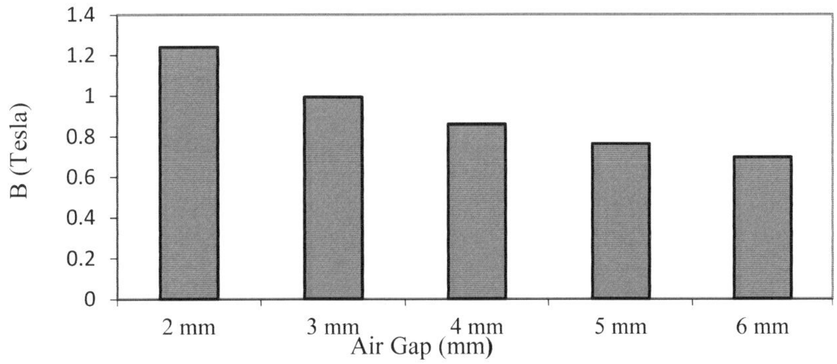

Figure 8. Magnetic flux density on the variation of air gap distance between the stator and the rotor.

Magnetic flux density, winding coil and air gap affected the value of output voltage generated from winding coil, figure 9 shows the relationship between the output voltage with rotation speed of rotor or permanent magnet, it demonstrates that the created model produced maximum voltage at 750 rpm axial speed. After 750 rpm, there was a reduction in the output voltage. This condition is in accordance with what has been done inside [13, 15-17], that the number of magnetic poles, magnetic strength and magnetic gap distance to the coil affect the voltage generated from the axial generator.

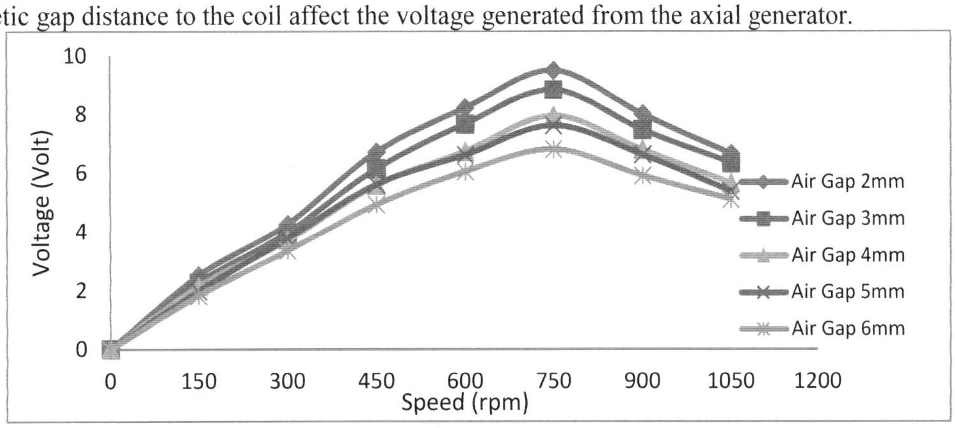

Figure 9. Relationship between output voltage and speed (rmp).

5. Conclusion

The evidence from this study suggests that magnetic flux density of NdFeB N52 on 1 mm of diameter trapezoidal shaped coil and coil wire density of 6.36 A /mm2, has a linear relationship to the air gap, which indicates that the bigger the gap the lower the magnetic flux density in the winding coil. However, the NdFeB N52 magnet has an equitable distribution of magnetic flux density on the winding coil. The uniform distribution will provide good quality outcome, however, it should be noted that the distance of permanent magnet placement on the rotor side and also the distance between the coils on the stator side must be precise, so that the magnetic distribution in the coil will be evenly distributed. Based on experiments with 2mm to 6mm of magnet to coils of gaps, the highest output voltage is at 2mm gaps at 750rpm. This condition is in accordance with what has been done inside, that the number of magnetic poles, magnetic strength and magnetic gap distance to the coil affect the voltage generated from the axial generator.

References

[1] Hiron N, Andang A and Setiawan H 2016 *Int. J. Futur. Comput. Commun.* **5** 163–166
[2] Hiron N and Andang A 2016 *2nd Int. Conf. on Science in Information Technology (Balikpapan-Indonesia)* (US: IEEE) p 250
[3] Yang Y *et al* 2016 *J. Sustain. Metall.* **3** 122–149
[4] Shewane P G, Gite M, Singh A and Narkhede A 2014 **2** 4056–59
[5] Sheridan R S, Williams A J, Harris I R and Walton A 2014 *J. Magn. Magn. Mater.* **350** 114–18
[6] Fraden J 2010 *Handbook of Modern Sensors: Physics,Designs, and Applications*, 4th ed. (New York: Springer)
[7] Hiron N, Andang A and Mubarok H 2015 *The 14th Int. Conf. on QiR (Lombok-Indonesia)* (US: IEEE)
[8] Tian J, Tang Z, Zuo Z, Pan D and Zhang S 2013 *Mater. Lett.* **105** 87–9
[9] HKCM Engineering 2018 *Neodymium (NdFeB)* (Eckernfoerde: HKCM Engineering)
[10] Magma Magnetic Technologies Ltd *Neodymium magnets magnetic characteristics* (Kibbutz Geser: Israel)
[11] Cipriani G, Corpora M, Di Dio V, Miceli R, Spataro C and Trapanese M 2015 *Int. Conf. Renew. Energy Res. Appl. (palermo-Italy)* (US: IEEE) p 1518
[12] Shaw S and Constantinides S 2012 *8th International Rare Earths Conference*
[13] Georgiev N 2017 *Electroteh. Electron. Autom.* **65** 90–96
[14] Wu Y Y 2014, "Design of NdFeB Permanent Magnet DC Generator," *Appl. Mech. Mater.* **496–500** 1113–16
[15] Minaz M R and Çelebi M 2017 *Results Phys.* **7** 183–88
[16] Zhao J, Shi G and Du L 2015 *Energies* **8** 11755–69
[17] Eklund P, Sjökvist S, Eriksson S and Leijon M 2014 *Machines* **2** 120–33

Acknowledgments

This paper was supported by personal foundation and dedicated for Siliwangi University. The author would like to thanks to Electrical departmen of Siliwangi University (UNSIL) for the facilities that presented in laboratorium of Electrical department. Thanks to Kemenristekdikti Republic of Indonesia.

IOP Conf. Series: Materials Science and Engineering **550** (2019) 012035 doi:10.1088/1757-899X/550/1/012035

Production and characterization of activated carbon from cashew nut shell using N_2 as activation agent

M Anas[1], M Jahiding[2], Ratna[3] and I N Sudiana[2]

[1]Department of Educational Physics, Faculty of Education, Halu Oleo University, Kampus Bumi Tridharma Anduonohu, Kendari 93232, Indonesia
[2]Laboratory of Material Physics, Department of Physics, Faculty of Mathematic and Natural Science, Halu Oleo University, Kampus Bumi Tridharma Anduonohu, Kendari 93232, Indonesia
[3]Department of Chemistry, Faculty of Education, Halu Oleo University, Kampus Bumi Tridharma Anduonohu, Kendari 93232, Indonesia

E-mail: anas.physedu.uho@gmail.com

Abstract. The activated carbon from cashew nut shell (CNS) has been prepared by physical activation. The cashew nut shell liquid (CNSL) is primarily extracted from a shell by using handed-rotary roaster kiln at the temperature around 170°C. The carbonization is done by augmenting temperature up to 350°C without taking out the CNS from kiln. The carbonized CNS is crushed and sieved of 60 mesh. While the activation process, nitrogen gas is flowed through the electric furnace with a constant flow rate and the activation temperature is varied from 400oC to 900oC. Sample then is characterized using ultimate analysis to determine composition of activated carbon including carbon (C), hydrogen (H), nitrogen (N), oxygen (O) and sulphur (S) and scanning electron microscope (SEM) and Energy Dispersive Spectroscopy (EDS) to know the surface morphology and element composition respectively and X-Ray diffraction (XRD) is used to know the structural properties. The ultimate analysis shows that carbon element in the sample increases as activation temperature increases up to 700°C of 79.80% and then decreases to 79.14% at 900°C. XRD pattern shows that it appears a broaden peak about $2\theta = 24°$ indicating an amorphous structure. The increasing of activation temperature tends to graphitization process of activated carbon.

1. Introduction

The research on activated carbon has attached much attention by many researchers because of its variety of applications such as an absorbent, an electric double layers capacitor, a filter, etc. So far, the utilization of biomass from agricultural waste as raw materials on the production of activated carbon has done by several researchers. The agricultural waste such as cashewnut shell [1], rice husk [2], durian shell [3], sugar cane and horn corn [4] have been used to producofe activated carbon. National production of raw cashew nut is about 95.000 ton per year and these amounts do not increase significantly among last 10 years. The main producer of raw cashew nut is Southeast Sulawesi (35 % of national production), South Sulawesi (25 %), Lombok, Flores and Sumbawa (30 %) and Jawa-Madura (10 %) [5]. About 35% weight of raw cashew nut consists of CNSL, 25% of nut and the rest is shell. So, it is approximately 38.000 ton CNS available in Southeast Sulawesi in a year. So far, cashew nut shell becomes an agricultural waste in this region. In general, there are two methods to fabricate activated carbon from agricultural waste such as chemical method and physical method. Chemical activation is done by immersing raw material in the proper chemical solution such as $ZnCl_2$, H_3PO_4. Physical activation is ideally carried out through 2 (two) steps include carbonization process of raw material and activation process by heating carbonized cashew nut to high temperature. For the physical method, researchers commonly use heat energy such as applied electric furnace or microwave. The use of microwave for material processing recently became popular because of its unique effect on the material such as selective heating and fast heating.

Many researchers have conducted the research on production, , and application of activated carbon, but it remains only a few researcher concerns on activated carbon from cashew nutshell. Therefore, this study aims to prepare and to characterize activated carbon from CNS. In our previous work, we prepared activated carbon from CNS only by physical activation without an activation agent [6, 7]. In this study, we prepare the activated carbon from CNS by flowing nitrogen gas into an activated chamber as activating agent while the activation process. Khezami et.al. (2007) reported that the common activation agent is steam, CO_2 and N_2 [8].

2. Methods

The CNS in this research is collected Lombe District in Buton Regency, Southeast Sulawesi. To remove the residual element in raw material, cashew nut shell is washed by aquadest, then it is dried in sunlight exposure. Carbonization process is done through 2 (two) stage. Firstly, the extraction of CNSL by using handed-rotary coaster kiln [9] in which the temperature is maintained about 170°C. Secondly, without taking CNS out, the heating is continued until the temperature reaches 350°C and it is maintained for 4 hours. Then, carbonized CNS is crushed and sieved in pieces of 60 meshes. The activation process is done by physical activation. Among the activation process, the nitrogen gas is flowed into the electric furnace with a constant flow rate. The activation temperature is varied from 500°C to 900°C with the interval of 150°C and activation time is 30 minutes. Proximate analysis is done to determine moisture content, ash content, volatile matter content and fixed carbon content, then, ultimate analysis is used to determine the element content such as C, O, H and N. ultimate analysis is done in Laboratorium Mineral dan Batubara Bandung and structural properties is characterized by XRD in Laboratorium Kritalografi IPB.

3. Experimental Results and Discussion

3.1. Ultimate analysis and proximate analysis

Table 1 shows the ultimate analysis result of activated carbon from cashew nut shell for different activation temperature. This table providesthe information about the effect of temperature on element composition such C, O, H, N and S of activated carbon from CNS. It seems that the C element is a dominant element in the activated carbon from CNS either before or after the activation process. Before activation, C element is about 70.25%. As activation temperature increase, the C becomes 79.80% in activation temperature of 700oC and decreases to 79.14% in activation temperature of 900oC. In relevance, the H and O element decreases as activation temperature increases. This result produced by ultimate analysis shows that the carbon content fulfills the quality standard of activated carbon issued by Standar Nasional Indonesia (SNI) 06-3730-1995 of 65%.

Table 1. The ultimate analysis result of activated carbon from CNS for different activation temperature.

	C (%)	H (%)	N (%)	S (%)	O (%)
carbonized CNS	70.25	4.61	0.94	0.01	20.13
T_a = 500°C	74.90	3.21	1.00	0.02	15.10
T_a = 700°C	79.80	1.93	1.03	0.01	9.22
T_a = 900°C	79.14	1.02	1.18	0.02	8.93

Table 2. The proxymate analysis result of activated carbon from cashew nut shell for different activation temperature.

	Moisture (%)	Ash (%)	Volatile Matter (%)	Fixed Carbon (%)
carbonized CNS	4.00	3.67	3.34	88.99
T_a = 500°C	1.40	1.85	1, 42	95.33
T_a= 700°C	2.40	1.04	1, 02	95.54
T_a= 900°C	1.60	0.82	1.02	96.56

Proximate analysis provides the information about moisture, ash, volatile matter and fixed carbon in activated carbon. Table 2 shows the proximate analysis result of activated carbon from CNS for different activation temperature. The fixed carbon content of activated carbon is determined by moisture, ash, volatile matter. The results show that the fixed carbon increases regularly as the activation temperature increase. As carbonized, the carbon content is 88.99%. It increases up to 96.56% in activation temperature of 900 °C. In general, the higher activation temperature the higher ash is. In contrast in this study, ash content decreases while activation temperature increases. It indicated that the use of nitrogen as activation agent is effective to prevent the formation of ash from the surface of char.

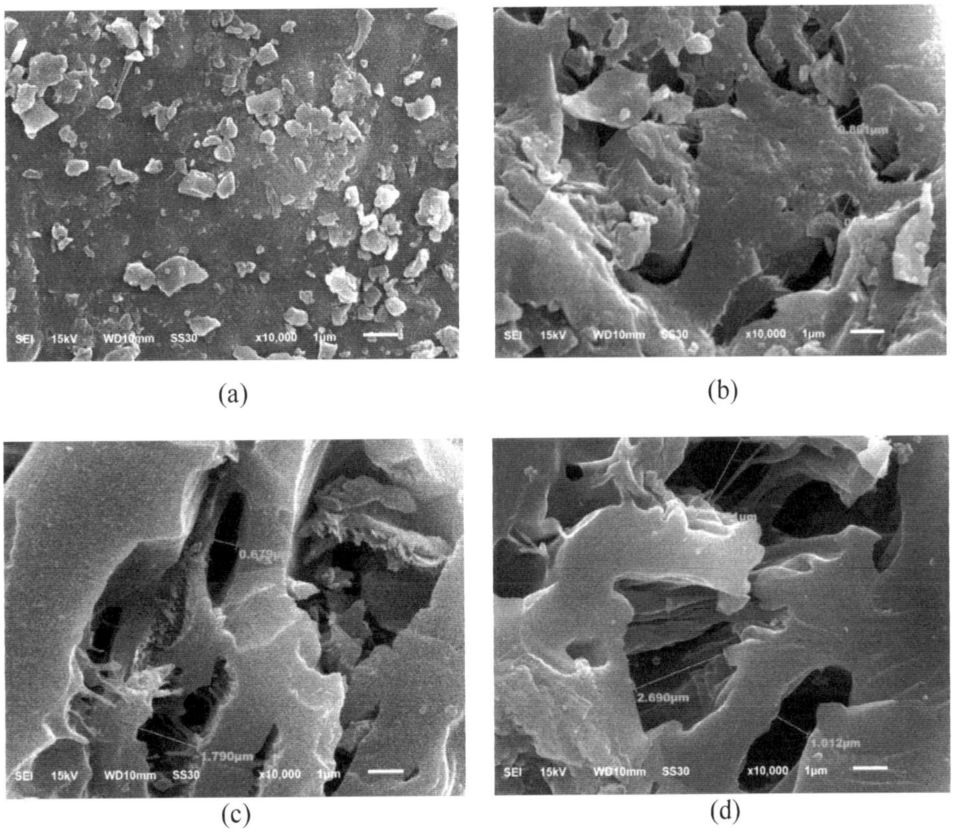

Figure 1(c-d). SEM micrograph of activated carbon from CNS (a) carbonized CNS (b) activation temperature of 500°C. (c) activation temperature of 700°C and (d) activation temperature of 900°C.

3.2. SEM and EDX characterization

The results for scanning electron micrograph (SEM) of activated carbon of CNS for the different activation temperature under 5000 times magnifications are shown in figure 1(a-d). The surface of the carbonized CNS was dense without any cracks and crevices. This surface pattern had poor and negligible adsorption capacity. In contrast, the activated carbon observed in the SEM was found in clear pore structure developed. Many cavities over the surface are well-developed pores. The size of surface pore increases as activation temperature increase.

3.3. Structural properties by X-ray diffraction

Figure 2 shows the X-ray diffraction pattern of activated carbon from cashew nutshell. As carbonized, XRD pattern shows a broaden peak located about $2\theta = 24°$ indicating an amorphous structure. The appearance this peak and the peak in $2\theta = 43°$ show the structure similar to graphite. The sharp peak on $2\theta = 38°$ is still unclear. It is convenient to relate to the characteristic of MgO (111) that can be confronted with EDX result. The peak on $2\theta = 43°$ is clearer after activation in which the increasing activation provoked the broader of peak located on $2\theta = 43°$ the appearance two new peaks about $2\theta = 29°$ and $2\theta = 32°$. This result seems to become the starting point toward the occurrence of the graphitization process. Zhao *et. al.* (2009) have reported that based on the evolution of peak intensity of XRD pattern, the graphitization process can be divided into three regions i.e. non-graphitization region, near graphitization region and graphitization region for activation below 900°C, 1000 – 1100°C and over 1200°C respectively [10].

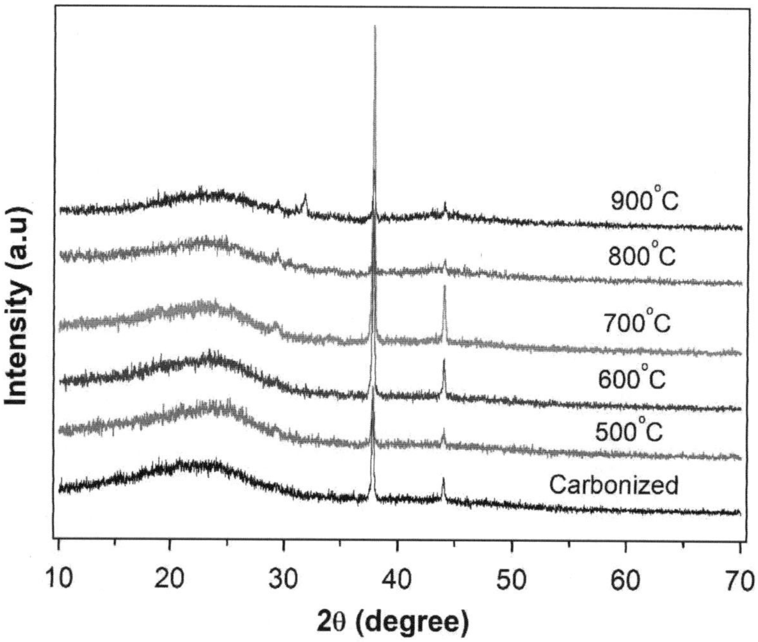

Figure 2. X-ray diffraction pattern of activated carbon from cashew nut shell with the different activation temperature.

4. Conclusion

This study results indicated that cashew nutshell as an inexpensive raw material to produce the activated carbon. Ultimate analysis shows that the activated carbon produced has optimum carbon content on the activation temperature of 700°C. The x-ray diffraction result provided the possibility of graphitization process by increasing the activation temperature.

References

[1] Tangjuank S, Insuk N, Udeyedan V and Tontrakoon J 2009 Chromium (III) Sorption From Aqueous Solutions Using Activated Carbon Prepared From Cashew Nut Shells *Int. J. of Phys. Sci.* **4**(8) 412 –7

[2] Yahayaa N K E M, Latiff M F P M, Abustan I and Ahmad M A 2010 Effect of Preparation Conditions of Activated Carbon Prepared from Rice Husk by CO2 Activation for Removal of Cu (II) from Aqueous Solution *IJET-IJENS* **10**(06)

[3] Jun T Y, Arumugam S D, Latip N H A, Abdullah A M and Latif P A 2010 Effect of activation temperature and heating duration on physical characteristics of activated carbon prepared from agriculture waste *EnvironmentAsia 3(special issue)* 143 –8

[4] Jadhav A S, Salwen S, M Tambedan and N H Shinde 2011 Chemical Characterisation of Biomass Waste by Proximate Analysis Method using Catalyst *IJAET* **II**(2) 1 – 19

[5] Metecelebes 2010 *Seluruh daerah Sultra memiliki perkebunan jambu mete yang potensial* http://metecelebes blogspot com/2010/01/seluruh-daerah-sultra-memiliki html

[6] Muzakkar M Z, Anas M and Ratna 2004 Kualitas karbon aktif dari kulit mete dan penentuaan konduktivitas listriknya *Laporan Penelitian Ilmu Dasar DIKTI* (Jakarta: DIKTI)

[7] Anas M Arifin dan R Baco 2003 Pemanfaatan Limbah Kulit Mete untuk Pembuatan Minyak Cashewnut (CNSL) *Laporan Penelitian FK8PT Universitas Halu Oleo* (Kendari: Universitas Haluoleo)

[8] Khezami L, AissaOuld-Dris and Richard C 2007 Activated Carbon fron Thermo-compressed Wood and Other Lignocellulosic Precursors *BioResources* **2**(2) 193 – 203

[9] Anas M M Z Muzakkar dan Ratna 2007 Aplikasi Arang Aktif dari Kulit Biji Mete untuk Penyerapan Logam Berat Fe Ni Cd Pb dan Hg *Laporan Penelitian Fundamental DIKTI* (Jakarta: DIKTI)

[10] Zhao J, Yang L , Li F , Yu R and Jin C 2009 Structural Evolution in The Graphitization Process of Activated Carbon by High-Pressure Sintering *Carbon* **47** 744 –51

Acknowledgments

This work has been partially supported by the Directorate of Higher Education, at the Ministry of Research Technology and Higher Education, Republic of Indonesia by means of Penelitian Unggulan Perguruan Tinggi (PUPT) grant.

Reduction behavior of medium grade manganese ore from Karangnunggal during a sintering process in methane gas

H Aripin[1], E Priatna[1], N Busaeri[1], N Hiron[1] and S Sabchevski[2]

[1]Department of Electrical Engineering, Faculty of Engineering, Siliwangi University, Tasikmalaya, Indonesia
[2]Lab. Plasma Physics and Engineering, Institute of Electronics of the Bulgarian Academy of Sciences, Bulgaria

Email: aripin@unsil.ac.id

Abstract. In this investigation, manganese has been produced from medium grade manganese ore from Karangnunggal mine (West Java, Indonesia). The effects of keeping it at a temperature of 1200°C in methane gas on the structural properties have been studied. The material's properties have been characterized on the basis of the experimental data obtained using X-ray fluorescence (XRF), X-ray diffraction (XRD), Scanning Electron Microscopy (SEM), and Fourier Transform Infrared (FTIR) spectroscopy. It has been found that MnO_2 is reduced to hausmannite (Mn_3O_4), manganosite (MnO) and manganese carbide (Mn_7C_3). An increase of the keeping time up to 45 minutes during the sintering process leads to an increase of the MnO concentration, while at 90 minutes, the concentration decreases. The analysis based on the interpretation of both XRD patterns and FTIR spectrum allows one to explain the blockage in the particle surface of the ore by an increase of the carbon deposition.

1. Introduction

The manganese is one of the most abundant minerals in nature and important engineering material. It is well-known that in the industrial processes and technical devices, the manganese is an environmentally friendly and cheap material. It has a long history as an electrode material for energy storage systems with high energy and power density. Many studies have been carried out in order to investigate the properties of manganese and to estimate its applicability to various energy storage systems e.g. fuel cells, electrochemical capacitors, and lithium-ion batteries. A promising and actively studied variety of the latter are namely the lithium batteries. The manganese has a porous microsphere morphology composed of MnO and Mn_3O_4 nanoparticles, which is very appropriate for its use as appropriate anode materials for lithium-ion batteries that exhibit a reversible capacity of 757 mAh/g [1]. When its morphology changes from microsphere to sponge-like porous, its capacity increases to 869 mAh/g [2]. From these studies, it has been found that the lithium batteries can reach an energy density up to 800 mAh/g and power density of 340 W/kg. Such technological and economic benefits (high number of charge-discharge cycles; faster charge-discharge; absence of disposable parts and environmentally harmful components; high efficiency up to 90%; low thermal load; safety and reliability) make this material a promising energy storage media for many modern electronic devices, such as digital cameras, cell phones, MP3 players and high-tech toys.

The rapid progress in the development and commercialization of battery systems in consumer electronics market of the modern society drives increasing demand for production of manganese. In order to meet the requirements of a rapid development of the electrical storage power industry and especially to provide an electrode material for batteries, the utilization of medium-grade manganese ore has become increasingly important. The manganese is usually obtained by reducing manganese oxide in manganese ores using a variety of different methods. In this respect, a promising approach is to introduce an appropriate gas medium in which the ore is immersed during the sintering process. In such a way, the manganese ore can be reduced to ferromanganese carbide keeping the sintering temperature of 1200°C for 45 minutes and exposing it to CH_4-H_2-Ar gas mixtures [3]. Sorensen et al., have studied the phase composition of the ore sintered at 800 – 1200°C in different media [4]. It has been found that MnO_2 in air is reduced to Mn_2O_3 and Mn_3O_4, in hydrogen is reduced to MnO_2 and MnO and iron oxides to metallic iron, while in CO gas it is reduced only to MnO. Furthermore, it has been found that in methane gas the manganese MnO_2 phase of the ore decomposes to form Mn_2O_3 phase and then transits to Mn_3O_4 at temperatures above 850°C and to MnO at 950°C [5]. Recently, by heating in the same medium, the manganese ore has been reduced to carbide Mn_7C_3 and Mn_5C_2 at temperatures under and above 1200°C, respectively [6]. From these studies, it has been found that manganese ores of different origin have different mineralogy which affects their reduction properties. In this paper, we present the results of our study on the structural properties of the medium grade manganese ore of Karangnunggal sintered at a temperature of 1200°C and exposed to the methane gas.

2. Experimental procedure

2.1 Preparation of manganese ore powder
The manganese ores were obtained from the artisanal mining area in Karangnunggal, Tasikmalaya, Indonesia. The materials for the samples were selected manually, cleaned by washing with water, and then dried in the sun. The dried manganese ores were crushed into small pieces with an area of approximately 1 cm2. Then, they were ground using a jar mill to pass through a 150 mesh sieve. The manganese ore powder was stored in a desiccator for further treatment.

2.2 Annealing the manganese ore
The powder samples were placed into a porcelain crucible and then inserted into a horizontal furnace. The powders were sintered with a controlled heating rate of 2°C/min up to the temperature of 1200°C in the air. Then the temperature was kept constant at 1200°C for 15, 30, 45 and 90 minutes in methane gas flow at a speed of 1000 mL/min. The cooling was performed by natural convection after turning off the electric furnace and leaving the samples inside.

2.3 The method used for characterizing the manganese ore
The XRF measurements were carried out on a Philips PW2400 wavelength dispersive spectrometer. The calibration was performed following the prescribed standard procedure. In addition, some correction factors for determining the influences of other elements were calculated. The measurement conditions were 24 kV – 100 mA, collimator mask 27 mm and Rh-X ray tube as an excitation source. A gas flow detector was used for controlling the rate of the gas flow. Furthermore, XRD was used to determine the change of the crystalline phase for samples sintered at 1200°C for 15, 30, 45 and 90 minutes and it was performed using a Smartlab X-ray diffractometer with filtered Cu Kα radiation at a wavelength of 0.15418 nm. The intensity of the diffracted X-ray was measured as a function of the diffraction angle 2θ ranging from 10 to 90 degree. We used scanning electron microscopy and energy dispersive spectrometry (SEM-EDS) in order to study the small-scale relationships between textures and to measure the elemental distributions and abundances. The images were taken at a magnification of 5000 x at 20 kV. The Fourier transform infrared spectroscopy (FTIR) was performed on a SHIMADZU-FTS 8400 FTIR using KBr pellet techniques in the 400 – 4000 cm^{-1} wavenumber range. The resolution and number of scans used in FTIR collection were 2.0 cm^{-1} and 16, respectively.

IOP Conf. Series: Materials Science and Engineering **550** (2019) 012036 doi:10.1088/1757-899X/550/1/012036

3. Result and discussion

Table 1 shows the chemical composition of oxide contents and the loss on ignition for medium-grade of Karangnunggal manganese ore. The ore contains MnO as the main constituents at about 46.80 wt.%. The content of such magnitude is classified as a medium-grade ore [7]. It is also accompanied by a significant amount of SiO_2, CaO, Fe_2O_3, and Al_2O_3. It can also be seen that the ore contains small amount of MnO. The loss on ignition is associated with the presence of hydroxides, organic matter and volatile components.

Table 1. Chemical composition for medium-grade manganese ore of Karangnunggal.

Chemical composition	Medium-grade manganese ore of Karangnunggal (wt.%)
MnO_2	46.80
SiO_2	7.29
CaO	5.95
Fe_2O_3	3.61
Al_2O_3	2.71
MgO	2.46
BaO	1.54
Na_2O	1.07
P_2O_5	0.80
SrO	0.66
K_2O	0.368
CuO	0.108
TiO_2	0.0777
Cl	0.0561
V_2O_5	0.0485
Co_3O_4	0.0256
SO_3	0.0207
Cr_2O_3	0.0178
NiO	0.0144
Y_2O_3	0.0057
ZnO	0.0051
Ga_2O_3	0.0027
LOI	26.07

Figure 1. SEM image of medium-grade manganese ores from Karangnunggal.

Figure 1 shows SEM image of medium-grade manganese ores from Karangnunggal. It can be observed that the ore particles are agglomerated forming dense grains with different texture and morphology. The EDS reveals that the irregular-shaped grains with dark contrast marked as A are associated with the presence of manganese [8]. The large brighter grains marked as B are associated with the presence of calcium, while the relatively smaller grains with bright texture and smooth surfaces marked as C are quartz [9].

Figure 2 shows some representative XRD patterns for the raw material samples of medium-grade manganese ore from Karangnunggal and samples sintered at a temperature of 1200°C with varying keeping time in methane gas. It is clear from figure 1 that the raw material sample consists of manganese oxide mainly in the form of pyrolusite (MnO_2) and quartz (SiO_2). For the sample sintered for 15 minutes, MnO_2 disappears and is replaced by three new phases, namely hausmannite (Mn_3O_4), manganosite (MnO) and manganese carbide (Mn_7C_3). The hausmannite (peak 1) appears at 2θ = 18.31°, 32.59°, 38.29°, 56.37° and 58.78° (JCPDS 24-0734), manganosite (peak 4) at 2θ = 36.39°, 40.69°, 60.33°, and 74.61° (JCPDS 00-001-1206), and manganese carbide (peak 5) at 2θ = 44.47°, 51.18°, and 54.12° (PDF 01-075-1498). At a temperature of 1200°C the keeping time does not have any effect on the formation of crystals and the crystalline phases remain almost unchanged in the time range from 15 to 90 minutes. The intensity of MnO at 2θ = 36,39° increases with the rise of the keeping time up to 45 minutes. The increase in the intensity can be explained by an increase of the degree of crystallinity of MnO. This crystallization affects significantly the grain growth and the sintering process provides energy, which is necessary for bonding the particles together and removing the porosity simultaneously [10]. For the sample sintered for 90 minutes, the intensity of MnO is reduced. At longer keeping times the excess reaction of the methane gas and manganese oxide in the ore causes an increase of carbon deposition [3]. As a result, it blocks the particle surface and then hinders the reduction process of manganese oxide.

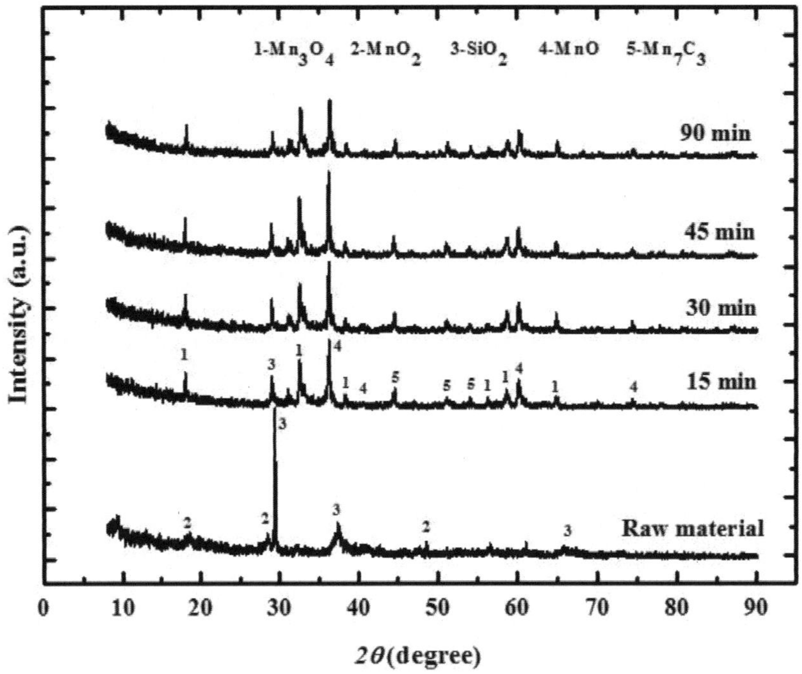

Figure 2. XRD patterns for medium-grade manganese ore from Karangnunggal with varying the keeping time at a temperature of 1200°C in methane gas.

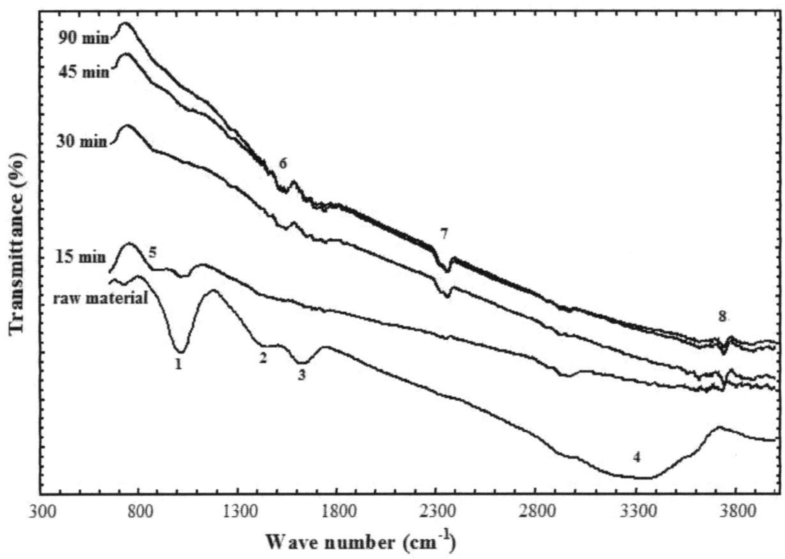

Figure 3. FTIR spectra for medium-grade manganese ore from Karangnunggal with varying the keeping time at a temperature of 1200°C in methane gas.

In figure 3, we show the Fourier transform infrared (FTIR) spectra for medium-grade manganese ore varying the keeping time at a temperature of 1200°C in methane gas. For the raw material sample, the band (1) identified at about 1020 cm^{-1} corresponds to pyrolusite (MnO_2). Additionally, the weak band

(2) at around 1450 cm^{-1} and the band (3) at 1640 cm^{-1} correspond to the existence of large numbers of residual hydroxyl groups, which imply the O-H vibrating mode of traces of adsorbed water [11-13]. In the sample sintered for 15 minutes, the characteristic bands of pyrolusite are gradually reduced as the keeping time increases and eventually disappear completely above time of 15 minutes. Then they are replaced by three new bands (5, 7 and 8) attributed to hausmannite [14], C=O stretching mode [15] and SiOH at about 895 cm^{-1}, 2370 cm^{-1} and 3750 cm^{-1} [16], respectively. The appearance of C=O stretching mode arises from absorption of atmospheric methane and manganese oxide of the ore. The intensity of this band increases with increasing the keeping time, indicating an increase of carbon deposition on the surface of manganese oxide. For the sample sintered from 30 to 90 minutes, there is an absorption band (6) at about 1552 cm^{-1} attributed to asymmetric O-C-O stretching [17].

4. Conclusion

Manganese was produced successfully from medium-grade ore of Karangnunggal by controlling the keeping time in methane gas flow during the sintering process. The results presented in this study show that the existing phases in raw material are manganese oxide in the form of pyrolusite (MnO_2) and quartz (SiO_2). In the sample sintered at 1200°C for 15 – 90 minutes, the crystalline phases are hausmannite (Mn_3O_4), manganosite (MnO) and manganese carbide (Mn_7C_3). An interesting phenomenon has been observed analyzing the XRD pattern and FTIR spectra of the sample sintered for 90 minutes. It manifests itself in the finding that for long times of sintering in methane gas the reduction process for manganese oxide ore is considerably hindered.

References

[1] Jiang X, Wang Y, Yang L, Li D, Xu H and Ding Y 2015 Dealloying to porous hybrid manganese oxides microspheres for high performance anodes in lithium ion batteries *J Power Sources* **274** 862
[2] Gao J, Lowe M A, and Abruña H D 2011 Spongelike Nanosized Mn_3O_4 as a High-Capacity Anode Material for Rechargeable Lithium Batteries. *Chem. Mater.* **23** 3223
[3] Ostrovski O, Anacleto N and Ganguly S 2004 Reduction of Manganese Ores by Methane Containing gas, *In Proceeding of Tenth International Ferroalloy Congress*, Cape Town South Africa, 173
[4] Sorensen B, Gaal S, Tangstad M, Ringdalen E, Kononov R and Ostrovski O 2010 Properties of Manganese Ores and Their Change in the Process of Calcination *In Proceeding of The Twelfth International Ferroallors Congress*, Helsinki, Finland 439
[5] Moradkhani D, Malekzadeh M and Ahmadi E 2013 Nanostructured MnO_2 synthesized via methane gas reduction of manganese ore and hydrothermal precipitation method *Trans. Nonferrous Met. Soc. China* **23** 134
[6] Bhalla A and Hurman E R 2015 Reduction Behavior of Manganese Ore Using Methane Gas, *In Proceeding of The Fourteenth International Ferroalloys Congress*, Energy efficiency and environmental friendliness are the future of the global Ferroalloy industry, May Kiev, Ukraine
[7] Fahim M S, Faramawy H E, Ahmed A M, Ghali S N and Kandil A T 2013 Characterization of Egyptian Manganese Ore for Production of High Carbon Ferromanganese *J. Miner. Mater. Charact. Eng.* **1** 68
[8] Miaha P P, Mohapatara B K and Mahanta K 2009 Upgradation of low-grade siliceous manganese ore from Bonai-Keonjhar belt, Orissa *Indian J Miner. Mater. Charact. Eng.* **8** 47
[9] Mehdilo A, Irannajad M and Hojjati-Rad M R 2013 Characterization and beneficiation of Iranian low grade manganese ore *Physicochem. Probl. Miner. Process.* **49** 725
[10] Aripin H, Mitsudo S, Nyoman Sudiana I, Tani S, Sako K, Fujii Y, Saito T, Idehara T and Sabchevski S 2011 Rapid Sintering of Silica Xerogel Ceramic Derived from Sago Waste Ash Using Sub-millimeter Wave Heating with a 300 GHz CW Gyrotron *J Infrared Mili. Terahz Waves* **32** 867

[11] Goreletsky V I, Mitichkin A I, Balenko L E and Rebrova T P 2001 IR Spectrocopy of KBr salt and crystals *Semicond. Phys. Quantum Electron Optoelectron* **4** 139
[12] Yang X, Makita Y, Liu Z, Sakane K and Ooi K 2004 Structural Characterization of Self-Assembled MnO_2 Nanosheets from Birnessite Manganese Oxide Single Crystals *Chem. Mater.* 16 5581
[13] Yang R, Wang Z, Dai L and Chen L 2005 Synthesis and characterization of single-crystalline nanorods of α-MnO_2 and γ-MnOOH *Mater. Chem. Phys.* **93** 149
[14] Swami M C, Jayasheela H M and Hulsure V C 2014 Infrared Spectra of Manganese ores from Deogiri of Sandur Schist Belt Area Karnataka India *Intern. J Research. in Adv. Technol.* **2** 341
[15] Christobel G J 2016 Vibrational Spectroscopy of ZnO-ZnS Nanoparticles *Intern. J Sci. Research* **5** 2228
[16] Aripin H, Mitsudo S, Prima E.S, Sudiana I.N, Kikuchi H, Sano S and Sabchevski S 2015 Crystalline mullite formation from mixtures of alumina and a novel material – silica xerogel converted from sago waste ash *Ceram. Intern.* **41** 6488
[17] Marwood M, Doepper R and Renken A 1997 In-situ surface and gas phase analysis for kinetic studies under transient conditions: The catalytic hydrogenation of CO_2 *Appl. Catal. A: Gen.* **151** 223

Acknowledgement
This research was supported by the fund of the Siliwangi University through the Project of Higher Education Innovation Research and collaboration with Nano Technology and Graphene Research Center (NTGRC), Padjadjaran University. The authors would like to thank the research team from the NTGRC for kindly helping to carry out the sintering for manganese oxide ore.

Synthesis and characterization of chitosan:SiO_2 nanocomposite by ultrasonic spray drying

S Y Azhary[1], D Purnama[2], F F Florena[3], M Vanitha[4], Muchtaridi[2], C Panatarani[4,5] and I M Joni[4,5]

[1]Dept. of Environmental Science, Graduate School, Universitas Padjadjaran, Jl. Raya Bandung-Sumedang Km 21, Jatinangor 45363, Jawa Barat, Indonesia
[2]Dept. of Pharmacy, faculty of pharmacy, Universitas Padjadjaran, Jl. Raya Bandung-Sumedang Km 21, Jatinangor, 45363, Jawa Barat, Indonesia
[3]Dept. of Physics, Graduate School, Institute of Technology Bandung, Jl. Ganesha No.10, Kota Bandung 40132, Jawa Barat, Indonesia
[4]Dept. of Physics, Faculty of Faculty of Mathematics and Natural Sciences, Universitas Padjadjaran, Jl. Raya Bandung-Sumedang Km 21, Jatinangor 45363, Jawa Barat, Indonesia
[5]Nanotechnology and Graphene Research Center, Universitas Padjadjaran, Jl. Raya Bandung-Sumedang, Jatinangor 45363, Jawa Barat, Indonesia

E-mail: imadejoni@phys.unpad.ac.id

Abstract. Chitosan:SiO_2 nanocomposite (Cs:SiO_2 NCs) is synthesized via aerosol processing. Ultrasonic spray drying apparatus was used to produce smaller sized composite particles with spherical morphology. Composite from shrimp shell and water glass (Na_2SiO_3) as silica precursor were used for the preparation. Cs:SiO_2 NCs were characterized by Fourier Transform Infrared (FTIR) spectroscopy and scanning electron microscopy (SEM) to know about the chemical bonding and morphology of the particles, respectively. Based on the FTIR investigation the absorption peak at 1090 cm^{-1} represents the reaction of silanol groups of silica with carbonyl groups of polymers leading to the formation of Si-O-C bonds. The morphology of Cs:SiO_2 NCs was more spherical than chitosan particles with relatively narrow size distribution. The smaller particle size of Cs:SiO_2 NCs is potentially applied to engineer the material for drug delivery system.

1. Introduction

Nowadays, Matrix formulation has become an essential step in the development of drug delivery systems. Chitosan is a polysaccharide obtained from the deacetylation of chitin, which is widely used as a drug delivery system for gene delivery, implants and nasal, oral, parental and transdermal administration [1]. Chitosan can be prepared by ionic cross-linking between positively charged amino groups of the chitosan chains and negatively charged polymers or ions such as tripolyphosphate (TPP) [2].

The addition of inorganic oxides to chitosan in order to impart resistance to microbial attack, higher mechanical strength, enhanced thermal stability, negligible swelling in organic solvents, and modulable textural properties, can be an alternative way to increase its chemical and mechanical stability [3]. Among the inorganic oxides, SiO_2 is an interesting choice for the formulation of an organic-inorganic composite carrier. Over the past years, the sol-gel process was widely used to create novel hybrid nanoscale materials based on organic and inorganic components [4]. This approach can produce particles at low temperature and pressure.

The advanced utility of chitosan:SiO_2 composite for the development drug carrier can be accomplished by controlling its morphology. The method proposed to obtain particles with a controlled morphology is spray drying, in particular, ultrasonic spray drying is one of the aerosol-assisted spray methods employing ultrasonic nebulizer [5]. By using ultrasonic nebulizer fine droplets in the range of 1-10 μm, particles can be produced which is smaller than particles attained by a nozzle or rotary atomizer [6].

In this work, chitosan:SiO_2 nanocomposite cross-linked with TPP was synthesized by ionotropic-sol gel method using water glass (Na_2SiO_3) as the silica precursor. The controlled morphology of particles was achieved by ultrasonic spray drying apparatus. The presence of functional groups, the morphology of the particles, and the size distribution of the obtained particles was investigated. In addition, the mechanism of inotropic-sol gel method was also proposed.

2. Materials and methods

2.1. Materials

Chitosan was received from commercially available flakes chitosan (>80% deacetylated) in Cirebon, West Java. The raw material of chitosan was pulverized using ball milling to obtain uniform sized powder by sieving with 100 mesh and subjected to further experiments and analysis. Sodium tripolyphosphate (TPP), water glass (Na_2SiO_3), acetic acid (CH_3COOH), was purchased from Brataco, Co., Ltd. (Bandung, Indonesia). Distilled water was used for all sample preparation.

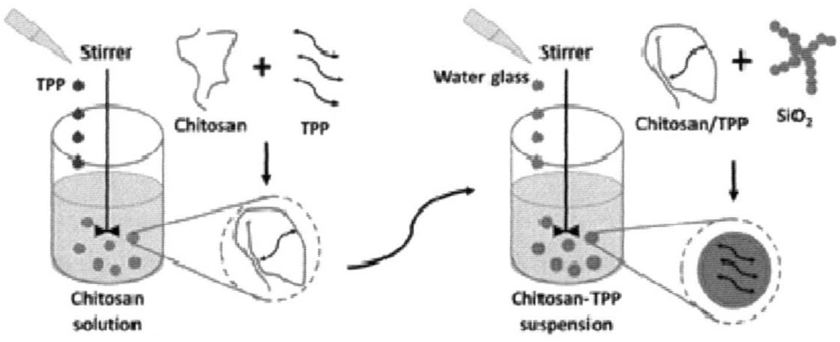

Figure 1. Schematic inotropic-sol gel process.

2.2. Synthesis and characterization

Schematic inotropic-sol gel process of the chitosan-SiO_2 nanocomposite is shown in figure 1. Briefly, 0.1 g of chitosan dissolved in 200 ml of 1% v/v acetic acid solution at room temperature. The solution was stirred at 600 rpm for 120 min to ensure complete dissolution of the polymer. Chitosan was cross-linked by adding 40 ml of 0.07% (w/v) TPP solution dropwise with a gravitational syringe, and then further stirred at 600 rpm for 60 min. The cross-linked chitosan was coded as Cs NPs. A 6 ml of water glass precursor was added dropwise to the chitosan-TPP solution to prepare nanocomposite. After the addition of water glass, the solution changes from transparent to opalescent. Opalescence occurs due to the hydrolysis of sodium silicate and polycondensation reaction in the chitosan-TPP solution to form an inorganic network in the presence of organic polymer, and the composite particles is coded as Cs:SiO_2 NCs.

To form structured Cs NPs and Cs:SiO_2 NCs, the colloidal suspension was spray-dried using ultrasonic nebulizer [7]. The temperature of the reaction tube was set at 80°C based on the drying temperature of the suspension, and also related to the residence time of the droplets inside the reactor. Air compressor was used as carrier gas with a flow rate of 5 L/min and the dry powder was collected

IOP Conf. Series: Materials Science and Engineering **550** (2019) 012037 doi:10.1088/1757-899X/550/1/012037

by an electrostatic precipitator. The functional groups of chitosan and chitosan-SiO_2 nanocomposite were characterized by Fourier-transform infrared (FTIR) analysis (Shimadzu prestige 21) operating in the wavenumber range of 400-4000 cm^{-1} with KBr as a beam splitter. The morphology of the particle was characterized by SEM Hitachi SU 3500 measurements.

3. Results and discussion

3.1. FTIR analysis

The functional groups present in the spray-dried Cs NPs and Cs:SiO_2 NCs are compared with raw Cs, TPP, and water glass by FTIR analysis. The FTIR spectrum of raw Cs (Figure 2 (a)), shows broadband at range 3200-3400 cm^{-1} which is contributed to different vibrations, namely hydrogen bonded O–H stretching at around 3400 cm^{-1} and NH stretching involved in hydrogen bonding at 3200 cm^{-1} [8]. The absorption band at 2873 cm^{-1} is related to the C-H stretching vibrations. The absorption peak at about 1658 cm^{-1} represents C=O stretching of the amide I band. The band at 1592 cm^{-1} corresponds to the vibrations of -NH (N-acetylated residues, amide II bands) and peaks at 1423 cm^{-1} and 1376 cm^{-1} are attributed to -CH_2 wagging coupled with -OH in plane deformation [8]. The smaller peak at 1318 cm^{-1} is ascribed C-N stretching of amide III.

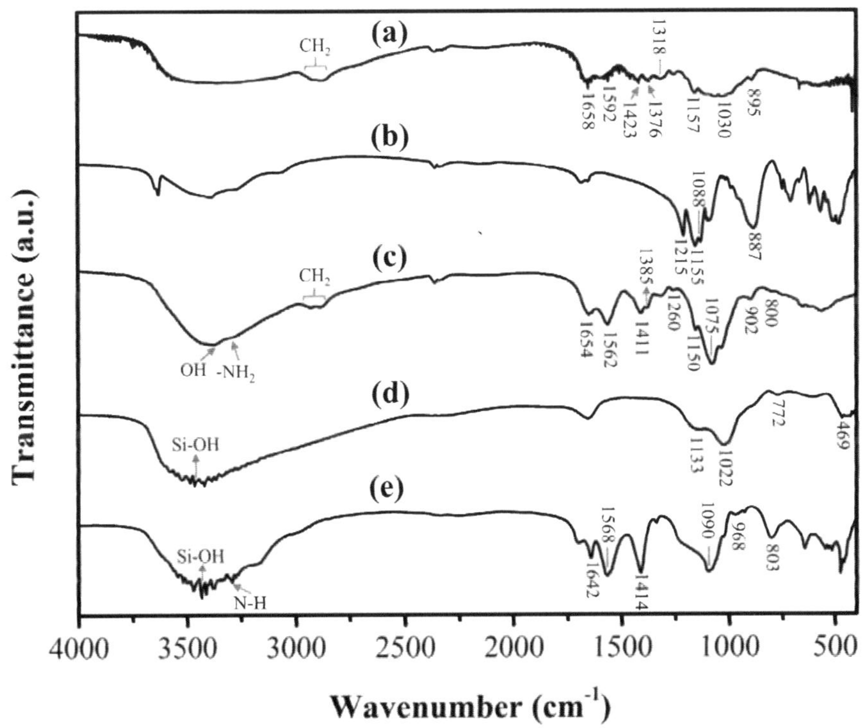

Figure 2. FTIR Spectra of (a) raw chitosan, (b) TPP, (c, e) Cs NPs and Cs:SiO_2 prepared by ultrasonic spray drying, (d) pure water glass.

The peak at 1157 cm^{-1} is assigned to asymmetric stretching vibrations of glycosidic linkage (C-O-C). The absorption band in the region of 1030 cm^{-1} is associated with the C-O stretching vibration and peak at 895 cm^{-1} corresponds to the saccharide structure. Figure 2 (b) represent the FTIR spectra of sodium tripolyphosphate (TPP) as a cross-linking agent that shows the absorption band at 1215 cm^{-1} corresponding to P=O stretching vibration. The peak at 1155 cm^{-1} is related to the symmetrical and asymmetric stretching vibration of the PO_2 groups. Two bands at 1088 cm^{-1} and 887 cm^{-1} are due to

the symmetric and asymmetric stretching vibration of the PO_3 groups and Stretching vibration of P–O–P bridge, respectively [9].
FTIR spectrum of spray-dried Cs NPs is shown in Figure 2 (c). The peak between 3372 cm^{-1} and 3290 cm^{-1} become sharper which is indicative of the enhanced O-H and N-H groups. The weak absorption band in the region of 2931-2882 cm^{-1} is due to the presence of C-H stretching vibration. The amide I band at 1658 cm^{-1} was shifted to 1654 cm^{-1} in the Cs NPs. The absorption peak at 1592 cm^{-1} was reduced and a new peak appeared at 1562 cm^{-1} confirming that amino groups were cross-linked by phosphate anion (N-O-P) [10]. The band at 1260 cm^{-1} which indicates P=O stretching and P-O bending, occurs at 800 cm^{-1} [11].
Figure 2 (d) and (e) shows the FTIR peaks for water glass (Na_2SiO_3) and spray-dried Cs:SiO_2 NCs, respectively. The broadband of water glass at 3420 cm^{-1} is attributed to Si-OH stretching vibration bounded with water molecules [12]. The two characteristic peaks at 1022 cm^{-1} and 772 cm^{-1} are assigned to Si-O bending vibration and asymmetric stretching vibration of the siloxane bonds (Si-O-Si). The absorption peak at 469 cm^{-1} corresponds to Na-O bond structure. After addition of water glass, the presence of amino groups in the Cs surface facilitates the hydrolysis of water glass and condensation of the created silanol groups (Si-OH) [13] as seen at 3372 cm^{-1} and 968 cm^{-1} (figure 2 (d)). Specific bands in the region of 1000–1250 cm^{-1} reveals the characteristic absorption for Si–O–Si bond after the complete condensation of water glass [14]. On the other hand, the silanol groups of silica can interact with carbonyl groups of polymer to form of Si-O-C bonds as assigned by a peak at 803 cm^{-1}.

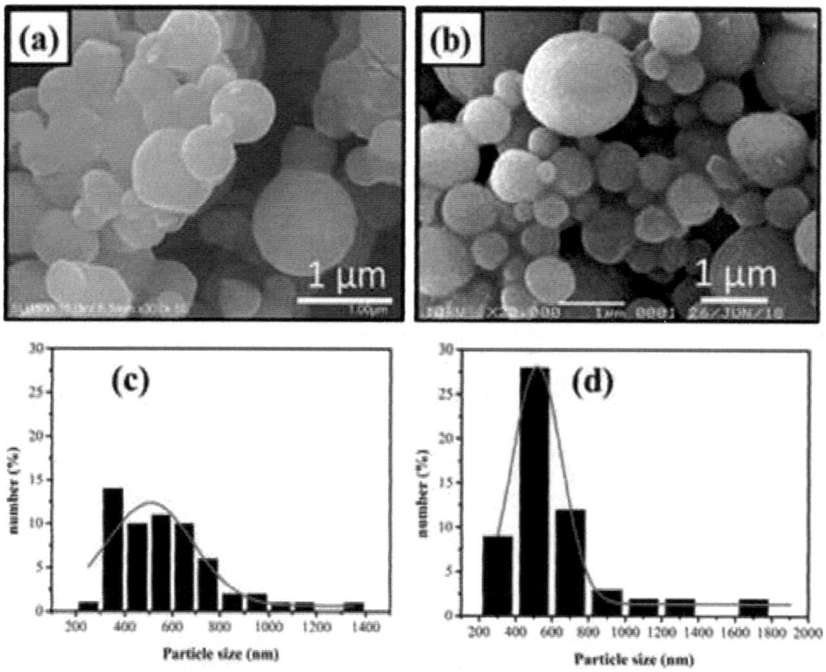

Figure 3. (a-b) SEM images of Cs NPs and Cs:SiO_2 (c-d) size distribution of CS NPs and Cs:SiO_2 NCs prepared by ultrasonic spray drying.

3.2. Morphology of particles

Scanning electron microscopy (SEM) was used to observe the morphology of the Cs NPs and Cs:SiO_2 NCs prepared by ultrasonic spray drying. Figure 3 (a) shows the Cs NPs have nearly spherical in shape with agglomerated particles indicated with a red arrow, while the Cs:SiO_2 NCs have a compact spherical shape. The agglomerated particles of Cs NPs after spray drying are due to the interaction of polyanions (TPP) and chitosan during the synthesis process. At the end of addition of polyanions, chitosan molecules are completely cross-linked and excess TPP got involved in intermolecular cross-linking resulting in larger particle size or agglomeration [15]. On the other hand, Cs:SiO_2 particles have a compact shape (figure 3 (b)) due to the hydrolysis and condensation of water glass. Therefore, it is suggested that the SiO_2 on the surface are bound with polymer forms some –Si–O–Si– chains covering the surface of Cs:SiO_2 NCs to prevent the excess intermolecular cross-linking by TPP anions. The particle size distribution calculated from SEM images using the Image-J free software is shown in figure 3 (c-d). Generally, the Cs NPs and Cs:SiO_2 have a mean size distribution at around 500 nm obtained by ultrasonic spray drying. However, the Cs NPs droplet produced by ultrasonic atomizer which have intermolecular interaction between chitosan particles tend to bring them together and generate broader distribution (figure (c)). In contrast, Cs:SiO_2 droplet containing –Si–O–Si– chains covering the surface particles achieved irregular to compact spherical shape with relatively narrow size distribution (figure (d)). Based on FTIR analysis and SEM observation, The possible formation of Cs NPs and Cs:SiO_2 NCs were obtained by ultrasonic spray drying as illustrated in figure 4.

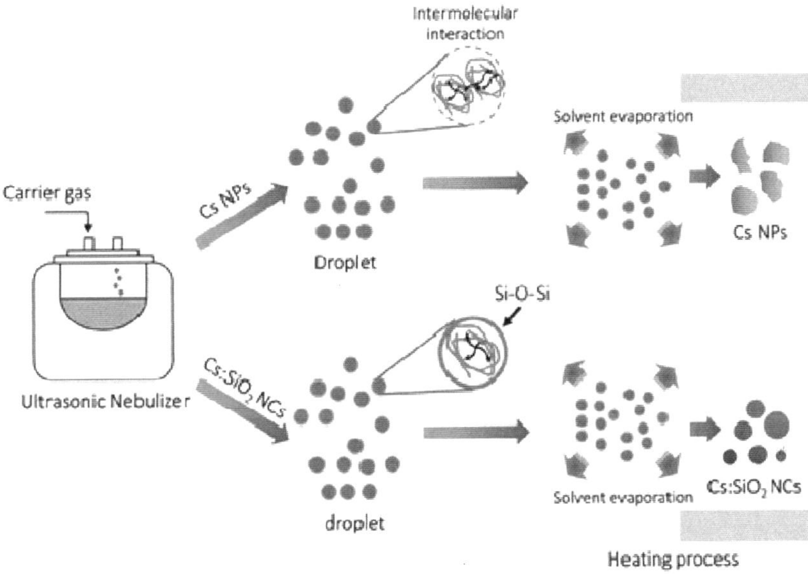

Figure 4. Schematic illustration of Cs NPs and Cs:SiO_2 NCs formation prepared by ultrasonic spray drying.

4. Conclusion

Cs:SiO_2 nanocomposite has been successfully prepared by ultrasonic spray drying with a compact spherical shape and relatively narrow distribution of particles. Furthermore, the TPP as anions play a crucial role in Cs NPs synthesis, it can be concluded that the intermolecular interaction between chitosan molecules along with the cross-linking tends to make the particle agglomerated or leads to a formation of large particles. It is established that Cs:SiO_2 NCs prepared by spray drying have greater potential as a drug delivery agent.

References

[1] Rinaudo M 2006 *Progress in Polymer Science* **31** 603
[2] Tokárová V, Kašpar O, Knejzlík Z, Ulbrich P and Štěpánek F 2013 *Powder Technology* **235** 797–805
[3] Budnyak T M, Pylypchuk I V, Tertykh V A, Yanovska E S and Kolodynska D 2015 *Nanoscale Research Letters* **10** 87
[4] Zou H, Wu S and Shen J 2008 *Chem Rev* **108** 3893–957

[5] Nandiyanto A B D and Okuyama K 2011 *Adv Powder Tech* **22** 1–19
[6] Joni IM, Panatarani C, Hidayat D, Setianto, Wibawa B W, Rianto A, and Thamrin H 2013 *AIP Conf Proc* **1554** 20-6
[7] Panatarani C, Muharam D G, Wibawa B M and Joni I M 2013 *Materials Science Forum* **737** 20–7
[8] Pati F, Adhikari B and Dhara S 2011 *Carbohydrate Research* **346** 2582–88
[9] Gierszewska, M and Ostrowska-Czubenko, J 2016 *Carbohyd Polym* **153**, 501–11
[10] Xu Y and Du Y 2003 *Int. J. Pharmaceutics* **250** 215–26
[11] Qi, L and Xu, Z 2004 *Colloids and Surfaces, A: Physiochemical and Engineering Aspects*, **251** 183–190
[12] Pignatelli I, Kumar A, Bauchy M and Sant G 2016 *Langmuir* **32** 4434–39.
[13] Sh. Al-Kandary, A. A. M. Ali and Z. Ahmad 2005 *Journal of Applied Polymer Science* **98** 2521–31
[14] Ahmad Z, Al-Sagheer F, Ali A. A. M, and Muslim S 2007 *Journal of Macromolecular Science* A **44**, 79–90
[15] Fan W, Yan W, Xu Z, and Ni H 2012 *Colloids Surf B: Biointerfaces* **90** 21–7

Acknowledgement
The authors would like to thank to Universitas Padjadjaran for the financial support under Academic Leadership Grant (ALG) 2018.

IOP Conf. Series: Materials Science and Engineering **550** (2019) 012038 doi:10.1088/1757-899X/550/1/012038

Startup of expanded granular sludge bed reactor treating undiluted palm oil mill effluent

A Ahmad, I Amri and R Fitrah

Department of Chemical Engineering, University of Riau
HR Subrantas Rd. km 12.5 Pekanbaru 28293

E-mail: adri@unri.ac.id

Abstract. A lab-scale Expanded Granular Sludge Bed (EGSB) Reactor was used to the treatment of undiluted Palm Oil Mill Effluent (POME) with the COD mean value of 53,333.33 mg/l. The startup was done at a mesophilic temperature of 35 ± 2^{0}C for 12 days with two days Hydraulic Retention Time (HRT) as a pre-study of enhanced biogas production by mesophilic and thermophilic anaerobic co-digestion of POME with Empty Fruit Bunches (EFB) in EGSB Reactor. The steady state was achieved at 8th to 12th day of continuous operation. The effluent COD at steady state was vary from 10,000 to 11,500 mg/l, COD removal 78.44 – 81.25 %, Methane content was 43.89 % and total biogas volume achieved was 2.760 ml (0.1486 – 0.1503 m^3 CH_4/kg COD removed). There were several operational problems met in the experiment and corresponding countermeasures were also discussed.

1. Introduction

Riau province was the biggest CPO producer in Indonesia, achieving 7.33 tons or equal to 34,904,761 tons of FFB in 2016. It was followed by a large amount of liquid waste or Palm Oil Mill Effluent (POME) estimated by 20,942,857 m^3 (0.6 m^3/tons FFB processed). POME was mainly consisted of processed sludge, condensate from sterilizing, processed water from clarification, Hydrocyclone /Clay bath and spent water from plant's cleaning activity. POME was relatively high in organic materials and non-toxic because no chemicals were used in the extraction process of palm oil.

POME's physical appearance was brown-colored, contain residual oil, dissolved solids, and suspended colloidal form with high temperature, high level of Biological Oxygen Demand (BOD) and Chemical Oxygen Demand (COD). Such characteristics would harm the environment when openly discarded into bodies of water. High level of BOD would reduce oxygen levels in the water thus threatened the aquatic ecosystem [1]. [10]. Most of CPO producer treated POME by using open ponds system which consists of a de-oiling tank, facultative aerobic and anaerobic ponds [2]. [11]. The number of ponds used varies that depends on the capacity of the palm oil mill.

Anaerobic pond for 54 tons FFB/hour Palm Oil Mill sized from 60 x 29.6 x 5.8 m (length x width x Depth) [3]. [12]. The size also depended on how many vacant areas could be used as open ponding system. Anaerobic ponds have the longest retention time among any other ponds in the ponding system, which vary from 20 to 200 days [2]. [11]. The drawbacks of this system was a long HRT, required vast areas, and the released of GHG in the form of Biomethane which cannot be collected thus harming the environment. Concerned researchers were trying to solve these problems by utilizing biogas from POME for power plants

IOP Conf. Series: Materials Science and Engineering **550** (2019) 012038 doi:10.1088/1757-899X/550/1/012038

To shorten the HRT, lessen the area required, and utilizing the biogas produced at the same time, researchers were studied various type of high-rate bioreactors such as Expanded Granular Sludge Bed (EGSB) [4] [5], tank digester [5] [19], anaerobic filter by [6] [13], anaerobic fluidized reactor [7] [14], anaerobic baffled reactor [8] [20], and upflow anaerobic sludge bed reactor (UASB)[9]. [15]. EGSB was a variant of the UASB reactor with a typical characteristic of a slender body. This design was aimed for improving the substrate-biomass contact time in granular sludge bed [4]. [5]. Upflow movement from the substrate would expand the sludge bed, releasing biogas bubbles that were trapped in between granular sludge as well as mixing the substrate with biomass.

Yejian *et al.* (2008) studied start-up of EGSB at 35^{0}C using a diluted POME with a constant level COD of 30,000 mg/l [4] [5]. The EGSB reactor used was 21,560 ml in total volume and 20.500 ml in effective volume. Highest COD removal was 91%, at two days HRT and 1.45 to 17.5 kg COD/m^3 of Organic Loading Rate (OLR). Yejian *et al.* (2011) studied another use of EGSB reactor in treating POME to generate biogas [10]. [4]. The reactor was operated continuously at a mesophilic temperature of 35^{0}C with the OLR was increasing in the range of 1.45 to 16.5 kg COD/m^3-day. The highest Biomethane Yield achieved was 46.00% at a temperature of 35^{0}C, HRT of 3 days, OLR of 10 kg COD/m^3, and pH 8.5 – 9.0.

Ohimain and Izah (2014) reviewed the biogas production and performance from many different types of anaerobic biodigester by using Nigeria's POME - Potential Biogas data from 2004 until 2013 [11]. [2]. The study revealed that EGSB reactor exhibits the highest Biomethane production in the shortest HRT compared to other types of the anaerobic biodigester.

This study was performed to review the startup of EGSB reactor treating undiluted POME at a mesophilic temperature of 35 ± 2^0 C and Hydraulic Retention Time (HRT) of 2 days. This study was also a pre-study of enhanced biogas production by mesophilic and thermophilic anaerobic co-digestion of POME with Empty Fruit Bunches (EFB) in EGSB Reactor.

2. Materials and methods

The materials used in this study were POME obtained from Sei Galuh Palm Oil Mill, PT. Perkebunan Nusantara V, granular sludge obtained from UASB reactors of PT. Insansandang Internusa treating textile wastewater, Nitrogen, and NaCl solution. Other supporting materials i.e.; $K_2Cr_2O_7$ 0.05 M; Ag_2SO_4 0.05 M; FAS; H_2SO_4 0.02 N; NaOH 0.1 N; indicator Ferroin, PP and MO.

The EGSB reactor series used in this research were presented in figure 1 below.

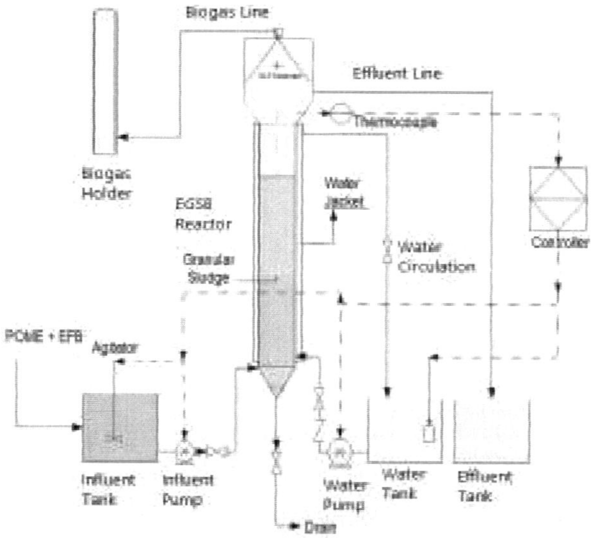

Figure 1. Schematic diagram of the EGSB reactor.

EGSB reactor used in this study was made of PVC pipe with 4 in ID and 6 in OD. The space between two pipes was used for a water jacket. The diameter to height ratio was 1:15 minimum, so the reactor's height (not including the head and bottom) is 190 cm. The reactor divided into three parts, namely the bottom, body, and head. All of these parts were joined by two pcs of 4 in PVC coupling. Inside the head section of the reactor, there was an inverted cone-shaped Gas-Liquid-Solid separator, which served as biogas capture. The lower part of the reactor served as influent distribution area. The total volume of the EGSB reactor was 30,570 ml and the effective volume was 19,820 ml.

All of the instruments used, i.e., influent pump, a water circulation pump for the water jacket, coil heater, and thermocouple were connected to a single control panel. The biogas outlet was connected to an inverted 2000 ml measuring glass, containing 10% NaCl solution to measure the amount of biogas generated. Other tools were test tubes, 250 ml Erlenmeyer, and a burette to perform the titration. Research variables were reactor's effective volume of 19,820 ml, HRT of 2 days, and temperature of 35 $\pm 2^0$ C [10, 12]. [4] [18].

2.1. *Material preparations:*
POME obtained from the bottom outlet of Fat Pit in Sei Galuh Palm Oil Mill, PTPN V. Granular Sludge took from UASB reactor treating textile wastewater in PT. Insansandang Internusa. Microbial consortia on the granular sludge mainly consisted of *Micrococcus sp Nitrosomonas sp, Nitrobacter sp, and Methanobacter sp.*

2.2. *POME characteristics*
POME's characteristics were analyzed in the Material Testing Lab of Dinas Bina Marga, Riau Province. The COD level measured using the titration method, while biogas content was analyzed using a portable gas analyzer (Biogas Geotech 5000).

3. Results and discussion
POME's characteristic was presented in table 1.

Table 1. Characteristic of POME obtained from Sei Galuh Palm Oil Mill, PTPN V.

Parameters	Value	Threshold*
pH	4.58	6.0 – 9.0
BOD5 (mg/l)	-	100
COD (mg/l)	55.000	350
TSS (mg/l)	-	250
Total Alkalinity (mg/l)	1.220	N/A
Volatile Fatty Acid (mg/l)	180	N/A
Total Nitrogen (mg/l)	-	50
Max. Debit (m^3/ton CPO)	-	2.5

**Indonesia's Ministry of Environment Regulation No. 5,2014*

3.1. *Influent's COD level*
Influent's COD level analysis was used as the basic calculation of EGSB reactor's COD removal. This value was the average values of influent's COD samples result per run with three repetitions. Influent's COD level analysis was presented in table 2 below:

Table 2. Influent's COD level in mg/L.

Sample 1	Sample 2	Sample 3	Average
55.000	50.000	55.000	53.333

3.2. Start-up preparation

3.2.1. Influent pump calibration. Before the startup was started, influent's pump was calibrated to ensure that HRT achieved in 2 days. Influent's pump capacity was set in the control panel. Once the pump ready, the step continued by calibration of the temperature.

3.2.2. EGSB reactor temperature calibration. Reactor's temperature was set to 35 ± 2^0 C by starting the heater coil to warm up the circulating water inside the water jacket. Once the temperature inside of the reactor reached 37^0C, the control panel will cut off the power to coil heater and water pump. The two instruments will back to work when the temperature of the reactor was 33^0C. This process repeats during continuous operation of the reactor.

3.2.3. Startup EGSB reactor. The startup begins with the mixing of 9 liters granular sludge and POME which has been prepared in advance. POME and granular sludge mixture were stirred and then fed into the reactor. Reactor's temperature was controlled at 35 ± 2^0C (Mesophilic). The use of freshly obtained granular sludge from another fully operated reactor aimed to shorten startup time [13, 14]. [16 - 17]. Make sure that all parts of the reactor, i.e., the hose inlet-outlet, pumps and instruments are correctly installedproperly, and then fed the Nitrogen gas into the reactor to ensure the reactor was in anaerobic condition. After this stage was completed, the reactor operated in a batch-mode for two days [4]. [5]. After the second day, Biogas bubbles was generated and attached to the inner side of the measuring glass containing 10% of NaCl solution. Formation of Biogas bubbles indicated the microbial consortia was digesting POME [4]. [5].
The next step was reactor operated in continuous mode. Influent's pump switched on to fed POME into the reactor. The continuous-mode was staged for 12 days. During the 4th day in continuous-mode, the effluent's hose and the GLS separator were clogged by scum. As a result, granular sludges were carried over the effluent's tank and Biogas outlet [4]. [5].
Some improvements were made in the head section and the Biogas outlet of the reactor to prevent the recurrence of similar incidents. Startup phase was restarted to ensure the reactor's operational excellence in the next step. The second startup attempt was also failed due to the limited volume of biogas holder. The biogas volume generated during the 4th day was much higher than the biogas holder volume. The third attempt of the start-up stage was quite a success. The batch-mode operation for two days, followed by continuous-mode for the next ten days went smoothly. During ten days of operation, the samples were taken from POME's preparation tank and effluent's outlet hose to determine the COD level.

3.3. COD removal

The effluent's samples were taken in 2 days of intervals. The COD removal during the startup stage was presented in figure 2:

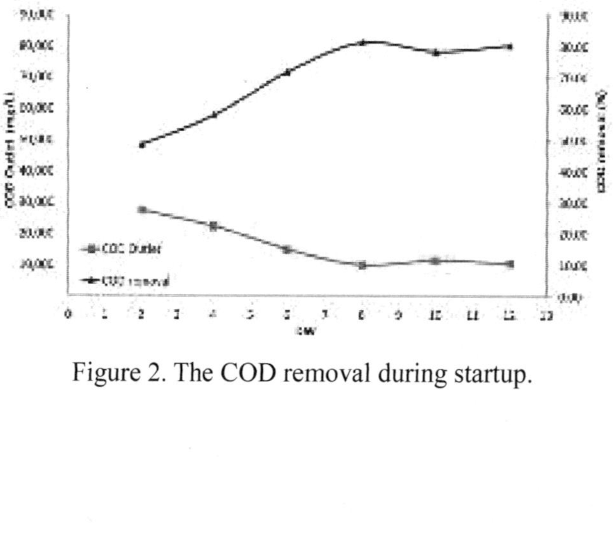

Figure 2. The COD removal during startup.

Figure 2. showed that the reactor was steady on day 8 to 12, where the COD level was relatively stable at 10,000 to 11,500 mg/l. The effluent's COD level on the second day was removed for 48.44% compared to the influent's level. Effluent's physical appearance still looks dark brown and there was a bit granular sludge carried away to the effluent outlet. The effluent's color tends to fade over the increase of COD removal during the start-up stage. The effluent's COD level was decreasing from 27,500 to 10,000 mg/l on day 2 to 8. The reactor was on steady condition starting from day 8 to 12. The effluent's dark brown color was faded, effluent's COD levels were stable at 10,000 to 11,500 mg/l, and COD removal vary from 78.44 to 81.25%.

3.4. The biogas volume

During the start-up stage, biogas volume was measured with inverted glass tube containing 10% NaCl solution. The volume of biogas during startup was presented in figure 3:

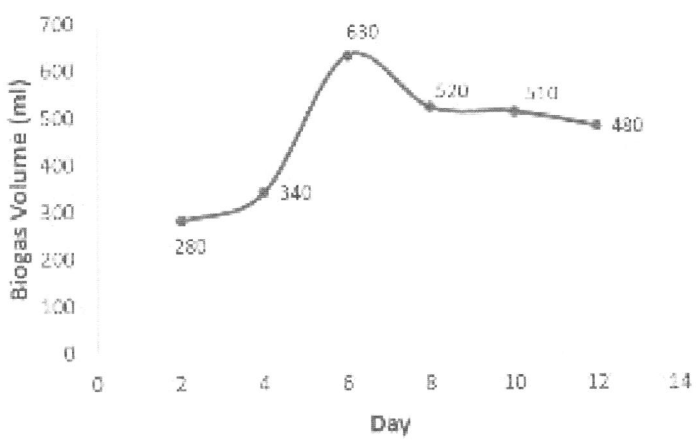

Figure 3. Biogas volume during the startup stage.

Figure 3. showed that biogas volume increased from 4th to 6th day, they tend to be stable in the range of 480 – 520 ml on day 8 to 12. Biogas was formed from the beginning of 2 days of batch-mode. Biogas volumes increased over days during the start-up stage. Observation during the startup stage showed that the biogas volume increased when the sludge bed was expanded by influent's pump operational pressure [14]. [17]. The biogas volume varied from 480 to 630 ml on day 6 to 12.
The total biogas volume generated during the start-up stage was 2,760 ml with 43.89% of Biomethane. The Biogas content was analyzed using Geotech Biogas 5000. The Biomethane yield during the startup stage was 0.1486 – 0.1503 m^3 CH_4/kg COD. The detailed experimental data showed in table 3.

Table 3. Detailed experimental data.

Parameters	Day					
	2nd	4th	6th	8th	10th	12th
COD influent (mg/l)				53.333		
COD effluent (mg/l)	27.500	22.500	15.000	10.000	11.500	10.500
COD removal (%)	48.44	57.81	71.88	81.25	78.44	80.31
Biogas volume (ml)	280	340	630	520	510	480
				2.760		

3.5. Scum formation

Undiluted POME which fed into EGSB reactor were taken directly from the bottom outlet of Fat Pit in Sei Galuh Palm Oil Mill. Scum formation occurred due to the high content of oil in POME digested by microbial consortia in granular sludge. The thickness of scum will continue to grow along with the high content of oil and fats in POME.

Scum layer could block the biogas. The thicker the scum layer, more difficult for biogas bubbles to pass through. In this study, a scum formed hinder the GLS separator outlet. To overcome this constraint, GLS separator can be modified, i.e., by adding a scum breaker. Development of GLS separator with scum breaker could be an interesting research topic in the near future.

3.6. Effluent's outlet blockage

The blockage of the effluent's outlet hose occurred due to scum carry over. The increased scum deposit in the outlet line could block the outlet flow. This blockage would failof GLS separator. The remaining scum could also stick to the inner wall of the hose, thereby reducing the effluent's outlet volume. In this study, the constraint was corrected by modified the GLS separator and increased the size of the effluent's outlet hose.

3.7. Granular sludge carry over

Granular sludge carry-over occurred on 1st to the 2nd day of continuous-mode operation. This was caused by the high pressure of influent's pump when it was run for the first time. Granular sludge was not fully settled yet, even after the two days of batch-mode operation. It was slowly disappeared after the 4th day of continuous-mode operation. On day 8, the carryover was stopped, granular sludge has already settled.

4. Conclusion

EGSB reactor can be used in the anaerobic digestion of raw and undiluted POME to generate Biogas at mesophilic condition (temperature 35 ± 2^0C). EGSB reactor start-up stage achieved the steady state on day 8 to 12 of continuous-mode operation with COD removal 78.44 to 81.25%. The Biogas volume generated during 12 days of start-up stage was 2,760 ml (0.1486- 0.1503 m^3 CH_4/kg COD).

References

[1] Ahmad A, Shah, Othman and Abdullah 2014 Enhanced palm oil mill effluent treatment and biomethane production by co-digestion of oil palm empty fruit bunches with Chlorella sp. *Can J Chem Eng* **92** 1636–42

[2] Chan and Choi 1984 Ponding system for palm oil mill effluent treatment. *Proceedings of the Regional Workshop on Palm Oil Mill Technology & Effluent Treatment* p 185–92

[3] Yacob, Hassan, Shirai, Subash, and Wakisaka 2006 Baseline study of methane emission from anaerobic ponds of palm oil mill effluent treatment *Sci Total Environ* **366** 187–96

[4] Yejian Z, Li Y, Lina C, Xiuhua L, Zhijian M and Zhenjia Z 2008 Startup and operation of anaerobic EGSB reactor treating palm oil mill effluent *J. Env. Sci.* **20** 658-63

[5] Ho C C and Tan Y K 1985 Anaerobic treatment of palm oil mill effluent by tank digesters *J. Chem. Technol. Biotechnol.* **35** 155–64

[6] Borja R, and Banks C J, 1994, Treatment of palm oil mill effluent by upflow anaerobic filtration, *J. Chem. Technol. Biotechnol.* **61** 103–09

[7] Borja R, and Banks C J 1995 Response of an anaerobic fluidized bed reactor treating ice-cream wastewater to organic, hydraulic, temperature and pH shocks *J. Biotechnol.* **39** 251–59

[8] Setiadi T, Husaini and Djajadiningrat A 1996 Palm oil mill effluent treatment by anaerobic baffled reactors:Recycle effects and biokinetic parameters *Water Sci. Technol.* **34**(11) 59–66

[9] Borja R and Banks C J 1996 Anaerobic treatment of palm oil mill effluent in a two-stage up-flow anaerobic sludge blanket (UASB) reactor *J. Biotechnol.* **45** 125–35

[10] Yejian Z, Hairen Y, Xiangyong Z, Zhenjia Z, and Li Y 2011 High-rate mesophilic anaerobic digestion of palm oil mill effluent (POME) in Expanded Granular Sludge Bed (EGSB) reactor *Int. Confer. Agricul. Natural Res. Engine. Adva. Biomed. Engine.* **3-5** 214-19

[11] Ohimain E I and Izah S C 2014 Potential of biogas production from palm oil mill effluent in

Nigeria *Sky Journal of Soil Science and Environmental Management* **3** 50–58
[12] Kim H S, Choi S, Ju H and Jung J 2013 Mesophilic co-digestion of palm oil mill effluent and empty fruit bunches, *Environmental Technology* **34**(13-14) 2163–70
[13] Goodwin, Forster, and Wase, 1992, Pre-granulated seeds for UASB reactors: how necessary are they? *Bioresour. Technol.* **41** 71–79
[14] Poh and Chong 2009 Development of anaerobic digestion methods for palm oil mill effluent (POME) treatment *Bioresour. Technol.* **100** 1–9

Acknowledgment
This study was supported by PT. Perkebunan Nusantara V.

The effect of different α and β gable on stainless steel and TMA segmental T-loop on the resulting force – an in vitro study

A Laviana

Orthodontic Department, Faculty of Dentistry, Universitas Padjadjaran
Jl. Sekeloa Selatan 1, Bandung 40132, Indonesia

E-mail: avi.laviana@fkg.unpad.ac.id

Abstract. Adding more angle to the segmental T-Loop gable can affect the amount of force created to retract the canine on an orthodontic treatment with extraction. In-vitro laboratory experimental design, performed at the Chemistry Laboratory of Faculty of Science, ITB. The samples are 60 segmental T-loop springs with α and β angle of 0°–0°; 0°-45°; 45°–0°; 22.5°–22.5°; 15°–30°; 30°–15°. Force was measured using autograph with retraction distance of 1, 2, and 3 mm, then analyzed using ANOVA statistics analysis with a factorial design of 2x3x6 and 5 replications for each cell. The results of the posthoc Student Newman Keuls and Tukey test show that the added angle to the α and β gables of SS and TMA T-loop will add more resulting force along with the increase of retraction distance. The added angle to the α and β gables of SS and TMA segmental T-loop will add more resulting force along with the increase of retraction distance. The force resulted by the SS T-loop with various α and β gables are significantly different, while the force resulted by the TMA T-loop are not (p-value <0.05), as long as α + β is the same. The force resulted by the SS T-loop with various α and β gables are significantly different. However, on the TMA segmental T-loop, as long as α + β is the same, the force resulted will not be different.

1. Introduction

The success rate of coverage in an orthodontic treatment with extraction highly determines the final position of the canine and incisor teeth, which are crucial to the stability of treatment result, the bite functions, and the smile aesthetics. The coverage of the extracted premolar by retracting the canine, followed by the retraction of the four incisors can be done in two ways, which are by using a continuous archwire that can create friction between the braces slot and the archwire during the teeth movement, and by segmental technique, which does not create friction [1,2].

Canine retraction by using the segmental technique can be done using loops with various configurations, materials, placing distance, archwire size, and additional gable to create various force and momentum [3]. The success rate of a force-controlling tool is marked by the type and direction of the teeth movement that are in accordance with expectations, fast teeth movement, minimal teeth or periodontal damage and pain – which correlates to the comfort level of the patient during the orthodontic treatment [1-6].

T-loop is one of the spring configurations that can control the applied force towards teeth at the time of activation. The control mechanism created by adding angle to the gables is aimed to overcome unwanted effects during canine retraction [1-6]. The α and β gable angles are commonly the same; however, few clinical operators often reshape them without measuring the symmetry of the two angles.

Since its introduction in the 1930's, stainless steel (SS) is still commonly used in dentistry, among others is as a material for orthodontic wirings. In the 1980's, titanium – commonly known as TMA

(Titanium Molybdenum Alloy) is developed with high elasticity, low level of stiffness, high formability, able to maintain its resilience during welding, corrosive-resistant, and can be flexed twice as large as SS wiring without permanent damage or change [3, 7-9]. Segmental T-loop retraction springs are now commonly made of TMA [3-5, 8-10].

Based on various α and β gable and different mechanical properties of SS and TMA, a research is conducted to discover the different force created by SS and TMA segmental T-loop retraction springs with various and different α and β gable angles on various retraction distances.

2. Methods

The materials of this research are 60 T-loops – 30 are made of SS and the other 30 are made of TMA archwire size 0.016 x 0.022 inches with 7 mm height, 10 mm width, and 18 mm length, with the combination of anterior (α) and posterior (β) angle of 0°–0°; 0°-45°; 45°–0°; 22.5°–22.5°; 15°–30°; 30°–15°, 5 of each angle. All T-loops are calibrated using a template and measured using glass lab for alignment (figure 1).

This research is a pure experimental research, done in-vitro at at the Chemistry Laboratory of Faculty of Science, ITB (Institut Teknologi Bandung). The pre-designed testing tool is made by the Physics Engineering Laboratory, LIPI (Lembaga Ilmu Pengetahuan Indonesia), from aluminum, consists of static part, movable retractor, and gyrating braces holder with edgewise standard size 0.018 inches. The spring is put on the testing tool and tied to the braces using ligature wires and set on distance 0 (zero). The quantity of force is noted using the unit gF on retraction distance 1 mm, 2 mm, and 3 mm.

The result of the experiment is tested using the ANOVA statistic test with experiment design 2x3x6 with 5 replications. The interaction between the three factors (material, gable, and retraction distance) is tested using the posthoc Student Newman Keuls and Tukey test on the level of 5%.

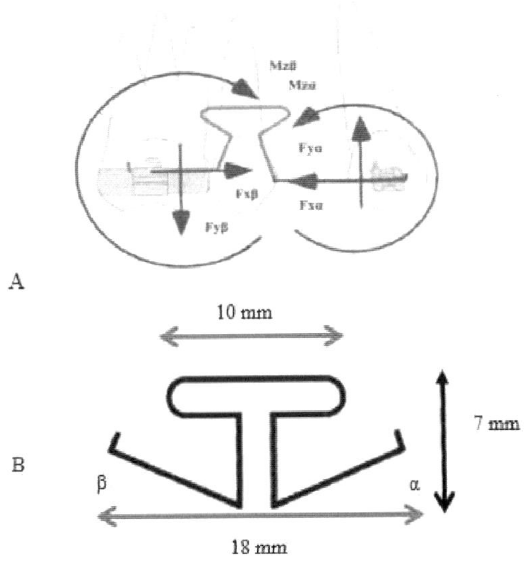

Figure 1. Force System of Segmental T-loop. (A) Fxα is the horizontal force which is applied to the anterior segment. Fyα is the vertical force which is applied to the anterior segment. Mzα is the momentum applied to the anterior segment. Fxβ is the horizontal force applied to the posterior segment. Fyβ is the vertical force applied to the posterior segment. Mzβ is the momentum applied to the posterior segment.10 (B) The configurations of the T-loops are made with α and β gable variations: 0°–0°; 0°-45°; 45°–0°; 22.5°–22.5°; 15°–30°; 30°–15°.

The sample size calculation for three-factor experiment design (SS and TMA, retraction distance, and gable angle) forms 2x3x6 treatments – to calculate the minimum required sample, the Federer sample formula is used:

$$(n-1)(k-1) \geq 15 \text{ or } n \geq 1 + \frac{15}{(k-1)}$$

With: n = minimum required sample, k = the number of treatments in the experiment design, the unit number of experiments is n x k. The research design with two materials, SS and TMA, 3 retraction distances, and 6 gables will receive 36 different treatments – this means that the minimum sample required is:

$$n \geq 1 + \frac{15}{(36-1)} \qquad n \geq 1.4285 \approx 2$$

In conclusion, the minimum amount of experiment or replication per treatment is 2, with the total of 2 x 36 = 72 required units. This research is done with 5 replications per treatment, with the total of 5 x 36 = 180 replication units, which meet the minimum required sample.

3. Results

Measurements have been conducted of the force created by 60 Stainless Steel and TMA segmental T-loop retraction springs with 0.016 x 0.022 inch, with anterior (α) and posterior (β) gables of 0°–0°; 0°-45°; 45°–0°; 22.5°–22.5°; 15°–30°; 30°–15°; 5 of each angles and materials. The springs are attached one by one to the force-test device and then attached to the Universal Strength Tester Shimadzu Corporation Autograph. Then, activation distance is applied with the measurement of 1, 2, and 3 mm. The measured force is shown on the monitor and then noted. The mean of the results can be seen in table 1.

Table 1. Mean value of the force created by 0.016 x 0.022 inc stainless steel and TMA segmental T-loop based on the α and β gable and retraction distance variations (gF)[a].

SS	Force (gF)			TMA	Force (gF)		
	Mean	Median	SD		Mean	Median	SD
A1B1C1	93.20	93.00	18.75	A2B1C1	52.40	60.00	17.61
A1B1C2	208.60	203.00	27.37	A2B1C2	94.00	90.00	11.04
A1B1C3	342.20	338.00	37.14	A2B1C3	149.20	149.00	14.20
A1B2C1	153.40	150.00	26.08	A2B2C1	84.80	83.00	19.82
A1B2C2	354.40	347.00	37.60	A2B2C2	164.80	170.00	16.66
A1B2C3	539.40	520.00	31.73	A2B2C3	245.00	240.00	22.07
A1B3C1	197.00	192.00	23.80	A2B3C1	72.80	65.00	31.25
A1B3C2	398.60	385.00	36.77	A2B3C2	144.80	150.00	30.34
A1B3C3	538.40	558.00	55.98	A2B3C3	220.40	220.00	31.48
A1B4C1	197.00	205.00	14.40	A2B4C1	89.60	75.00	31.08
A1B4C2	384.40	362.00	37.51	A2B4C2	171.00	162.00	23.65
A1B4C3	554.20	542.00	50.27	A2B4C3	246.60	242.00	13.22
A1B5C1	165.20	168.00	36.16	A2B5C1	82.00	75.00	22.80
A1B5C2	342.60	355.00	74.28	A2B5C2	159.40	160.00	20.58
A1B5C3	530.40	542.00	49.42	A2B5C3	232.00	237.00	23.16
A1B6C1	188.20	198.00	23.17	A2B6C1	93.40	87.00	20.12
A1B6C2	403.40	412.00	41.44	A2B6C2	174.80	167.00	21.58
A1B6C3	583.40	590.00	50.89	A2B6C3	250.00	240.00	25.73

[a]A1=SS, A2=TMA, B1-B6 = gable 0°–0°; 0°-45°; 45°–0°; 22.5°–22.5°; 15°–30°; 30°–15°, C1-C3= retraction distance 1, 2, and 3 mm.

Table 1 shows the mean value of the force created by the stainless steel and TMA segmental T-loop retraction spring with different gables and activation distances. Based on the table, it is discovered that on stainless steel segmental T-loop, the least amount of force created is 93 gF, created by gable 0°–0° and activated by the distance of 1 mm, and the most amount of force is 583 gF, created by gable 30°-15° and activated by the distance of 3mm. On the TMA segmental T-loop, the least amount of force created is 52.40 gF, created by gable 0°–0° and activated by the distance of 1 mm, and the most amount of force is 250.00 gF, created by gable 30°-15° and activated by the distance of 3mm.

Based on the analysis from ANOVA from SPPS ver.19 with experiment design 2x3x6 with 5 replications, it is discovered that all treatment – whether it is wire material, gable, retraction distance, material's interaction with gable and retraction distance – gave the most significant effect towards the created force (p-value <0.05), shown with different letterings.

Table 2. Amount of force based on the material: Stainless Steel (SS) and TMA (Titanium Molybdenum Alloy)[a].

	Force (gF)					
Material	Mean	Median	SD	Minimum	Maximum	Posthoc Test*
SS	343.00	352.50	160.86	70.00	622.00	a
TMA	151.50	151.00	68.39	33.00	287.00	b

[a] SS = Stainless steel, TMA = Titanium Molybdenum Aloy, SD = Standard Deviation, *S-N-K and Tukey test.

Table 2 shows that based on the factor of material, which are SS and TMA, it is discovered that the force created by SS is greater than TMA. The force created significantly differs on the 5% level by using the posthoc Student Newman Keuls and Tukey test.

Table 3. Amount of force based on the interaction between the material and angle (AB)[a].

AB	Force (gF)					
	Mean	Median	SD	Minimum	Maximum	posthoc test*
A1B1	214.67	203.00	108.63	70.00	393.00	c
A1B2	349.07	347.00	165.85	127.00	580.00	d
A1B3	378.00	385.00	149.95	177.00	603.00	e
A1B4	378.53	362.00	154.87	175.00	620.00	e
A1B5	346.07	355.00	162.70	105.00	580.00	d
A1B6	391.67	412.00	171.31	150.00	622.00	e
A2B1	98.53	90.00	43.19	33.00	172.00	a
A2B2	164.87	170.00	70.10	60.00	270.00	b
A2B3	146.00	150.00	68.67	37.00	265.00	b
A2B4	169.07	162.00	69.93	73.00	270.00	b
A2B5	157.80	160.00	66.65	65.00	263.00	b
A2B6	172.73	167.00	69.42	75.00	287.00	b

[a] A1=SS, A2=TMA, B1-B6= gable 0°–0°; 0°-45°; 45°–0°; 22.5°–22.5°; 15°–30°; 30°–15°, SS = Stainless steel, TMA = Titanium Molybdenum Aloy, SD = Standard Deviation.

Table 3 shows that based on the interaction between wire material (A) and gable angle (B), the force created is greater along with the addition of gable angle on the same material. On SS segmental T-loop, the measurements are significantly greater on the 5% level by using the posthoc Student Newman Keuls and Tukey test – the same goes to the TMA segmental T-loop 0°–0° compared to other gables. However, in the group of TMA T-loops with asymmetric α and β gables with $\alpha + \beta = 45°$ (0°-45°; 45°–0°; 22.5°–22.5°; 15°–30°; 30°–15°), the results show no significant difference.

Table 4. Amount of force based on the interaction between the material and retraction distance (AC)[a].

AC	Force (gF)					
	Mean	Median	SD	Minimum	Maximum	posthoc test*
A1C1	165.67	177.50	43.14	70.00	235.00	b
A1C2	348.67	361.00	78.98	175.00	455.00	d
A1C3	514.67	531.00	90.89	292.00	622.00	e
A2C1	79.17	75.00	26.14	33.00	145.00	a
A2C2	151.47	160.00	34.06	82.00	213.00	b
A2C3	223.87	236.00	40.98	135.00	287.00	c

[a] A1=SS, A2=TMA; C1-C3= retraction distance 1, 2, dan 3 mm; SS = Stainless steel; TMA = Titanium Molybdenum Aloy; SD = Standard Deviation; *S-N-K and Tukey test.

Table 4 shows that based on the interaction of the material (A) and the retraction distance (C), it is discovered that the force created becomes greater the further the retraction distance is for the same wire material. The force created significantly differ on the 5% level using the posthoc Student Newman Keuls and Tukey test.

Table 5. Amount of force based on the interaction between gable (B) and retraction distance (C)[a].

BC Interaction	Force (gF)					
	Mean	Med	SD	Min	Max	Posthoc test*
B1C1	72.80	70.00	27.50	33.00	115.00	a
B1C2	151.30	142.50	63.52	82.00	238.00	b
B1C3	245.70	232.00	105.12	13500	393.00	c
B2C1	119.10	117.00	42.24	60.00	185.00	b
B2C2	259.60	248.50	103.62	143.00	395.00	c,d
B2C3	392.20	390.00	157.29	218.00	580.00	e
B3C1	134.90	149.50	70.50	37.00	235.00	b
B3C2	271.70	271.00	137.49	105.00	455.00	c,d
B3C3	379.40	363.50	172.98	187.00	603.00	e
B4C1	143.30	160.00	61.04	73.00	210.00	b
B4C2	277.70	281.50	116.29	155.00	430.00	c,d
B4C3	400.40	387.50	165.78	238.00	620.00	e
B5C1	123.60	112.50	52.30	65.00	198.00	b
B5C2	251.00	201.00	109.38	132.00	398.00	c,d
B5C3	381.20	361.50	161.43	200.00	580.00	e
B6C1	140.80	138.50	53.99	7500	205.00	b
B6C2	289.10	272.50	124.44	157.00	447.00	d
B6C3	416.70	392.00	179.78	223.00	622.00	e

[a] B1-B6= gable 0°–0°; 0°-45°; 45°–0°; 22.5°–22.5°; 15°–30°; 30°–15°, C1-C3= retraction distance 1, 2, dan 3 mm; SD = Standard Deviation; *S-N-K and Tukey test.

Table 5 shows that based on the interaction between the gable angle (B) and the retraction distance (C), it is discovered that the force created becomes greater the further the retraction distance is for the same gable angle, as well as the angle of the gable. The force created significantly differ on the 5% level by using the posthoc Student Newman Keuls and Tukey test, with the exception of the group with asymmetric α and β gables (but α + β =45o) which show no significant difference.

4. Discussion

As much as 60 SS and TMA Segmental T-loop retraction springs with diameter of 0.016 x 0.022 inches with different anterior (α) and posterior (β) gable angles of 0°–0°; 0°-45°; 45°–0°; 22.5°–22.5°; 15°–30°; 30°–15° are retracted within a distance that represents the common clinical activation distance, which are 1, 2, and 3 mm. The research was conducted in-vitro in a laboratory to measure the amount of force created in unit of gF.

Table 1 shows the mean value of the force created by the stainless steel and TMA segmental T-loop retraction spring with different gables and activation distances. Based on the table, it is discovered that on stainless steel segmental T-loop, the least amount of force created is 93 gF, created by gable 0°–0° and activated by the distance of 1 mm, and the most amount of force is 583 gF, created by gable 30°-15° and activated by the distance of 3 mm. On the TMA segmental T-loop, the least amount of force created is 52.40 gF, created by gable 0°–0° and activated by the distance of 1 mm, and the most amount of force is 250.00 gF, created by gable 30°-15° and activated by the distance of 3 mm.

The force differences created by the TMA T-loop spring with gables of 0°-45°; 45°–0°; 22.5°–22.5°; 15°–30°; 30°–15°on each retraction distance are not significant; this shows that the angle difference of gable α and β does not affect the force created, as long as α + β are the same (45°). The results are different when using stainless steel T-loops – despite gable angle α + β = 45°, the force created have significant differences. This is caused by the stainless steel wire's lower formability property compared to TMA wire 8, so that the wire is damaged when bent to a significantly large angle and affects the force created [10].

This research takes thought, planning, and a complicated preparation in order to imitate the real situation. However, because the observation is only done two-dimensionally, what actually happen inside the mouth cannot be simulated completely on the contraption designed for the in-vitro laboratory research. As stated by Chen [12], Gajda [13], and Katona [14], due to the multiple three-dimensional systems applied to the tool, there should be an adjustment on the material choice and T-loop configuration on the use of T-loop spring for anterior teeth retraction in clinics.

5. Conclusion

The conclusion of this research is that there is significant difference on the force created by the SS segmental T-loop sized 0.016 x 0.022 inches compared to the ones made of TMA with various gable angles and activation distance, where the force created by the SS segmental T-loop is significantly larger compared to TMA. Stainless steel segmental T-loop springs with gable angle α and β produce different force, while TMA T-loops create the same amount of force, as long as α + β is the same.
The force created by SS segmental T-Loop with the same gable angles and retraction distance is twice the force compared to the TMA – this should be a concern for orthodontic practitioners when determining the retraction distance during tooth retraction.

References

[1] Bisol G K and Rocha R 2010 Laboratorial study of the cuspid's retraction timing and tipping effects during space closure, using the segmented arch technique *Dental Press J Orthod* **15**(1) 53–64
[2] Felemban N H, Al Sulaiman F F, Murshid Z A, and Hassan A H 2013 En masse retraction versus two step retraction of anterior teeth in extraction treatment of bimaxillary protrusion *J Orthod Science* **2**(1) 28–37
[3] Blaya M B G, Westphalen G H, Guimaraes M B and Hirakata L M 2009 Evaluation of tensile strength of different configuration of orthodontic retraction loops for obtaining optimized forces, *Stomatologija Baltic Dental and Maxillofacial Journal* **11**(2) 66–9
[4] Lim Y, Quick A, Swain M and Herbison P 2008 Temperature effects on the force, moments and moment to force ratio of nickel-titanium and TMA symmetrical T-loops *Angle Orthod* **78**(6) 1035–42
[5] Martins R P, Buschang P H, Viecilli R and Santos-Pinto A 2008 Curvature versus V-bends in a group B titanium T-loop spring *Angle Orthod* **78**(3) 517–23

[6] Choi Y J, Chung C D, Choy K and Kim K H 2010 Absolute anchorage with universal T-loop mechanics for severe deepbite and maxillary anterior protrusion and its 10-year stability *Angle Orthod* **80**(4) 771–82
[7] Krishnan V and Kumar K J 2004 Mechanical properties and surface characteristics of three archwire alloys *Angle Orthod* **74**(6) 825–31
[8] Vijayalakshmi R D, Nagachandran K S, Kummi P and Jayakumar P A 2009 Comparative evaluation of metallurgical properties of stainless steel and TMA archwire with timolium and titanoum niobium archwire - an in vitro study *Indian J Dent Res* **20**(4) 448–52
[9] Gurgel J A, Pinzan-Vercelino C R M and Powers J M 2011 Mechanical properties of beta-titanium wires *Angle Orthod* **81**(3) 478–83
[10] Laviana A, Hambali T S, Thahar B and Mardiati E 2015 Pengaruh heat treatment untuk mengembalikan sifat mekanik kawat t-loop segmental stainless steel terhadap besaran gaya yang dihasilkan, *Bandung Medical Journal* **47**(3) 167–73
[11] Proffit W R, Fields H W Jr, Sarver D M and Ackerman J L 2013 *Contemporary Orthodontics* ed. 5 (St Louis, Missouri: Mosby)
[12] Chen J, Isikbay S C and Brizendine E J 2010 Quantification of three-dimensional orthodontic force systems of T-loop archwires *Angle Orthod* **80**(4) 754–8
[13] Gajda S and Chen J 2012 Comparison of three-dimensional orthodontic load systems of different commercial archwire for space closure *Angle Orthod* **82**(2) 333–9
[14] Katona T R, Isikbay S C and Chen J 2014 Effects of first- and second-order gable bends on the orthodontic load systems produced by T-loop archwires *Angle Orthod J* **84**(2) 350–7

IOP Conf. Series: Materials Science and Engineering **550** (2019) 012040 doi:10.1088/1757-899X/550/1/012040

Comparison of shear bond strength of light-cured resin-modified glass ionomer and moist insensitive primer on contaminated enamel

I A Evangelina, T Hambali, B Thahar and J Salim

Department of Orthodontics, Faculty of Dentistry, Universitas Padjadjaran, Jl. Sekeloa Selatan 1, Bandung, Indonesia

E-mail: ida.evangelina@fkg.unpad.ac.id

Abstract. Despite rapid developments in adhesive technology, contamination of bonding surfaces remains a major problem. The aim of this study was to evaluate the influence of contamination on shear bond strength of two adhesive systems: resin-modified glass ionomer cement (Fuji Ortho LC; GC America Corp) and Moist Insensitive Primer (Transbond MIP, 3M Unitek) combination with Transbond XT (light-activated), under different enamel conditions. Metal brackets were bonded to the buccal surfaces of 42 human premolars. The total sample was divided into 6 groups (n = 7 teeth) and brackets were bonded with one of the 2 adhesives under 3 enamel surface conditions: (1) dry (2) saliva-contaminated (3) saliva and blood-contaminated. The shear bond strength was tested using Universal Strength Tester Autograph Shimadzu Corp. There was no significant difference in the shear bond strength between dry, saliva, and saliva-blood contamination bonded with MIP combined Transbond XT. There were significant differences in the shear bond strengths between dry and saliva -blood contaminated enamel bonded with Fuji Ortho LC. In conclusion, Fuji Ortho LC is not recommended for clinical bonding application under saliva-blood contaminated enamel conditions.

1. Introduction

Orthodontists and oral surgeons often collaborate in the exposure and orthodontic treatment of unerupted ectopic teeth, since it is difficult to work under ideal conditions [1,2]. In this situation, the treatment options are either exposure of the tooth only or exposure of the tooth and direct bonding of attachment for orthodontic traction. Delaying the bonding procedure until healing occurs will reduce the risk of blood contamination and moisture. However, the soft tissues that cover the tooth must be excised or repositioned to expose the crown, which will cause an undesirable situation, a poor gingival margin. Another option is removing the mucosa to expose the impacted teeth. In this situation, the only strategy is to bond the attachment at the time of operation. However, saliva and blood contamination can result in failure of the attachment bonding procedures [3]. The new orthodontic cements, adhesive resins, and resin-cement hybrid offer improved physical properties and clinical advantages but there are indications and contra-indications for each material.

Glass Ionomer Cement (GIC) began to be used widely in the field of restorative dentistry due to its ability to release fluoride, adherence to the tooth structure, and biocompatibility. However, this material also has some shortcomings, such as long hardening time, sensitivity to moisture during the initial hardening, and rough surfaces which will reduce the mechanical resistance. Resin-modified glass ionomer cement (RMGIC) was developed to overcome the problem of sensitivity to moisture and increase the strength during hardening while maintaining the clinical advantages of conventional glass ionomer that include a chemical bond to enamel, adhesion in a wet field, and fluoride release [4-6]. The addition of 10 % - 20 % resin monomers to the GIC result in accelerated hardening, through polymerization of the monomer. However, either light or chemical activators are required. RMGIC is adhesive cement with improved physical properties and a hydrogel phase that is more stable than GIC. Although the amount of resin monomer added to a solution of polyalkenoic acid is very limited, the polymerization of the resin monomers accelerates the initial hardening of RMGIC without significantly interfering the acid-base setting reaction, and the release of fluoride or the attachment of carboxyl groups to metal and tooth surfaces. In addition to the chemical bonding of RMGIC, resin monomers penetrate surface irregularities to produce a micromechanical interlock (bond) after polymerization. Not only is there the advantage of operator-controlling setting, light-activated polymerization also proceeds significantly faster than acid-base (cement forming) reactions, resulting in improved early physical properties, especially fracture resistance. However, despite the favorable characteristics of such materials, the bond strength in contaminated enamel still has to be tested and proven [5,6].

The conventional method for bonding orthodontic attachments to the enamel surface requires three different agents: acid etchant, liquid primer, and adhesive resins [7,8]. Bonding with the use of conventional materials require a dry tooth surface to get adequate bonding strength to withstand the orthodontic force and the load of chewing. However, various clinical conditions do not permit ideal isolation for common orthodontic bonding adhesive use and protocols. To address this, some manufacturer have introduced hydrophilic bonding material and suggested that it may be possible to obtain successful orthodontic bonding to a contaminated enamel surface with this material. New hydrophilic enamel primers for orthodontic treatment use have been formulated with alcohol and/or acetone as ingredients to displace moisture from contaminated enamel [3,9,10]. This formulated differently as reflected in the manufacturers' protocols for use and may play a role in the successful bonding to a contaminated enamel surface. This primer is called Moist Insensitive Primer, commercially available as Transbond MIP (3M Unitek Dental Products, Monrovia, Calif.). The manufacturer recommends that this primer can be used on both wet and dry enamel surfaces and combined with a resin adhesive that is activated either chemically or light. MIP is chemically identical to containing ethanol dentin bonding materials (Single Bond; 3M, St. Paul, Minn.). In this study, bracket bonding was conducted by using Transbond XT adhesive resin. The composition of Transbond XT, Transbond Moist Insensitive Primer, and Fuji Ortho LC is shown in Table 1 [4,11,12].

Table 1. Composition Transbond XT (paste) and TMIP.

Adhesives	Composition	% by Wt
Transbond	Paste :	
XT	Silane-treated quartz	
	Bisphenol A diglycidyl ether dimethacylate	70-80
	Bisphenol A bis(2-hydrocyethyl ether) dimethacylate	10-20
	Dichlorodimethylsilane reaction product with silica	5-10
		<2
	Ethyl alcohol	30-40
TMIP	Bisphenol A diglycidyl ether dimethacylate	10-30
	2-Hydroxyethyl metacrylate (HEMA)	10-30
	2-Hydroxy-1,3-dimethacryloxypropane	7-13
	Copolimer itaconic and acrylic acid	7-13
	Diurethane dimethacrylate	3-7
	Water	3-7

Although literature has been published regarding the bond strengths of RMGIC and MIP, to our knowledge there is no reported study yet available about the effect of the mixture of saliva and blood contamination on these adhesive systems.

2. Materials and method

This research was a laboratory experiment using 42 extracted maxillary premolar teeth. The criteria for tooth selection included intact buccal enamel and no caries. The teeth were cleaned and then polished with pumice, then embedded in a resin block. The teeth were divided into 3 groups (A, B, C) each group consisting of two sub-groups, each sub-group consisting of 7 teeth. The bonding procedure was done as follows:

2.1. Group A: Non-contaminated enamel

A1: The tooth was cleaned, then dentin conditioner was applied for 30 seconds, then blotted away with a cotton pellet. Next the capsule containing Fuji Ortho LC was triturated for 10 seconds, bonding material was applied to the bracket base, the bracket was bonded to buccal surface 4 mm from the occlusal. Excess adhesive was carefully removed from the tooth using a scaler. The tooth was immersed in artificial saliva for 24 hours, then the shear bond strength was measured.
A2: The tooth was cleaned, then etched with 37 % phosphoric acid for 30 seconds, rinsed with water for 15 seconds, blown dry with air, MIP was then applied to the teeth and bracket base, Transbond XT paste was applied to the bracket base. The bracket was bonded to buccal surface 4 mm from the occlusal. Excess adhesive was carefully removed from the tooth using a scaler. The tooth was immersed in artificial saliva for 24 hours, then the shear bond strength was measured.

2.2. Group B: Saliva-contaminated enamel

B1: The tooth was cleaned, then dentin conditioner was applied for 30 seconds, blotted away with a cotton pellet. Next, artificial saliva was applied to the tooth's surface, the capsule containing Fuji Ortho LC was triturated for 10 seconds, bonding material was applied to the bracket base, and the bracket was bonded to buccal surface 4 mm from the occlusal. Excess adhesive was carefully removed from the tooth using a scaler. The tooth was immersed in artificial saliva for 24 hours, then the shear bond strength was measured.
B2: The tooth was cleaned, then etched with 37 % phosphoric acid for 30 seconds, rinsed with water for 15 seconds, and blown dry with air then artificial saliva was applied to the etched enamel, MIP was applied to the teeth and bracket base, and Transbond XT paste was applied to the bracket base. The bracket was bonded to the buccal surface 4 mm from the occlusal. Excess adhesive was carefully removed from the tooth using a scaler. The tooth was immersed in artificial saliva for 24 hours, then the shear bond strength was measured.

2.3. Group C: Saliva + blood-contaminated enamel

C1: The tooth was cleaned, then dentin conditioner was applied for 30 seconds, blotted away with cotton pellet. Artificial saliva + fresh human blood was applied to the tooth's surface, the capsule containing Fuji Ortho LC was triturated for 10 seconds, and bonding material was applied to the bracket base. The bracket was bonded to the buccal surface 4 mm from the occlusal Excess adhesive was carefully removed from the tooth using a scaler. The tooth was immersed in artificial saliva for 24 hours, then the shear bond strength brackets was measured.
C2: The tooth was cleaned, then etched with 37 % phosphoric acid for 30 seconds, rinsed with water for 15 seconds, and blown dry with air. Artificial saliva + fresh human blood was applied to the etched enamel, MIP was applied to the teeth and the bracket base, and Transbond XT paste was applied to the bracket base. The bracket was bonded to the buccal surface 4 mm from the occlusal. Excess adhesive was carefully removed from the tooth using a scaler. The tooth was immersed in artificial saliva for 24 hours, then the shear bond strength was measured.

The shear bond strength measurements were carried out in the Laboratory of Department of Chemical Engineering, ITB (Bandung Institute of Technology). The instrument used was the Universal Strength Tester Autograph Shimadzu Corp. set at a speed of1mm/minute. Measurements were taken after the sample had been immersed in artificial saliva for 24 hours. The embedded tooth and the adhesively fixed bracket were positioned in the testing apparatus. Shear bond strength measurement begins by providing a tensile load 0 kgf and continued until shear tested to failure.
The results obtained from Autograph were converted to MPa by using the following formula in equation (1).

$$Bond\ strength\ (mPa) = \frac{Forces\ (Newtons)}{Surface\ area\ of\ brackets\ (mm^2)}\ (1)$$

To determine statistical differences, a Student-Newman-Keuls test was performed. The means and standard deviations were calculated. The level of significance was set at $\alpha = 0.05$.

3. Results and discussion

The mean of shear bond strength of Fuji Ortho LC adhesive material and Transbond MIP in dry enamel, saliva-contaminated and saliva + blood contaminated enamel shown in table 2. The results of the ANOVA showed that under dry conditions, the bond strengths of all adhesives systems were comparable. Under saliva contamination, both Fuji Ortho LC and MIP showed acceptable and comparable bond strengths; however, the bond strength displayed by Fuji Ortho LC on saliva+ blood contamination was not adequate. (Figure 1)
There are significant differences of shear bond strength on the different enamel conditions ($p<0.05$). When the two adhesives systems were compared, nor significant differences on shear bond strength and enamel conditions ($p>0.05$).Descriptive statistics for shear bond strength are given in Table 3.

Table 2. Mean shear bond strength of Fuji Ortho LC adhesive material and Transbond MIP in dry enamel, saliva-contaminated and saliva + blood contaminated enamel.

Adhesives	Dry	Saliva	Saliva + blood
Fuji Orto LC	12.79	13.05	6.76
	10.08	9.56	4.51
	8.08	8.18	8.95
	10	8.96	5.36
	8.96	9.52	6.125
	11.9	9.87	5.11
	13.78	7.66	4.32
Transbond MIP	4.95	5.42	4.32
	4.79	5.7	8.23
	11.44	6.5	14.91
	12.375	11.15	4.3
	9.875	11.46	9.97
	4.32	10.44	7.13
	12.5	6.5	5.34

The highest shear bond strengths produce on dry enamel either with Fuji Ortho LC and Transbond MIP. There are reductions in shear bond strength for all contaminated enamel groups. Fuji Ortho LC revealed a distinctive and significant decrease in shear bond strength after saliva + blood contamination. For Transbond MIP no significant differences in shear bond strength values were detected between dry and contaminated enamel. For Fuji Ortho LC there is significant differences between dry enamel and saliva+blood contaminated enamel.

In saliva-contaminated enamel there are reductions in shear bond strength for Fuji Ortho LC and Transbond MIP. The lowest shear bond strength value found in saliva + blood contamination for Fuji Ortho LC, even below than clinically required for adhesive bonding (6Mpa).

Table 3. The shear bond strength of FOLC and Transbond MIP under different enamel conditions.

Interaction		Statistic				
		Mean	Median	Std. Deviation	Min	Max
Shear bond strength	Fuji Ortho LC-Dry	10.7986	10.0800	2.08138	8.08	13.78
	Transbond MIP-Dry	8.6071	9.8750	3.77069	4.32	12.50
,	Fuji Ortho LC-Saliva	9.5429	9.5200	1.73926	7.66	13.05
	Transbond MIP-Saliva	8.1671	6.5000	2.71104	5.42	11.46
	Fuji Ortho LC-Saliva and blood	5.8764	5.3600	1.60380	4.32	8.95
	Transbond MIP-Saliva Transbond Transbond MIP-Saliva and blood	7.7429	7.1300	3.79218	4.30	14.91

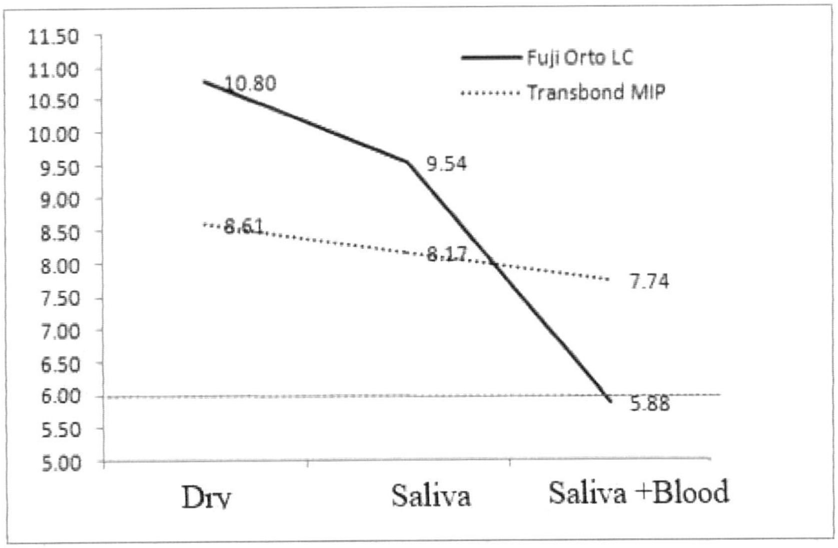

Figure 1. Mean shear bond strength of Fuji Ortho LC and Transbond MIP under 3 testing conditions (dry, saliva contamination, and saliva + blood contamination).

The most common contaminant of enamel during bonding procedures are saliva and blood. Contamination causes plugging of porosities caused by acid etching and a reduction in surface energy [13]. Whereas saliva occurs in all bonding situations, blood is mainly a problem on operation procedure to expose the impacted teeth. Saliva consists mostly of water (99.4%), with 0.6% solids. The solid is composed of macromolecules like proteins, glycoprotein sugars and amylase, inorganic particles like urea, amino acids, fatty acids and free glucose [14]. It seems that within seconds, an organic smear layer is formed, covering the etched porous surface [10]. It has been reported that the minimum bond strength of 6–8 Mpa was adequate for most clinical orthodontic needs [7].

Our study showed differences in the bonding characteristics of the 2 adhesives systems on contaminated surfaces. In this study, the shear bond strength of the group tested ranged from 5.88 to 10.80 Mpa. The Fuji Ortho LC is light-cured resin-modified glass ionomer cements and formulated to bond orthodontic brackets in wet environment. This eliminates the need to maintain the teeth in a completely dry condition during the bonding procedure. This cement not only provides enamel bonding but also have a mechanism for releasing and reloading fluoride so that the white lesions around the orthodontic brackets and bands can be reduced. In addition, this material facilitates the debonding procedures without injuring the enamel surface [15]. The result of this study indicate that the Fuji Ortho LC adhesive when used on dry and saliva-contaminated enamel, has an adequate bond strength: 10.80 Mpa and 9.54 Mpa respectively. Fuji Ortho LC had the best results in a humid environment whether by saliva contamination or water. The effective bond produced by Fuji Ortho LC in the presence of water could have three plausible explanations: (1) The cementing agent may be able to displace a sufficient amount of water so that the chemical bonding between the Fuji Ortho LC and the calcium in the tooth is not impeded [2]. The water present is simply incorporated into the cement because water is the carrier for the acidic component in this reaction.[3] Another explanation might be the presence of HEMA as a major constituent of the resin component in Fuji Ortho LC; this water-soluble hydrophilic monomer is an essential ingredient to inducing wetting and penetration.

In contrast to the contamination with saliva tolerated by Fuji Ortho LC, contamination with blood led to highly decreased bonding forces. It seems that high amounts of organic substances impede the binding between the primer and adhesive. Fuji Ortho LC showed significantly lower values for the saliva + blood contamination group (5.88 Mpa). This is in agreement with previous reports [1,3,10,11,14,16]

Previous studies that evaluated the effect of blood contamination on the bond strengths of light-cured composites showed a significant reduction in bond strength values [3,9,11]. To address this, manufacturer have introduced hydrophilic bonding material, suggesting the possibility of obtaining successful orthodontic bonding to a moisture-contaminated enamel surface. A different behavior was observed in the present study with Transbond MIP. There is no significant difference on the shear bond strength of Transbond MIP on 3 conditions of enamel. For saliva + blood contaminated enamel, the hydrophilic primer showed the shear bond strength value higher than Fuji Ortho LC. It seems that ethanol, which is responsible for humidity tolerance, and addition HEMA can enhance the adhesive strength on saliva-contaminated and saliva + blood contaminated enamel [4,12]. By applying a layer of Transbond MIP to acid-conditioned enamel, in addition to micromechanical retention, a reversible hydrolytic bond mechanism can be established by breaking and reforming of carboxylate salt complexes formed between the ionized carboxyl groups of the methacrylate functionalized-polyalkenoic acid copolymer and residual enamel calcium. These results agree with other study and contrary to some studies [2-4,9]. This difference might be explained by the use of anticoagulant in blood on experiment material.

In conclusion, there are significant differences of shear bond strengths on different enamel conditions. Statistically there is no significant difference between the shear bond strength between Light-Cured Resin-Modified Glass Ionomer Cement and Moist Insensitive Primer. The shear bond strength generated by Light-Cured Resin-Modified Glass Ionomer Cement and Moist Insensitive Primer on contaminated enamel surface lower than uncontaminated enamel. Light-Cured Resin-Modified Glass Ionomer Cement and Moist Insensitive Primer on blood + saliva-contaminated enamel produced the lowest shear bond strength. Using of light-cured resin-modified glass ionomer on blood + saliva-contaminated enamel is not recommended because it produces the shear bond strength lower than required for orthodontic application.

4. Conclusion

There was no significant difference in the shear bond strength between dry, saliva, and saliva-blood contamination bonded with Moist Insensitive Primer combined Transbond XT. There were significant differences in the shear bond strengths between dry and saliva -blood contaminated enamel bonded with Fuji Ortho LC. In conclusion, Fuji Ortho LC is not recommended for clinical bonding application under saliva-blood contaminated enamel conditions.

References

[1] Sfondrini M F, Cacciafesta V, Scribante A, De Angelis M and Klersy C 2004 *A. J. O. D. O.* **125**(3) 357–60
[2] Faltermeier A, Behr M, Rosentritt M, Reicheneder C and Müßig D 2007 *E. J. O.* **29**(6) 559–63
[3] Cacciafesta V, Francesca Sfondrini M, Scribante A, De Angelis M and Klersy C 2004 *A. J. O. D. O.* **126**(6) 703–8
[4] Vicente A, Toledano M, Bravo L A, Romeo A, la Higuera De B and Osorio R 2010 *Med. Oral Patol. Oral Cir. Bucal.* **15**(5) 820–6
[5] Bala O, Arisu H D, Kivanc B H and Samur S 2009 *J.D.O.H.* **1(2)** 17–21
[6] Al Shamsi A, Cunningham J L, Lamey P J and Lynch E 2006 *The Angle Orthodontist.* **76**(4) 694–9
[7] Turk T, Elekdag-Turk S, Isci D, Cakmak F and Ozkalayci N 2007 *The Angle Orthodontist* **77**(5) 901–6
[8] Endo T, Ozoe R, Sanpei S, Shinkai K, Katoh Y and Shimooka S 2008 *Springer* **96**(1) 50–4
[9] Patel D P, Gupta B and Thakkar S A *Measuring The Shear Bond Strength Of Orthodontic Brackets Using Self Etch Primer, Moisture Insensitive Primer & Conventional Primer (Light & Self Cure): A Comparative Study* (Editorial Review Board) p 24
[10] Brauchli L, Eichenberger M, Steineck M and Wichelhaus A 2010 *A.J.O.D.O* **138**(4) 435–41
[11] Öztoprak M O, Isik F, Sayınsu K, Arun T and Aydemir B 2007 *A.J.O.D.O* **131**(2) 238–42
[12] Vicente A, Mena A, Ortiz A J and Bravo L A 2009 *The Angle Orthodontist* **79**(1) 127–32
[13] Rajagopal R, Padmanabhan S and Gnanamani J 2004 *The Angle Orthodontist* **74**(2) 264–8
[14] Eiriksson S O, Pereira P N, Swift E J Jr, Heymann H O and Sigurdsson A 2004 *Dental Materials* **20**(2) 184–90
[15] Pithon MM, Santos dos RL, de Oliveira M V, Ruellas ACO and Romano FL 2006 *The Angle Orthodontist* **76**(4) 700–4
[16] Sayinsu K, Isik F, Sezen S and Aydemir B 2007 *A.J.O.D.O* **131**(3) 391–4

Acknowledgment

I would like to express my gratitude to my promotor who have given me fruitful guidance to do this study and to my Dean who have given me opportunity to take the postgraduate programme.

Institute of Physics
Dirac House, Temple Back
Bristol BS1 6BE UK

ISSN: 1757-8981
ISBN 978-1-5108-9384-9